JOHN DEERE'S COMPANY

JOHN DEERE'S COMPANY

The Golden Age; From Johnny Popper to the Iron Horses 1928–1982

VOLUME 2

WAYNE G. BROEHL JR.

Octane Press, Edition 2.0, November, 2023
Copyright © 1984 by Wayne G. Broehl Jr.

On the cover: John Deere Model R owned by the late Charles Klein. *Lee Klancher*
A farmer hauling oats with a John Deere Model B on August 6, 1948. *National Archives*

All rights reserved. With the exception of quoting brief passages for the purposes of review, no part of this publication may be reproduced without prior written permission from the publisher.

The author is grateful to the following for kind permission to reprint material as noted below.

Two cartoons and photograph from *Farm Implement News* (August 4, 1910; September 23, 1920; November 21, 1929). Copyright © Intertec Publishing Corporation, Overland Park, Kansas, reprinted by permission.

Table from Lester Larson, *Farm Tractors: 1950–1975*.
Copyright © American Society of Agricultural Engineers, reprinted by permission.

ISBN: 978-1-64234-135-5
ePub ISBN 978-1-64234-148-5

LCCN: 2023947456

Design by Tom Heffron
Copyedited by Dana Henricks
Proofread by Faith Garcia

octanepress.com

Octane Press is based in Austin, Texas

Printed in the United States

ACKNOWLEDGMENTS

Any study as complex as this book involves major debts to others. A host of them is owed here, and I wish to directly acknowledge a few, recognizing that there were many other significant contributions.

A business history typically involves a single firm, in this case Deere & Company. Cooperation of living executives, line employees, and retirees can be fundamentally useful in preparation of such a book. Deere & Company recognized from the start the wisdom of engaging an outside scholar and allowing him to view the company at arm's length, for only with such independence can an objective study be accomplished. Deere management exhibited a keen sense of this requirement, their single caveat being that, given the highly competitive nature of the agricultural machinery industry, no future product information prematurely be made available. In order to retain this arm's length relationship, it was agreed that there would be no financial relationship between me and the company. Instead, Deere made a grant to Dartmouth College, and the research and writing of this study was a formal project of the Research Program at the Amos Tuck School of Business Administration. For their perception of, and enthusiasm for, this basic principle of a sound independent business history, I owe particular thanks to the three members of the Deere chairman's office at that time—William Hewitt, chairman and chief executive officer; Elwood Curtis, vice chairman; and Robert Hanson, president. All three were profoundly helpful in my research; the objectivity of the manuscript would not have been the same without their unstinting cooperation.

Several other Deere personnel read the entire manuscript, in the process helping to ensure that it was as free of factual errors as possible. Two people were particularly helpful in this regard: Dr. Leslie Stegh, the archivist, and Elizabeth Denkhoff, corporate secretary for the company. Walter Vogel, Boyd Bartlett, Robert Boeke, Robert Weeks, Curtis Tarr, and Joseph K. Hanson Jr., also reviewed the entire manuscript. Commenting on key sections were James Davis, Thomas Gildehaus, Gordon Millar, Neel Hall, Joseph Dain, Clifford Peterson, George Neiley, and Roy Harrington. Several

ACKNOWLEDGMENTS

retirees of the company provided valuable insights; particularly to be mentioned are George French, the late Edmond Cook, and William Bennett. Ruth Moll, curator for the William Butterworth Foundation, contributed important insights from her personal relations with William Butterworth and his family, and Kay Vogel, curator of the Deere-Wiman House was also helpful. Dr. Herbert Morton and Barbara A. Standley ably edited the manuscript, and the transition from manuscript to book was aided measurably by Kenneth McCormick, Chester Lasell, and Wayne Burkart. The entire staff of the archives department of the company was involved importantly in the project; the contributions of Vicki Eller, Ann Lee, and Nancy Swanson were particularly helpful. Several executives in other companies in the industry, as well as in the industry association, the Farm and Industrial Equipment Institute, contributed valuable insights; I particularly want to recognize the help of Greg Lennes at the International Harvester Company Archives, and David Crippen at the Greenfield Village and Henry Ford Museum. Pat Greathouse, international vice president at the United Automobile, Aircraft and Machine Workers (AFL-CIO) was invaluable to my gaining perspective on the labor relations of the farm equipment industry.

The list of research librarians who have helped me at one stage or another of this project is very long indeed. I especially wish to thank those at the Vermont Historical Society; the Sheldon Museum in Middlebury, Vermont; the University of Vermont; the Illinois State Historical Society; the Chicago Historical Society; the State Historical Society of Iowa; the Minnesota Historical Society; the State Historical Society of Missouri; and the F. Hal Higgins Library of Agricultural Technology at the University of California, Davis. Significant research materials were also studied at the Worcester Polytechnic Institute. Finally, my debts to the Baker Library of Harvard Business School and to my own colleagues in the Baker Library of Dartmouth College and the Tuck School of Business at Dartmouth Library.

Several of my academic colleagues read significant sections of this book. Professor Alfred D. Chandler Jr., of Harvard Business School, reviewed a number of chapters, and his perceptive suggestions aided the manuscript at an important point. Dr. Glenn Leggett, president emeritus of Grinnell College, was most influential at the early stages of the book. Professor Fred Carstensen of the University of Connecticut also made significant contributions. My own Tuck School colleague, Dr. David Bradley, likewise helped at a critical point in the manuscript. My research assistant, Barbara Morin, was indispensable to me throughout the project. I would also like to thank Mrs. William Hewitt for her insights about her father, Charles Deere Wiman, and for aiding my understanding of the Deere family relationship to the business.

Finally, a profound thanks for those people who loyally aided the preparation of the many drafts of the manuscript: Eleanor Lackey, Joan Adams, Frances Moffitt, and Suzanne Sweet.

CONTENTS

PART IV
FROM RETRENCHMENT TO EXPANSION: THE CHARLES WIMAN YEARS (1928–1954)

CHAPTER 10
INNOVATION AND NEW LEADERSHIP 3
The Combine Comes to the Midwest 7
Deere Builds a Combine 9
The General-Purpose Tractor 13
Deere's Search for an All-Crop Tractor 15
Research & Development in the Organization 22
Deere and International Harvester Pull Ahead 25
Company Stores and Price Maintenance 27
A New Leader 32

CHAPTER 11
DEPRESSION YEARS 37
Financing the Farmers 40
Soviet Debts are Paid 42
The Argentine Trade Turns Sour 46
John Deere of Canada 48
Employee Traumas 51
Innovating in a Depression 57
The Death of William Butterworth 60
Management (and Family) Transition 61
Recovery for the Farmer and His Suppliers 65
Labor Relations in the 1930s 66
Henry Dreyfuss Styles the Tractor 69
Deere and Caterpillar Join Forces 72
California and the Caterpillar Link 76
Product Development in the Late 1930s 79

CONTENTS

CHAPTER 12
WORLD WAR II AND THE POSTWAR DECADE *85*
Deere's World War II Efforts *88*
Wiman's Washington Years *89*
Unions Come to Deere *94*
Postwar Anatomy of the Company *97*
Charles Wiman's "Postwar Reconversion" *101*
Defining Decentralization, Once Again *103*
Financing Expansion *108*
Postwar Labor Conflicts *112*
Antitrust Charges and Deere's Defense *121*
The Government Tries the J. I. Case Company *124*
Innovations in Engine Design *127*
Combine and Cultivating Product Development *134*
Manufacturing Abroad: The Scotland Proposal *135*
Deere's Fertilizer Venture *139*
The Scotland Project Falls Through *142*
Management Succession *142*

PART V
WORLDWIDE CORPORATION: THE WILLIAM HEWITT YEARS (1955–1982)

CHAPTER 13
RISE TO INDUSTRY LEADERSHIP *153*
Exploring Opportunities Abroad *157*
The Massey-Harris-Ferguson Proposal *159*
The Organizational Shake-up *161*
Moving into Mexico and Germany *166*
The Deere Administrative Center *175*
A New Generation of Products *180*
Deere Day in Dallas *183*
Miscalculations in Germany *187*
Deere Moves into France *193*
Mexico: Local Content, Local Ownership *196*
Argentina, South Africa, Japan *198*
The "Worldwide" Tractor *202*
Hewitt's First Decade in Retrospect *205*

CHAPTER 14
OBSTACLES ABROAD *215*
Reorganizing Foreign Operations *216*
Seeking a Partner: Talks with Deutz *219*
Second Effort: The Fiat Venture *223*

Going It Alone *228*
Economic Nationalism in Mexico *231*
"Yo-Yo" Nationalism in Argentina and Venezuela *234*
Contrasts: Iran, Turkey, Brazil, Australia *238*
Ideological Encounters in South Africa *242*
A Little Trade with the Soviet Union *248*
Mainland China: Deere's "Friendship Farm" Experiment *251*
Foreign Reprise *258*

CHAPTER 15
CHALLENGES OF THE 1980s *263*
World Agricultural Markets *264*
Changes in US Agricultural Practices *265*
Shifting Market Shares and Recessionary Fallout *267*
Deere's Non-Agricultural Divisions *269*
Deere Increases Its Market Share *274*
Deere Tractor Strategy *275*
The Small Tractor: The Deere-Yanmar Collaboration *278*
Global Tractor Strategy *284*
Dilemmas in Combines and Tillage and Haying Equipment *285*
The Nurturing of Innovation *286*
Should Deere Sell Its Technology to Competitors? *291*
Further Implications of Centralization *293*
The "Automatic Factory" *296*
"Give Backs" in Industrial Relations *301*
Financial Performance *303*
Preoccupation With the Recession *305*

CONCLUSION

CHAPTER 16
CHARACTER OF A CORPORATION *313*
As Seen by Others *314*
Instilling Initiative *317*
The Search for Excellence *319*
The Transition in Leadership *321*
Looking Ahead *326*

List of Appendix Exhibits *331*
Appendix Exhibits *333*
Bibliographic Notes *353*
Index *361*
About the Author *368*

PART IV

FROM RETRENCHMENT TO EXPANSION: THE CHARLES WIMAN YEARS (1928–1954)

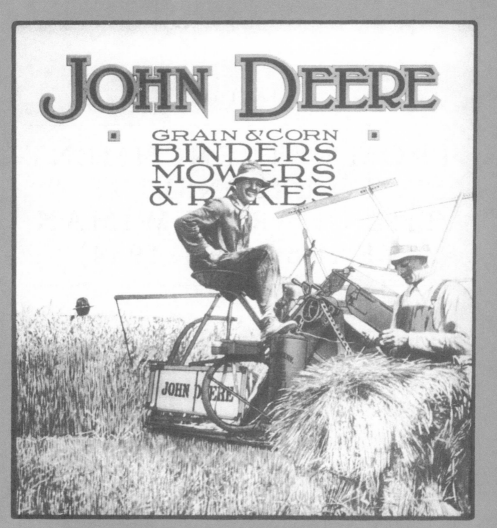

CHAPTER 10

INNOVATION AND NEW LEADERSHIP

Never believed in a two-row cultivator, as a man had all he could do to run and watch a single row . . . could see the tractor and planter but not the cultivator . . . the boys could try it, but did not expect to let them cultivate a whole row but would have to tell them to take it out . . . had horses ready to go and do the job. However, the boys cultivated one row and then another, and then told the boys to go ahead and cultivate with a three-row . . . frankly surprised to find out how well the three-row did.

C. C. Veech, 1927

The early 1920s, especially the depressed years of 1920–1921, had been very distressing for William Butterworth. At numerous critical junctures in his presidency he had articulated a conservative view of the business—urging caution in taking on a harvester line in the early 1910s when faced with International Harvester's aggressiveness and consistently opposing greater centralization of authority until he was forced to act in the 1920–1921 downturn because purchasing and inventory decisions had gotten out of control. He had been faulted often by Willard Velie for dividing his loyalty between the Deere trust and the company; he had repeatedly opposed the entry into tractors, despite the enthusiasm of most of the rest of the

◀ Advertising brochure for Harvester Works products, 1920. *Deere Archives* board. He seemed less a chief executive than a careful, financially oriented lawyer who felt strongly his fiduciary responsibility to the major shareholders of the company and who believed implicitly that the company should be preserved intact, even if one were to forego some entrepreneurial opportunity. Butterworth's thinking seemed so often to reinforce the status quo.

Butterworth's lack of entrepreneurial spirit was accompanied, however, by an abiding concern for the employees of the company and a view of employee relations that stressed loyalty, trust, and service. The pitched arguments of the World War I period about incentive bonus plans poignantly illustrated this. Butterworth advocated a broad sharing of ownership among all employees, based primarily on length of service, while others, particularly C. C. Webber, George Mixter, and George Peek, adhered to a selective, subjective plan designed to reward management performance.

The 1920–1921 downturn heightened, if anything, Butterworth's strongly held views. He warned over and over of the necessity of purchasing carefully—almost hand-to-mouth—and urged the cutting of expense at every turn. In his personal diary are many remarks during the 1920–1924 period urging "intelligent saving" and "careful spending of the Company's money." Even after business began to recover in 1923, Butterworth still discouraged building of inventory, writing Charles Wiman from Europe: "Let us not be frightened into buying a lot of high-priced material. . . . I notice our inventory is growing . . . this is a bad sign and should not continue." A roundtable discussion in the board in late 1925 recorded a "spirit of optimism" and a "bright outlook," and C. C. Webber expected "a very large increase in their volume for 1926." Yet Butterworth still raised cautionary caveats about "watching schedules closely," with inventory at a minimum.

Butterworth spent much time with Charles Wiman in the early 1920s, encouraging him and explaining his philosophy of family and company responsibility. In January 1924, Butterworth wrote Wiman about a managers' meeting that Wiman had chaired, complimenting him on "splendid success." He continued: "The fellow we want to keep out of our organization is he who says 'it can't be done.' We'll do it or bust and we're not going to 'bust,' is our homely motto." In this letter Butterworth reiterated his view of keeping inventory in control and concluded: "I am writing you somewhat at length, not because I think it is necessary, for I know you have these things in mind, but I'm doing so to emphasize their importance and to indicate to you that the Company officers and directors are thinking along the same lines and that we are all thinking about you and boosting for you."

As business did indeed get better in 1925, Butterworth finally seemed to relax a bit, confiding to his diary: "I feel more comfortable about the future of our Company than I have since its reorganization in 1910. In fact, this is the first time that I really have felt comfortable and contented with our policy and program. We are going ahead now with everybody from top to bottom

Exhibit 10-1. Top, Trolley leaves for Campbell's Island picnic ground on "Deere Picnic Day," ca. 1905; bottom, Deere & Company headquarters, 1919. *Deere Archives*

working for just one thing and that is the success of Deere & Co." At this point, Butterworth decided to share with Wiman, Frank Silloway, and T. F. Wharton a memorandum about his views toward the company that he had written during World War I but had never made public. ("I offer to you the following statements, which I wrote more than seven years ago and which indicates my ideas as to the policies we should pursue.") It is not surprising that he had been reluctant about advancing these views until this moment, feeling that the company was still not stable enough to warrant his being so definitive. The statement again emphasized his concern of employees. "I want to keep in close touch with you and your department and know your successes and your troubles," Butterworth told the three younger men. Each and every employee from the lowest to the highest should have from the management a "real, live, personal interest" and, in turn, each employee should "likewise understand that the Company expects and wants him and her to take and show a real, live interest in the welfare and growth of the Company." The earnings of employees should be watched carefully so as to give "the average man a chance to live comfortably and lay aside something for the rainy day." Thrift was to be encouraged among the employees, with a particular emphasis on purchasing of the company's preferred stock. Young men were to be encouraged to join the business, for which they would be educated "along definite and prescribed lines." College graduates were to be employed, "with preference for the sons of *our men*." (This particular plan, though, met with some resistance. "Not all of the branch houses had been convinced that it was a desirable thing to do," R. B. Lourie reported.) Board meetings should be held weekly, Butterworth continued, to provide a clearinghouse of ideas and to keep all informed, and there should be at least a once-a-year meeting of all branch managers and factory managers with top management, hopefully meeting together. Management should make periodic trips to the factories "to keep our relations close and the shop policies as uniform as possible." He stressed the importance of quality and careful workmanship and reiterated his profound concern for proper management of inventory. Finally, "Watch the expense account, both large and small, keep them at a minimum. A small reduction of each account will help our showing."

This is a document that rings true throughout, epitomizing the benign spirit and cautious ways that Butterworth had brought to the organization and to his personal relations over the previous two decades. Butterworth had become the highly respected—yes, loved—father figure for the whole company. It seems likely that he was now thinking of new leadership for the company, for he had become increasingly active in state and national organizations, particularly the US Chamber of Commerce.

Charles Wiman had moved quickly into a major position in top management after his appointment as director of manufacturing in 1924. As the ranking member of the next generation of the family, it was obviously important for him to prepare for an eventual leadership role in the company.

Wiman moved into this role self-assuredly and well, bringing a verve and drive that had been earlier filled in part by George Mixter, George Peek, and Leon Clausen. Wiman had started in the shop in 1915 as a line employee (salary, $.15 an hour); over the ensuing years he had built a strong following among shop employees, engineers, and production supervisors. By 1926 Wiman also had clearly won the respect and cooperation of his older fellow board members; at a board meeting in July of that year, Floyd Todd noted "the excellent feeling and cooperation existing not only in the factory organization but between the branches and the factories as well. This condition . . . was due largely to the efforts of Mr. Wiman."

As business expanded after 1925, it was Wiman who constantly was urging new product development. He seemed to have an instinctive feeling for the larger picture—the needs for coordination, the balancing of units, and so on. The tractor was center stage, he constantly reminded the board, for "it is more important that the tractor factory operate at a uniform pace . . . because it manufactures but one tool, and when they have a given schedule all of the equipment is utilized every day." In effecting this, though, Wiman advocated an aggressive strategy: "We must remember we should be willing to spend considerably more money in promoting tractor sales, either by increasing our field force or through advertising, or both." At one point in the tractor expansion, he even asked the board for a discretionary appropriation of $25,000, to be used for bonus payments to hasten the construction of a building.

Wiman was also willing to move decisively if any manager or employee was not reasonably fulfilling expectations. A few months later, for example, the Waterloo tractor factory foundry and mill room were falling short of quotas and reporting high scrap rates. Wiman relieved the foundry superintendent without delay and sent in two troubleshooters from Moline. "The Waterloo management has had the seriousness of the situation explained to them and has been shown the necessity of correcting the foundry difficulties at the earliest practical date," Wiman told the board. The Waterloo men got the message!

Wiman's influence on product development was particularly felt in two key innovations—the adaptation of the combine to Midwest farming conditions, and the development of the "general-purpose" tractor.[1]

THE COMBINE COMES TO THE MIDWEST

A persistent dream of farmers had been to harvest grain and thresh it in one operation. The first "combine," the machine that could do both jobs in a single operation, antedated even Cyrus McCormick's reaper—a Maine inventor had patented a traveling thresher with a harvesting attachment in 1828. In the early period of commercial development, in the 1870s and '80s, the combined harvesters were commonly known as "travellers." A farmer, asked how he was going to harvest his crop, would either reply "by stationary" (a stationary thresher) or "by traveller" (the combined harvester). After the 1880s the machine was generally known as a "harvester."

Midwest agriculture was unsuited to these early combines, for they would only work well under ideal conditions in which the crop was ripe and dry, which was rare in the Midwest. But the combine was ideal for the great wheat farms of California, which by 1880 had become the number one wheat-growing state of the country. Enormous combines were constructed by competing inventors to harvest for these big spreads—combines that could each harvest as much as fifty acres a day, two thousand acres a season. Many were variations of Hiram Moore's Michigan combine, first perfected in the 1830s, and soon other names were jostling each other for fame—the Shippee, the Berry, and, later, the Holt and the Best. (The Holts had bought out Shippee, and later, in 1925, Holt and Best merged as the Caterpillar Tractor Company to become a prominent producer on the West Coast.) By 1910, Massey-Harris had entered the field with its first combine, developed initially for the Australian market.

Australian inventors had experimented with combines as early as the 1850s, and South Australia sponsored a major international competition in 1877. The Australian versions were usually smaller, generally being light, horse-drawn, ground-driven machines of the stripper type with an eight-foot cut. The California machines were more often heavy, horse- or power-drawn, motor-driven machines with complete cutting, threshing, separating, and cleaning mechanisms cutting swaths of about twenty-four feet.

The Midwestern wheat farmers knew the cost savings of combining were great. But would the combine work for sections east of those regions where the wheat dried out thoroughly in the head? The year 1924 was the pivotal one in answering this question. In November of that year a bellwether demonstration was made at a Stonington, Illinois, farm of a Massey-Harris combine, in this case harvesting and threshing soybeans. More than two thousand people were there, including farmers, implement dealers, county agents, and experiment station men. A *Farm Implement News* editor attended and reported "entire satisfaction"—the thirty acres of beans had been harvested, threshed, and carried to a granary at a cost of about $.05 a bushel. The Massey-Harris representative running the test averred that the cost of harvesting with a binder, tying in bundles, and threshing in the old way was $.30 a bushel. The editor of *Farm Implement News* made a prophetic statement at the end of his article: "One of the drawbacks to raising soybeans has been the difficulty in harvesting and threshing. It is believed that the success of this demonstration will lead to a big increase in the acreage of soybeans in central Illinois."

Over this same harvest season, the combine was also tried with wheat and oats, as well as occasionally on crops of sweet clover, red clover, and timothy.

Again the results were most promising. The question remained of just how effective the new machine would be in humid conditions. The crop season of 1925 in Illinois was humid, an ideal condition for tests of the combine. The trade press soon trumpeted its continuing success and touted its potential as a Midwestern harvesting addition. The conditions of 1926 in Illinois and other Midwestern states were not quite as favorable, yet the combine again registered great records in Kansas, Texas, and Oklahoma, as well as in Iowa and Illinois. The US Department of Agriculture also conducted extensive tests during the crop season of 1926. The crop still had to be as dry as possible— users recommended delaying the commencement of harvest from five to nine days later than if headers or binders were used. If the farmer was patient for this time, the results were excellent and the costs remarkably lower. By 1927, *Farm Implement News* was headlining a major editorial, "It's a Combine Year." The demand for combines quickly outstripped the production capacity of the factories, forcing the branch houses of the manufacturers to ration the coveted machines. "The future of the farm equipment trade never looked brighter," commented *Farm Implement News*. There had been just over 3,600 combines produced in 1920; the number for 1927 jumped to 18,307, up over 6,000 from the previous year of 1926.[2]

DEERE BUILDS A COMBINE

Deere had been building harvesting machinery for about fifteen years, the product line consisting of a grain binder and a corn binder, together with several sizes of mowers and rakes. But it was not a leader in the field. Sales of harvest machinery in 1926 were just over $6 million, about one-seventh of the total sales of the company of $47.8 million, and for the first time the harvester division showed a profit.

Deere had shown little interest in entering the thresher field; companies there were well entrenched. The combine business, on the other hand, seemed more promising. But a catalyst was now needed to shake the company out of its lethargy in harvester product development, and it came in the person of Charles Wiman. The Harvester Works was headed by Charles N. Stone, who had succeeded W. R. Morgan after the latter's death in 1923. The Kansas City and Portland branches had already been pressuring Moline to consider combine possibilities; now Wiman and Stone wanted to move ahead fast, and they came to the board with an appropriations request for experimental work. Their memorandum was blunt: "We believe that the combine harvester and thresher will almost completely replace headers . . . and will probably be used almost exclusively in the sections where it can be used."

C. C. Webber was surprisingly negative at the start about the combine, at least as a tool to be sold by the Minnesota branch: "Here is the combination machine. You remember a few years ago that that was talked about everywhere and we did not know but what we would have to make it. Well, we believe it is being demonstrated that the combination machine is useful in a territory where the grain gets dead ripe and it's fit to cut and thresh at the same time, and to use it in any other territory is dangerous. So it seems to me that we have done well to stay out of the combination trade."

The harvesting situation in Minnesota was challenging. There often was an uneven ripening and sometimes a nagging weed problem, so for a long time after other parts of the country turned to combines, many farmers in the Webber territory used windrowers and grain binders. Still, it was incongruous for Webber to be so provincial in this instance, for he generally espoused a broad management perspective.

There was always a tradeoff between new product development and additional product engineering on present products. Webber wanted the latter: "It would be a pioneer work and an uphill job, and it does not seem to the writer that we are in shape at your factory to undertake that kind of a job. Your situation is such that you do not want to take on a machine that you have got to put a lot of money into for several years before it commences to yield any return. . . . I have never been in favor of the undertaking to make a header in the Harvester Works and I have held to that position constantly because I knew you were at work and had to work on a new mower and a new grain binder and that when they had been perfected you would have to equip your factory to make them economically and I felt that you had your hands full, and that all your energy should be devoted to the present line and perfecting that and enabling you to produce it comparatively before you undertake anything else." Given Webber's opposition to the combine, it is all the more remarkable that Charles Wiman stuck to his guns. He got the appropriation.

Next, another dilemma had to be faced. Should some outside machine, already perfected, be purchased or licensed, or should the company itself take on the laborious, expensive process of developing its own product from scratch? Wiman and Stone hedged their bets. There was, indeed, an existing combine available for sale to the company—the Universal, made by a Kansas City company of the same name. Stone and Wiman were uneasy about the company: "The makeup of the Universal Harvester Co. does not inspire confidence. The men of the company have been in bad company, having been with . . . Savage Harvester Company and the Standing Grain Thresher Company of Wichita, Kansas. Both of these companies had checquered careers and farmers lost money through them after investing as a result of more or less false representation." The Universal machine used a rapidly revolving cylinder with a system of square pin teeth instead of the cutter bar used by most other combines; there were some advantages in terms of simplicity of manufacture, as well as a reputedly high recovery of grain from the field, more than a standard combine. "Statements on this point vary greatly," noted Wiman and Stone, "and some competent observers say that the machine wasted too much grain."

Wiman and Stone finally decided to buy one Universal machine, together with a Massey-Harris machine, and to take these two into the field for testing, along with whatever prototype machine could be developed by Stone's own product development engineers. By the season of 1925, the company had its own prototype ready for tryout, and the three machines were sent for testing near LaCrosse, Kansas. The new Deere machine competed well with the Massey-Harris machine and performed considerably better than the Universal, which was dropped from consideration forthwith.

At the same time that the company was experimenting with the combine, its Harvester Works also had taken on the task of designing and manufacturing a corn picker. Commercially successful pickers had been sold by other companies since before the turn of the century; now Deere decided to develop its own. Twenty-three prototype machines were assembled for the crop season of 1925 and tested in the Minneapolis, Omaha, and Moline branches, again with considerable success "in a wet, difficult picking season."

After both the combine and the corn picker were tested again in the field during the crop season of 1926, the board met in April 1927 to decide what to do about both. The corn picker was a simpler machine and was put right into the product line for the selling season of 1927. As for the combine, the decision was to build forty machines for the crop season of 1927. Stone was blunt about what he felt was needed: "We must act promptly now and move faster in this development than is usually necessary in our business . . . but we should not go at it in a way that is too amateurish. It seems likely that real competition in combines will develop shortly and we would be better off not to get into the business at all than to enter it in a half-hearted way." Wiman proposed that they make 500 machines in 1928 and move quickly to 1,500 the following year. "A new building will be needed," Wiman stated, and the directors promptly gave Wiman and Stone the green light.

At this critical juncture, a surprising new development surfaced that sidetracked everyone for a moment. Representations had been made to Deere that a unit of the old Holt Combine Thresher Company in Stockton, California, one of the pioneer companies in the field, was available for purchase. The Holt company had just been merged with the C. L. Best Tractor Company as the Caterpillar Tractor Company, with its corporate headquarters at San Leandro, California (later at Peoria, Illinois). In 1916 Holt had formed a subsidiary, Western Harvester, and it was this entity that was being offered to Deere. The asking price for the Western Harvester organization, however, was $1,256,000; this seemed too steep an entry price, even for a successful member of the combine field, and the board finally rejected the Caterpillar overture.

By 1927, the John Deere No. 2 combine was available to the branches in quantity and a year later the No. 1 was also in the catalog. The No. 2 was available in 12- and 16-foot platforms, the No. 1 with 8-, 10-, and 12-foot platforms. (The larger No. 2 had a grain tank of sixty-five bushels, the No. 1

Exhibit 10-2. Top, Shipping No. 1 and No. 2 combines to Kansas City by rail, 1929; middle, Combining, Sioux Falls, SD, 1930; bottom, Mississippi River barge shipment of combines, 1940. *Deere Archives*

had a forty bushel capacity). Four-cylinder Hercules engines were purchased by the company for the machines. Stone and the others first had felt that the larger No. 2 machine would be the more popular, but the results in the field soon convinced everyone that greater numbers of farmers wanted the smaller machine. The No. 2 was heavier than it should have been and difficult for a Model D tractor to pull in sandy or muddy soil (so it was better adapted for the Minneapolis territory than, say, the Kansas City regions). Further, because Deere had made these first two combines at high quality levels, Stone also wanted a lower-priced machine, somewhat cheaper in construction and lighter in order to meet price competition. By 1929 the No. 1 and No. 2 had been replaced by two new versions, the No. 5, designed as a smaller, lighter-weight, popularly-priced combine, and the No. 3, a larger-sized machine similar to the No. 2, but also lighter in weight.

The combine clearly was going to make obsolete the old-fashioned thresher that had served so well over so many years. Despite the pessimistic outlook for the thresher, Deere surprised the industry in 1929 by adding a thresher "in order to complete our line"; so said Frank Silloway in presenting the matter to the board. Silloway and Charles Stone had investigated a nearly defunct company in Minneapolis, the Wagner-Langemo Company, which had been manufacturing the Grain Saver threshing machine. The owners of the faltering concern were willing to dispose of their patents, dies, tools, patterns, and so forth, as well as twenty-three complete machines, for a total of $42,500. Stone felt that these machines could be readily built at the Harvester Works, without adding any buildings or even very much equipment. Silloway and Stone anticipated a maximum volume of approximately five hundred machines a year, not an enormous potential. Nevertheless, the company voted the purchase and the John Deere Thresher made its debut, albeit very late in the game! The machine stayed in the line until 1936, selling only in modest quantities throughout.[3]

THE GENERAL-PURPOSE TRACTOR

There had been one significant shortcoming of all the gasoline tractors built before 1924—a lack of versatility. They could not be used for many farm tasks that agriculturalists might wish to convert to tractor power. Farmers had been using their tractors to prepare seedbeds (plowing, disking, harrowing) and probably for some harvesting (especially the cutting of small grains). Tractors were also enormously helpful in driving other pieces of machinery—in threshing, silo filling, feed grinding, wood sawing, baling, shelling, shredding. But these tasks were only part of a farmer's spectrum of horsepower needs, for with all of the row crops—corn, hay, cotton, kafir, potatoes, tobacco, peas, peanuts, beans, sugar beets—not only the cultivation but generally the planting was done by horse. In the way tractors were constructed, it was not very easy to drive them down a row to plant or cultivate. Similarly, many functions in the harvest were more readily accomplished by horse—mowing hay, cutting corn, digging potatoes, and so on. One industry executive in 1924 maintained that the tractors in use replaced on an average only 2.8 horses per tractor (though he did admit that owners of larger tractors often disposed of six to eight horses upon purchase of the tractor). A US Department of Agriculture bulletin of 1925 graphically showed the relatively small part of the total horsepower need assumed by the gasoline tractor in that period—only 1.6 billion of 16 billion horsepower hours, work animals providing 9.7 billion horsepower hours (Appendix exhibit 21).

The incredible potential for the tractor cultivation of row crops was just waiting for an innovative idea. Indeed, it seemed anomalous that after more than two decades of gasoline-tractor development, no one had been able to perfect such a row-crop tractor. But 1924 brought a solution, a very successful new four-cylinder tractor from the industry leader, International Harvester. Its name ideally described the essence of the need—the Farmall.

In construction, the Farmall differed radically from other tractors, its rear axle was built high, with a clearance of thirty inches above the ground, and the larger rear drive wheels were wide apart (seventy-four inches between the centers of the wheels); it could clear corn or cotton as well as the ordinary riding cultivator, and the wheels were wide enough apart to straddle two rows and thus cultivate two rows at a time. The front wheels, some twenty-five inches in diameter, were located close together in order to run between the two rows. A particularly attractive feature to the farmer was the placement of the International Harvester cultivator, out in front of the tractor where the operator could watch the work. The Farmall also filled a real need in planting, being able to plant either two or four rows at a time.

It is difficult to overemphasize the breakthrough in farming technology brought by this one new tractor. The response from the field was instantaneous—the Farmall became an abiding success. It was patently clear to the farmer that the Farmall could do things that the Fordson could not do, that the Deere Model D could not do.

The remarkable dominance of the Fordson tractor in the early 1920s—70 percent of all gasoline tractors in 1922, 76 percent in 1923, 71 percent in 1924—now began to decline. Though Henry Ford still commanded more than 50 percent of all gasoline tractors produced in 1926, the onrush of the International Harvester Farmall that year soon reversed the figures. By 1928 there were only 12,500 Fordsons produced, and late in that year Henry Ford gave up making Fordsons in the United States (although he did continue to make them in Cork, Ireland, and Dagenham, England). The Fordson was not adaptable to row-crop farming and that must have been one factor, among many, that persuaded Ford to retire from domestic competition in tractors, a surprising decision in retrospect, given his preeminence just a few years earlier. By 1928 International Harvester had moved into a commanding position, building more than 47 percent of all farm tractors that year and almost 60 percent in 1929.

The row-crop tractors did not replace their larger counterparts, of course, and International Harvester continued to make its famous Titan 10-20 tractor and its larger version, the 15-30. Nevertheless, the advent of the Farmall gave International Harvester a competitive edge that brought its dominance in the tractor field to an all-time high. It was an epochal step for the industry.

DEERE'S SEARCH FOR AN ALL-CROP TRACTOR

Deere & Company had entered the tractor field quite late and did so over the grudging opposition of many of its management—William Butterworth, George Crampton, and others. The precipitous decline of the Waterloo Boy right after World War I had been offset by the development at Waterloo of the first commercially viable Deere tractor, the Model D. This machine was an instant market success. Leon Clausen had almost single-handedly ensured that the Model D would be made in enough quantity to give it a good field test; at this point he left the company, thus placing the tractor's future in the hands of Charles Wiman.

Wiman had no doubts about the Model D's viability, and in 1925 he pushed production to more than 3,900 tractors. Further, he quickly committed the company to an all-crop tractor. Theo Brown was given the responsibility in 1925 for designing the machine with the help of a number of Waterloo engineers, particularly H. E. McCray and J. E. Cade, aided by H. B. McKahin, the Planter Works manager.

Right at the start, Brown made an important design decision that was to prove highly controversial—to enable Deere's all-crop tractor to handle a three-row cultivator. McKahin, meanwhile, had gone to Texas to test an International Harvester Farmall that Deere had purchased. He reported back that it could handle both two-row cultivators and four-row versions, and he commented somberly, "A three-row is not going to be popular there as the rows have to be watched to make sure that the old stalks do not drag."

By July 1926, three prototype tractors had been built that could pull a two-bottom plow (both the earlier Waterloo Boy and the Model D were designed for three-bottom plows). There was one especially striking innovation on the tractor—Brown and his colleagues had developed a power lift mechanism that would raise the integral tools of the cultivator or planter by engine power rather than mechanically by operator levers. Thus it had four forms of power—drawbar, belt, power take-off, and power lift. The latter became an "industry first," soon widely adopted in similar forms by competitors.

Within a few weeks the new Deere cultivator arrived to be tested on the all-crop tractor. Brown felt encouraged: "In the afternoon we had the outfit in the field and there is no question but what the outfit followed the ground in fine shape. The swinging frame is fine and is necessary. The lost motion has been taken out of the steering gear, which helps. We had Farmall out too, and in crossing corn we have the advantage of quicker dodge and rather better view ahead. On straight-way Farmall may be a little better, but McCray thinks we are nearly equal and in respect to crossing corn, better.... It really seems as though the idea of our cultivator is right and that we are on the right track. It is really encouraging."

By this time the Farmall was in the International Harvester dealers' shops and there was mounting pressure from Deere branches to "get the new tractor into our hands." Wiman feared a misfire by premature release of the prototypes (à la George Peek's Universal tractor fiasco), and he cautioned the board: "Development work of this kind is rather a slow process.... This was the best way to proceed, rather than to place the machine on the market before proper field and experimental work had been taken care of."

So further tinkering was done on a number of features of the tractor and it was once more tested early in the crop season of 1927. The test site was Mercedes, Texas, and again the Farmall and the All-Crop competed side by side. Brown reported the tester's reaction: "Hornberg said that he never thought corn could be cultivated with a tractor, but now he thinks it can be, and much faster. However, he is not yet convinced that the three-row idea is right."

Exhibit 10-2a. Three-row cultivator on prototype Model C / GP being tested on August 6, 1926, by Theo Brown and the Deere engineering team. Brown wrote, "McCray reports that the shorter the wheelbase, the less drift there is to [the] side." *Theo Brown Collection*

Exhibit 10-2b. This is the prototype Model C / GP in the field being tested in August 1926. Theo Brown referred to the model as the "All-Crop" in his notes, writing that "we had the basis of a real cultivator." *Theo Brown Collection*

The prototypes were also tested by a Decatur, Illinois, farm family, C. C. Veech and his sons. The former put the problem of the three-row cultivator bluntly: "Never believed in a two-row cultivator as a man had all he could do to run and watch a single row . . . could see the tractor and planter but not the cultivator . . . the boys could try it, but did not expect to let them cultivate a whole row but would have to tell them to take it out . . . had horses ready to go and do the job. However, the boys cultivated one row and then another, and then told the boys to go ahead and cultivate with a three-row . . . frankly surprised to find out how well the three-row did . . . would say, as did sons, that the cultivator seemed to be okay."

Wiman agonized over what was obviously a calculated risk—that the concept of three-row cultivation would not be effective. The alternative—to scrap the prototypes and begin afresh—was unpalatable. Wiman warned the board of the opportunity costs of this: "Re-design of the Model 'C' to a so-called little Model 'D' tractor would keep us two years away in point of time from production on such a cheaper two-plow tractor. . . . It seems that our best opportunity, considering everything such as time, economy of manufacture of one model only, supplying to our dealers a reasonably full line of tractors, would dictate that our best procedure now was to manufacture the Model 'C' as a two-plow tractor and should the all-crop feature fail, then proceed to lay out a little Model 'D.'"

If the new model was to be built, space would be needed either at Waterloo or elsewhere. If only existing space was used at Waterloo, the 100-tractor-a-day schedule for the Model D would have to be cut back—and just at the point when demand for the D was going up. If the Model D schedules were to be maintained, probably even expanded, then separate additional factory space would be needed for the new model. Wiman estimated the cost of additional space and of its tooling to be more than $3 million, but he pointed out that "even though we should go so far as to abandon the All-Crop idea and go to a small D machine, all of the machinery and tools would be useful, except possibly 10 percent of the original appropriation." It was the time loss in such an abandonment that would be costly.

After extensive testing over the spring and summer of 1927, considerable questions still remained about whether the three-row cultivator was the right configuration. C. C. Webber wrote from Minnesota: "It worked better than he expected it would work, and you assure us that the men who are running the tractor find it easier than driving a two-row cultivator." Webber then added an upsetting observation: "You have not the power at the present time to pull this three-row cultivator under all of conditions—that we presume you realize."

The question of the tractor's power had made everyone uneasy right from the start, and doubly so when McCray reported in late 1927 about a test on hilly ground. "Straight up on side hills Model C was okay, but sideways not as good as IHC 10-20." Theo Brown confided to his diary: "At meeting in afternoon it was decided to go ahead with Model C, having in mind that it was not equal in power on hills to IHC 10-20. Also, it was decided to design on paper a two-bottom tractor, simply as a plowing job that could be sold for $775 to the farmer. Silloway and Lourie are strong for the General Purpose tractor. Minneapolis wants a two-bottom plowing tractor only."

Frank Silloway, speaking from the sales manager's perspective, argued persuasively for building the full component of the Ds and Cs, and doing it in Waterloo. Others on the board, notably the often-conservative George Crampton, argued against it: "I am very much averse to putting any more money in Waterloo for brick and mortar."

C. C. Webber had the final word: "The tractor business is so definitely connected with the other tools in our line that I think we must protect the interests of stockholders and be prepared to supply the demand. We have suffered very much by not having sufficient tractors this year. We must meet the situation and I would favor the program." Wiman, still harboring many personal misgivings about the new tractor, moved that $3.9 million be expended for new buildings at Waterloo and the board passed the measure without further argument. The die had been cast.

Silloway now decided to rename the tractor and at first considered a popularized name—two suggestions were "Powerfarmer" and "Farmrite." He worried about detracting from the Model D, however: "If we were to name either tractor 'Powerfarmer,' we would give that name to our real power farmer tractor, the Model D. . . . We do not care to popularize with a name such as 'Powerfarmer,' the smaller tractor on which we make little profit, as against the Model D which is a profitable tractor for both the dealer and ourselves to sell and the one that most farmers should buy."

Silloway finally decided against a trade name and told the branches why: "Some of the branch houses have objected to Model 'C'. . . . 'D' and 'C' both have the sound of ee. When the dealer orders a tractor over the telephone, it would be very easy for a misunderstanding to arise because of the similarity in sound." Silloway chose the initials "GP," to stand for "general purpose." In the process, the term "all-crop" was discarded.

By October 1928 the production of the GP tractors was up to about twenty-five per day. But at this point Wiman reported to the board about an unsettling development: "For some unknown reason the horsepower had not been up to expectations. It was hoped that the tractor would develop at least 25-horsepower in order to give proper feel of performance but the dynamometer test has indicated that some of the tractors have developed only from 20 to 22 horsepower. The last 300 tractors show on dynamometer tests, 22.6 HP (belt)." Brown was with Wiman in that week and reported: "He is *very blue* because there has been complaint that the GP tractor is short of power. It seems to me this should be rectified as soon as possible. He is discouraged (too easily) and thinks we should build a regular two-plow tractor. All this is disturbing." Wiman even brought the issue to the board, telling them that he had "given some thought to the proposition of making a straight two-plow tractor to be known as the Model B to either replace the GP or be an addition to the line. Such a tractor would have many parts in common with the D but would only pull a two-bottom plow."

Wiman had been the hell-for-leather advocate for the combine, and at that time Webber had demurred. Now the roles seemed reversed. Webber urged the board: "Regardless of the consideration being given to the Model B or two-plow tractor, no efforts should be spared towards completing the job with a GP tractor as this . . . [is] absolutely necessary before going to the two-plow machine." Wiman finally deferred to Webber's judgment.

As if this was not enough, horsepower concerns relating to the Model D now surfaced. The D's horsepower had been improved a bit in 1928 with the addition of a 1/4-inch increase in the cylinder bore and an improved carburetor, resulting in an increase in the drawbar horsepower to about 28, and 36 horsepower on the belt. Unfortunately, Wiman reported, "No more horsepower could be built into our present Model D tractor without a pretty thorough redesign throughout. . . . [T]his tractor is now up to its limit of strength and stability with the horsepower delivered to the drawbar as at present. . . . If the I.H.C. is able to beat us in the field on field performance . . . we will have to rely on our selling of these machines in volume on the greater simplicity of the Model D tractors, its two-cylinder construction and its reputation as a low-cost machine from an economy standpoint and from the non-use of a large number of repairs."

CHAPTER 10

The horsepower fears were exacerbated when in January 1929 Deere learned that International Harvester had also incorporated a 1/4-inch increase in the bore of the cylinder of their 15-30 tractor, which it was rumored would add anywhere from five to twelve horsepower to the machine. This worried Deere not only because of the disappointing horsepower of the GP, but also as a direct assault on the Model D. Inasmuch as the D was the large-production star of the Deere line, any threat to its sales was even more serious. Wiman reported to the board: "It is unfortunate that this vicious circle or race for horsepower has again been started, but it is not strange when you consider the fact that for the last two years we have been able to outperform the International 15-30 machine, that they should increase the speed and power on their machines to equal or better our own. To sell their old 15-30's on the territory, the International Company have cut the price of this machine $100 and we believe this will have to be cut further yet."

With the renewed threat from the 15-30's increased horsepower, Silloway decided to reemphasize the competitive advantages of the Model D and issued a bulletin headed, "John Deere, a two-cylinder tractor." The memorandum bluntly denied that Deere was thinking of shifting to a four-cylinder tractor. "Why should we?" Silloway wrote, "The John Deere two-cylinder tractors will do plowing and other field work at the lowest possible cost per acre and, after all, that is what the farmer is interested in." Silloway emphasized the light weight that led to fuel economy, the extreme simplicity, with its low cost of upkeep, and the fact that the owner could do his own servicing.

Detractors of the two-cylinder concept often used as their argument the excessive vibration of a two-cylinder motor. But Deere counteracted this attack by sending a Model D tractor to fairs, mounting the machine on four pop bottles, and putting the tractor in operation with the rear wheels turning. There was not enough rhythmic vibration to shake the machine off the mouths of the bottles (exhibit 10-3).

Wiman's skepticism about the GP was soon confirmed by reactions from farmers. To start with, there were a number of breakdowns in operation, requiring rebuilding of some of the tractors in the field. Even more important was the farmer reaction to the three-row cultivating feature. While many farmers in the cornbelt seemed to accept the notion reasonably well, the cotton growers in the South clearly preferred two- and four-row operations. There were also inherent design problems; one was visibility for the driver on the GP. "It seems to me that the view is the real problem now. The Farmall has the best of us there," noted Theo Brown. Wayne H. Worthington, in his definitive study of the agricultural tractor for the Society of Automotive Engineers, succinctly summed up the reputation of the GP: "Unfortunately, the three-row idea failed in its acceptance and the tractor fell far short of Model D performance and durability."

Within a year, Brown and his development team were back at the drawing boards trying to rectify the GP's problems. A crash program soon

Exhibit 10-3. "Pop bottles bear witness to balance." *Farm Implement News, November 21, 1929*

brought into production a variation of the GP—the GP Wide-Tread. In many respects it was similar to its predecessor except that it had a long rear axle that allowed the machine to straddle two rows and the first John Deere tricycle front to run between two rows, just as the Farmall did. By the crop season of 1929, the GP Wide-Tread was available to buy and its acceptance not only in the South but in the Midwest was more gratifying. In 1931 the GP itself was modified to raise its horsepower ratings to 15.52 drawbar, 24.30 belt-pulley. A year later, in 1932, the GP Wide-Tread was modified to provide a tapered hood to increase the visibility of the driver and to provide changes in the steering apparatus that prevented the tendency of the front wheels to whip on rough ground. With the second model of the GP, and the modifications and additional features, the GP finally made a respectable showing in the Deere line. Charles Wiman always remembered his experience with the GP, however, as "an outstanding failure" and often mentioned: "Well do I recall how much tractor business was lost by our company due to bad design of the 'GP' line."[4]

RESEARCH AND DEVELOPMENT IN THE ORGANIZATION

In the process of masterminding the new product developments of the mid-1920s, Charles Wiman came into his own within the company's board of directors. His engineering training at Yale University's Sheffield School had given him the substantive knowledge to understand the complexities of technology all across the spectrum of products, from combines and corn pickers to tractors. Wiman spent a great deal of time with his product engineers during this period, and he was constantly out in the field at the test sites watching performance ("Driving his Stutz back and forth," Theo Brown often reported in his diary).

Driven by a sense of urgency about these new products, he sometimes leaned quite hard on his staff for results. Brown, who as head of the experimental department was himself working on a number of innovations in cultivating tools, seemed sometimes to resent this pressure: "These are busy days and I have to be careful to keep my feet on the ground. There seems to be a tendency for others to want to do redesigning of the cultivator to some extent. Also Charley Wiman is pushing hard to have this new outfit shipped Wednesday. We must do things right and then fast, but right fast." At several points in Brown's diary he railed at Wiman's intrusions into the development process; "all this is very annoying," he commented at several points. Yet Wiman also believed in high standards of quality and, while exhorting his design people for faster and faster work, still was willing to test the products until they were truly ready. "We have had bitter enough experience in the past," Wiman wrote Charles Stone during the high-tension days of the combine development, "in getting into production too soon without a settled, proven-up, practical, from a field standpoint model."

In the early years of the modern company, product development at Deere had been quite decentralized, and this remained the policy during the 1920s. Theo Brown, heading the experimental department, presumably had legitimate interests across all of the product lines. The department had indeed been accorded heightened prestige in the 1920s; its budget had increased markedly throughout this decade (jumping 30 percent in one year alone, in 1923). Yet, the experimental department was never really in charge of a complete product—its engineers were more concerned with new breakthroughs in components, the kind that could be patented. The factory product engineers sometimes sneered at the experimental people as "never having to innovate under budget constraints."

But the product development people in the Harvester Works, in the John Deere Tractor Works, and in the other factories knew where the real power lay—it was right in their own factories, where they made all the new completed machines. And they jealously guarded their prerogatives. When

Stone was just beginning the combine push, in 1925, Brown went to the harvester plant to see if he could give some help. "It is a rather delicate task," Brown wrote in his diary, "and so I want to make the right start. . . . Charley Wiman wants me [to] take on gradually the general company job of being a sort of consulting engineer for the various factories in the line of experimental work."

Even Wiman appeared not to be able to force the factories to share their experimental and product development efforts very widely. The factory managers were a strong, independent cadre, brooking no interference from within or without. The 1920s and '30s became an era of long-tenured, dominant personalities in those roles, with the seniors passing reins to equally strong juniors, thus perpetuating the mystique of power. In some of the smaller factories, a single individual could hold sway for many years—Herman Moschel at the Dain plant in Ottumwa from 1919 to 1947; F. H. Clausen heading the Van Brunt plant at Horicon from 1911 to 1943. Virgil Bozeman led the company's Wagon Works from 1924 to 1949; Carl H. Gamble ran the Spreader Works from 1919 to 1947.

It was at the mammoth bread-and-butter factories—plow, planter, harvester, tractor—that the pattern of powerful leaders was most apparent. There a coterie of works managers passed along their tricks of the trade to their handpicked successors, in the process creating legends about the "power of Harvester," the "intransigence of Waterloo," the "tight control at the Planter Works," and so forth. Charles Stone headed the Harvester Works from 1923 to 1934, going on to become the vice president for all tractor and harvester production. In the process, Stone was probably the closest confidant of Charles Wiman among any of the factory and branch managers. When Stone stepped up to corporate-wide responsibilities in 1934, L. A. Paradise, another able leader, assumed the Harvester Works post for another twelve years. At the Waterloo Tractor Works, A. H. Head held the reins for a full eighteen years (1919–1936), during a seminal period when seven basic models of John Deere tractors were spawned. His successor was L. A. "Duke" Rowland, one of the legendary figures among manufacturing men in the company; when Charles Stone relinquished his line assignments in 1944, Rowland then headed all tractor manufacturing. At the Plow Works, where company machines had been built continuously since John Deere brought his shop from Grand Detour in 1848, Benjamin Kough headed the operation from 1919 to 1936, and he, too, was succeeded by a company legend, L. A. "Pat" Murphy, who served until 1942 and then took over responsibility for all implements. Finally, at the Planter Works, H. B. McKahin held the number-one spot from 1917 to 1937; he was succeeded by Harold White, the manager who was consistently the lowest-cost producer (or near to it) all through his tenure. White held his post there until 1949.

Right in this group of people lay great abilities, great prestige and power, and great egos. They guarded every piece of their company "turf" jealously, nowhere more assiduously than in product development. They were secretive about their plans, often even to the board, and certainly to any other factory or branch. This high degree of decentralization in product development had many positive assets—it encouraged responsible engineering, for the factory had to build the dreams of the development engineers. But it often made the factories recalcitrant and unbending when presented with a new product challenge from one of the sales branches. The latter, of course, did not always realize the need of the factories for steady production of long runs.

A further problem began to intrude more and more in factory and branch relations in the 1920s, triggered in particular by the advent of the GP tractor. In the past, an implement factory in the company would assume that it could proceed independently, with little need to check other units—after all, most of the implements would be pulled by a horse, with tractors mostly for plowing. When the general-purpose tractor achieved almost instantaneous success, however, it became considerably more important to coordinate tractor and implement. So, too, with the combine, and indeed with many aspects of so-called "power farming" (the term that became popular in the 1920s). It was a sobering experience for Charles Wiman and his colleagues to realize that the GP tractor was now "driving" many other product developments that had formerly been almost autonomous. Moreover, they saw that the high costs of tractor development and tooling made it imperative that good decisions were made at the start and that everyone involved—implement makers as well—participated in the decisions. This realization led the company to convene all factory managers for the Power Farming Conference of February 1929. Over his years as chief executive officer, Wiman often used simple adages to cut through the maze of detail characteristic of large organizations, and one of his favorites—"Every horse looks fast running by itself"—had particular relevance here. Wiman was blunt in his keynote speech at the conference—he wanted everyone to "thoroughly understand" the need for coordination, and he insisted on the establishment of a "clearing house" of ideas. He stressed the ties between manufacturing and sales, making clear in the process that "sales are made in the designing rooms of the factories. . . . You gentlemen here are really the force behind the sales force."

The conference still left intact just about all of the personal power of the factory managers—this was not an assault on decentralization. Product development was still a close-to-the-belt decision, held by the individual managers probably to excess. Indeed, it was not until the early 1970s that certain research and development, as well as product engineering decisions and activities, were centralized. Yet the need for greater coordination was clear and the Power Farming Conference made the first halting step in that direction.[5]

DEERE AND INTERNATIONAL HARVESTER PULL AHEAD

The decade of the 1920s was a watershed for the agricultural machinery industry. First, there were major consolidations. In 1928 the Canadian-based Massey-Harris Company took over the J. I. Case Plow Works in Racine, Wisconsin; Massey immediately sold its rights to the "Case" name to the J. I. Case Threshing Machine Company, also in Racine, and kept for itself similar rights to the prestigious Wallis tractor. In turn, the J. I. Case Threshing Machine Company bought the farm implement business of the Emerson-Brantingham Corporation—the old-line Rockford, Illinois, implement manufacturer that had been in continuous operation since 1852—and then changed the combined company's corporate name to J. I. Case Company, Inc. The next year, 1929, saw another new company organized—the Oliver Farm Equipment Company, a merger of the Oliver Chilled Plow Works of South Bend, Indiana; the Hart-Parr Company of Charles City, Iowa; and the American Seeding Machinery Company and the Nichols & Shepard Company of Battle Creek, Michigan. Oliver, continuing to make its own chilled plow (of which it was a pioneer), now added grain separators and farm tractors. In this same year yet another new long-line company came into being, the Minneapolis-Moline Power Implement Company—a consolidation of the Moline Implement Company, the Minneapolis Steel and Machinery Company (builder of the Twin City tractor, the machine that briefly had been sold by Deere in the 1910s), and the Minneapolis Threshing Machine Company (also a builder of tractors). In this same period, Allis-Chalmers Manufacturing Company of Milwaukee, Wisconsin, added to its tractor line by purchasing the business of Monarch Tractors Corporation of Springfield, Illinois, manufacturer of a track-laying tractor, and then becoming a full-line manufacturer with the acquisition in 1931 of the Advance-Rumeley Thresher Co. of LaPorte, Indiana.

In 1921, there had been 186 tractor companies; by 1930, the number had been reduced to 38. Whereas there had been many dozens of tractor companies and many hundreds of implement companies in earlier years, now there were just seven main companies offering full lines—the "long-line" companies were Deere, International Harvester, Case, Oliver, Allis-Chalmers, Minneapolis-Moline, and Massey-Harris.

A second stage of realignment came when every one of these newly combined companies embarked on the quest for its own general-purpose tractor: in 1930 Oliver introduced its new tricycle Row Crop, Massey-Harris came out with a four-wheel-drive general-purpose model, and Allis-Chalmers announced its All Crop general-purpose tractor. The playing field of tractor competition was fast becoming crowded!

CHAPTER 10

Deere, meanwhile, had not only bought two small companies (the potato harvesting equipment of the Hoover Manufacturing Company and a threshing line, purchased from the Wagner-Langemo Company), but had added more than $7.6 million to existing capacity in its factories and branches. By 1930, the company had opened new branches in Fort Wayne, Indiana; Amarillo, Texas; and Edmonton, Alberta, and had also constructed warehouses in Lansing, Michigan; Sidney, Nebraska; Aurora, Illinois; and Swift Current, Saskatchewan—quite a change from Charles Deere's earlier marketing philosophy of a few large branch houses.

The 1920s witnessed major shifts in market position. The long-line companies picked up increasing percentages of the market for most of the key products. Deere and International Harvester were particularly strong; together they dominated most product categories, with Deere moving up faster in many of these. A Federal Trade Commission report made these public (Appendix exhibit 22).

The reasons for dominance by the larger, long-line companies, in the opinion of both industry and outside analysts, lay particularly in the strong marketing structures of these leaders, with the branch house as the linchpin. Warren Shearer's comprehensive study of the industry puts particular emphasis on the 1919–1929 period as "the formative decade." Shearer broke down the extensive documentation in the FTC report (cited above) to show that there were indeed sharp differences among the long-line manufacturers in regard to both manufacturing costs and marketing costs, with the latter particularly striking. He used 1929 for comparative purposes, as shown in Appendix exhibit 23.

Shearer's study confirmed the striking strengths of Deere and International Harvester in the 1920s. He concluded, "It has never been charged that IH and Deere had power to restrict production or set limits on the output of their competitors. It is equally true that their pricing policies were not of a sort to discourage competition except in the sense that such competition was unable to meet Deere and IH prices and still make a profit. If we are to discover any control over the market exercised by these two acknowledged leaders it must be found in the distribution system which characterized the industry. . . . If we neglect the possibility that these two concerns produced a markedly superior product, and it is fair to assume that this was not the case, the great strength of IH and Deere must have been their dealer organizations or the control they exercised over these organizations."

There were technological product breakthroughs for certain products, for example, the general-purpose tractor of International Harvester. Still, Shearer's judgment corroborates the findings of a number of other analysts: it was the large, strong long-line companies that were able to develop comprehensive distribution systems and use the branch-house concept to its fullest, and thus to develop a marketing relationship with both dealers and farmers that gave them greater strength than the smaller long-line companies. Most of the short-line companies were not able to mount extensive dealer networks that included branches (though some were able to maintain their market share by virtue of patent protection).[6]

The strikingly low selling costs of the three industry leaders (Deere, International Harvester, and Allis-Chalmers) shown in Appendix exhibit 23, testify to the substantial economies of scale in the larger branch-house systems. Yet, these three were also serving their dealers, and the latter serving their customers, the farmers, in a more effective way than the other long-line companies. Deere, in particular, never forgot the lesson that John Deere had learned so well—it was the farmer who would be the ultimate arbiter of success for the company. Deere had always been keenly attuned to the needs of the farmer and had been able to inculcate this in its dealers. The company's quality products were assiduously advertised and promoted, always with the customer's need uppermost. Deere had pioneered with a farmers' magazine, *The Furrow*, first published in 1895 and by the 1920s one of the country's preeminent farmers' publications. At the end of the war, its supervision had been assumed by H. M. Railsback (who had joined the company in 1917 to head advertising and publicity). Deere's catalogs were themselves straightforwardly service oriented, chock full of the kinds of information the farmer demanded; several of the company's service manuals themselves became classics (one, *The Operation, Care and Repair of Farm Machinery*, went through twenty-eight editions). All of these printed sales efforts were complemented by a group of sales-force personnel in both the branches and the dealerships, knowledgeable about their products and thoroughly dedicated to the longstanding Deere pride in its close relationship with the farmer.

COMPANY STORES AND PRICE MAINTENANCE

There were tensions between manufacturer and dealer in this period, and they focused particularly on two issues—the number of competing dealerships (with the exasperating issue of the company-owned store), and the maintenance of price levels. When the court decree in the International Harvester case in 1920 limited the firm to one dealer per town, it became much more likely that Harvester would exert pressure on its dealers to carry the complete line and give it "adequate" representation. Deere, too, knew it had to limit its own dealerships to one for each town, but Frank Silloway saw no problem; he felt that this would lead to strengthened dealers. International Harvester and Deere (as well as the other long-line companies) did not explicitly require exclusive dealerships, but, as Shearer pointed out: "There can be little question that this injunction was and, to some extent, is ignored by salesmen in efforts to increase their volume in business." The concern by dealers that they were being pressed to accept exclusivity remained high on their agenda of complaints.

Whether there were too many dealers in a locality was another issue that antagonized the dealers. Particularly threatening was the sharp expansion by the manufacturers of the company-owned store. The major long-line companies all had owned retail store properties from time to time; sometimes the only way to get an operation started in a new area was to purchase a property and lease it to an independent dealer. Deere had opened a small number of these before the establishment of the modern company in 1912. (Indeed, Deere was reputed to be the first agricultural machinery manufacturer to use them.)

But what if the manufacturer chose to continue operation of the store itself? The dealers held a simplistic view that they could not compete with a manufacturer's outlet because they themselves had to use local capital, whereas the factory-owned outlet was primarily concerned with volume for the factory and therefore would not worry too much about return on capital invested in the local outlet. Manufacturers were also accused of using excessive trade-in allowances, extended payment dates, and below-market interest rates as devices for maintaining volume, all subsidized by the factory itself. A John Deere dealer in Oklahoma complained: "I believe the implement business with Hi-Power factory salesmen and sales-promotional policies, with excessive trade-ins, extended payments dates and low interest rates will break any independent dealer. I expect to see implements sold through factory-branch stores only in the next few years." Another Deere dealer, in the state of Washington, reiterated the same theme: "If the farmer is smart and plays one [factory-owned outlet] against the other [independent outlet] he can get his machinery for practically his own price. They extend long credit terms to people who do not have any credit standing in the community. I suppose they can do this because of their legal help and collectors they employ in the field."

The manufacturers argued that a few company-owned stores gave them a better feeling for problems at the retail level and provided an excellent training ground for younger executives. It was left unsaid that the potential for opening new stores served as an implied threat to independent concerns that were themselves not too efficient. R. B. Lourie, Deere's Moline branch manager, cleverly put this point across in a speech before an implement dealers' convention in Illinois in 1927. "You dealers have a slogan, 'To the Retail Implement Dealer Belongs the Retail Trade.' I agree in principle with this doctrine, but thinking straight on this subject, does it not predispose that the dealer should be efficient and that he is to give proper service to his customers and to his suppliers?" Lourie hinted, not too subtly, that if the dealers themselves did not prove efficient, the companies would take over for them: "In business there is always a survival of the fittest. The inefficient man is going to be eliminated and the unneeded man is going to be eliminated.

Exhibit 10-4. Top, Road mowers for shipment, Harvester Works, ca. 1929; bottom, Twelve farmers ready to drive grain drills from dealer's store, Casselton, ND, ca. 1930. *Deere Archives*

CHAPTER 10

The implement jobber, for example, has been very largely eliminated and for reasons that you gentlemen perfectly well know. There are going to be some implement dealers eliminated. The retail implement establishment of the future is going to be one that will have a suitable organization and proper management, so that it will render prompt and efficient service and on a basis that will produce profit for the owners."

By the time of the next Federal Trade Commission investigation of the industry in 1938, the issue of factory-owned stores was a major one. By then, Deere had more than one hundred such stores. The company made a formal statement to the FTC about this issue, noting that they had always attempted to put independent dealers in such stores but in cases where they could not find one, "and the community was too important to leave without a dealer representation, Deere & Company, instead of permitting the liquidation of the business, bought it out and continued its operation as a John Deere retail store." Even by this time, though, the company was not enthusiastic about expanding the concept of retail store management and vowed to the FTC: "We are not interested in retail stores and prefer to transact our business through independent local dealers. All these retail stores now owned by us as listed in Section A are for sale as soon as we can find a suitable independent dealer to purchase same."

The dealers also wanted protection from each other when competition seemed too stiff, particularly when price cutting became endemic. The early 1920s, with so much farmer resistance to price, revived price-cutting tactics, and the dealers appealed to the manufacturers for resale price maintenance or, to use the terminology of that day, a "list and discount" method of doing business. The dealers wanted manufacturers to establish a fixed retail price, quoted widely, and then police its maintenance in order that there not be "unfair" competition on the part of dealers shaving the price. The manufacturers, while interested in maintaining retail prices congruent with their own set prices to their branch-house and jobber price lists, were reluctant to see fixed prices on most implements. Only in the tractor market was the list price common (mirroring the approach of the automobile industry).

Both the federal government and the states seemed genuinely ambivalent as to whether to allow or perhaps even support resale price maintenance. One state, New Jersey, had legislation on the books supporting the concept; a number of other states had statutes, usually generally worded, prohibiting price fixing in general terms. At the federal level, as there was no specific legislation relating to resale price maintenance, the law was the product of judicial interpretations under the common law and under the Sherman and Clayton acts and the administrative rulings of the Federal Trade Commission. These administrative rulings of the FTC became the key government constraint. In the 1920s, the FTC promulgated a set of administrative orders against resale price maintenance, based upon an earlier landmark case, a suit between the Great Atlantic & Pacific Tea Company and the Cream of Wheat Company (1915). The FTC views were subsequently modified by the courts, again leaving the issue nebulous. Implement manufacturers were to be able to "request" their dealers not to resell at less than a stated minimum price and were even to be allowed to refuse to sell to a customer if that customer resold on a basis below the requested minimum prices. Further, manufacturers could use "reasonable," though not "cooperative" or "coercive," methods in collecting information about price cutters and in working with them. On the other hand, they could not demand a contract or written agreement about resale prices, nor could they use such methods as group action, licensing, or other cooperative action to coerce dealers. So assiduous persuasion of dealers for price maintenance was countenanced, provided it was not coercive, secret, or involving group sanctions.

Deere now felt in limbo. For years it had been mailing its dealers copies of a retail price schedule, prepared so that dealers could quote such retail prices to farmers who happened to call directly at a branch house. Frank Silloway and the others in the sales organization were careful to point out that these were suggested prices, but not required by the company. Earlier, other manufacturers had not followed such a practice, but when International Harvester decided to do so in 1928, Silloway issued a major memorandum, reiterating Deere's policy: "Of course, according to existing laws we cannot insist upon dealers adhering to any Retail Price Schedule which we make. We, however, can take a fatherly interest in the retail price at which our goods are sold. We believe that from now on all implement manufacturers will take a greater interest in the welfare of the local implement dealers and that in the future we will have cooperation in this endeavor, which for many years we have been obliged to carry on alone. We believe that this cooperative effort will make the retail implement business more profitable, which will have a beneficial and salutary reaction all along the line." The implement business had recovered very markedly by this year, 1928, and many competitive pressures, such as price cutting, were not as pronounced. Thus retail price quotations could be more readily maintained. Later, in the period of the Great Depression, the situation became more difficult again.

The company did have a brush with the government on pricing in 1926, when the Federal Trade Commission brought suit against the National Malleable Castings Association, also involving one hundred of its members (all of whom were made defendants in the case). Union Malleable Iron Company was one of these, and its manager, John Simmon, was also made a defendant. The case was to come to trial in September of that year, but a number of the members of the association decided to plead nolo contendere and take fines ranging from $3,000 to $6,000. Burton Peek found no evidence that the subsidiary had participated in any of the alleged price fixing. "If this were a case against the Union Malleable Iron Company alone, I would advise contesting it." As a number of Malleable's sister companies were involved, however, Peek felt that "our situation is affected by our relations to the other defendants" and he advised that Deere also enter such a plea and submit to the fine of $3,000.⁷

A NEW LEADER

In June 1928, Butterworth was elected president of the United States Chamber of Commerce. (He was also mentioned as a possible presidential candidate to succeed Calvin Coolidge, just as he had been mentioned in 1924 as a possible running mate for Coolidge.) The Chamber of Commerce post was to be a demanding one in time and travel, and Butterworth decided to transfer the mantle of operating management to Charles Wiman. In October 1928 Butterworth proposed the creation of a new position of chairman of the board, which he himself would assume. In turn, Charles Deere Wiman was elected the fourth president of the company, thus keeping family leadership intact at Deere & Company. Butterworth told the board that "this would give him an opportunity to work with Mr. Wiman in his position as President, and at the same time get him acquainted with outside work which has to do with the financial side of the business." The bylaws of the company were amended to incorporate the new post; the role of chairman was to be as "ranking officer of the Company" and it was to have "the duty of formulating the policy and of directing the operation of the Company." The president, in turn, was to be "the executive officer of the Company" and was to act for the chairman when the latter was not available. Though the designation "chief executive officer" was not given to either of the two men, it was obvious from that time forward that Charles Wiman was taking the major role in decision making for the company.

The decade had opened for Deere & Company with two years of heavy losses and considerable difficulties in recovering momentum. The decade closed, however, with three years of resounding success—the last year and a half under Wiman. US agriculture as a whole had had an excellent year in 1925, a mild depression in 1926, and then three more excellent years; most of the machinery manufacturers had done well in these years. The net sales at Deere had climbed steadily from just over $28 million in 1923 to more than $76 million in 1929. The years 1927–1929 had highly profitable margins, particularly striking in comparison with the rest of the industry; in 1929 Deere had a return on investment of 26.7 percent, with International Harvester at 17.9 percent and the rest of the industry much below (Appendix exhibit 24).

Deere's spectacular figures for 1929 were published in the annual report dated October 31, 1929—just two days after the Tuesday when 16 million shares were traded at the New York Stock Exchange in a frenzied sag in the stock market that was one of the triggers for the Great Depression. Survival during an economic crisis became the challenge facing Charles Wiman, just over a year after he had assumed the presidency of Deere & Company.

Endnotes

1. William Butterworth to Charles Wiman, January 28, 1924, DA, 4153; ibid., March 21, 1923, Deere Scrapbook; Butterworth diary, December 31, 1924; Butterworth to Frank Silloway, Charles Wiman, and T. F. Wharton, January 1, 1925, Deere & Company *Minutes*, November 3, 1925. For the Floyd Todd quotation, see ibid., July 27, 1926; for the election of Charles Wiman as president of the company, see ibid., October 30, 1928; for the quotation on evening the production of tractors, see ibid., July 28, 1925; for contractors' bonus, see ibid., July 31, 1928. For William Butterworth's mention as a possible presidential candidate, see *Manufacturers News*, January 1928.
2. For nineteenth-century history of the combine, see Graeme Quick and Wesley Buchele, *The Grain Harvesters* (St. Joseph, MI: American Society of Agricultural Engineers, 1978), 86–90. See also Lillian Church, *Partial History of the Development of Grain Harvesting Equipment*, United States Department of Agriculture, Agricultural Research Administration, Bureau of Plant Industry, Soils and Agricultural Engineering, Information Bulletin 72 (October 1947), 45–52; F. Hal Higgins, "The Moore-Hascall Harvester Centennial Approaches," *Michigan History* 14 (1930): 415–37. For the soybean test on the Garwood farm in Stonington, Illinois, see *Farm Implement News*, November 13, 1924; ibid., February 12, 1925; ibid., July 23, 1925. The question of humid-condition constraints on the combine is discussed in ibid., August 26, 1926; ibid., March 29, 1928; and ibid., April 18, 1929. See also ibid., July 28, 1927; ibid., March 31, 1927.
3. The initial Charles Stone memorandum on the harvester project is dated October 23, 1947; see Deere & Company *Minutes*, October 28, 1924. For Stone's memorandum on plans for 1927, see ibid., April 26, 1927. See C. C. Webber to C. Stone, October 1, 1923, and ibid., March 11, 1925, DA, 34818, 34827. Original contact with the Caterpillar Tractor Company had been made in 1925; for rejection of the offer of $1,256,000 by the Deere board, see Deere & Company *Minutes*, June 7, 1927; for Silloway's report on the acquisition of the Wagner-Langemo plant, see ibid., October 29, 1929.
4. For the decline of the Fordson tractor, see Nevins and Hall, *Ford: Expansion and Challenge;* Wik, *Henry Ford and Grass-roots America*, 97; A. C. Seyforth, "Tractor History," International Harvester Company, DA, 13864 and 14238. The need for a "general purpose tractor" is discussed by F. A. Wirt in *Farm Implement News*, April 17, 1924; International Harvester's Farmall was introduced in ibid., April 29, 1926; see also ibid., February 10, 1927. For Theo Brown's involvement and McKahin's quotations, see "Theo Brown Diaries," February 9, 1926, and ibid., June 2, 1926, DA. For Wiman's quotations, see Deere & Company *Minutes*, July 27, 1929, ibid., April 26, 1927. For Veech's quotation on trials, see "Theo Brown Diaries," July 7, 1927. See also ibid., July 5, 1927; ibid., October 12 and 20, 1927. For Webber's quotations, see Deere & Company *Minutes*, February 20, 1928; ibid., March 26, 1928; and ibid., April 24, 1928. For Silloway's description of nomenclature, see branch-house bulletin 650, June 20, 1928. For Webber's "absolutely necessary" quotation, see Deere & Company *Minutes*, October 30, 1928; see also "Theo Brown Diaries," October 23, 1928. For the discussion of Model D horsepower, see Deere & Company *Minutes*, January 29, 1929; branch-house bulletin 665, August 28, 1929. Wayne H. Worthington, *50 Years of Agricultural Tractor Development* (Farm, Construction and Industrial Machinery Meeting, Society of Automotive Engineers, September 1966; republished by American Society of Agricultural Engineers: St. Joseph, Michigan). For Wiman's quotation on bad design, see his speech to the Deere Factory Managers Conference, April 17, 1953.
5. Theo Brown quotations are from "Theo Brown Diaries," January 28, 1925; ibid., January 30, 1928; ibid., March 12 and 16, 1928. For quotation on "bitter experience," see C. Wiman to C. Stone, October 10, 1925, DA. The first Power Farming Conference was held on February 7–8, 1929; the Wiman quotations are from its "Proceedings," DA, 663.
6. See Federal Trade Commission, *Report on the Agricultural Implement and Machinery Industry* (1938); Warren Wright Shearer, "Competition through Merger: An Economic Analysis of

CHAPTER 10

the Farm Machinery Industry" (PhD diss., Harvard University, 1951), 122–34. The mergers involving Massey-Harris, Allis-Chalmers, Case, Oliver, and Minneapolis Moline are also chronicled in Gray, *The Agricultural Tractor*, 15–21.

7. See Frank Silloway's report on the single dealer per town, Deere & Company *Minutes*, September 21, 1920. International Harvester's retail stores are discussed in *Farm Implement News*, August 3, 1927, and ibid., September 14, 1927. For the quotation on Deere's policy regarding company-owned stores, see Federal Trade Commission, *Report on the Agricultural Implement and Machinery Industry* (1938), 120. R. B. Lourie's speech is reported in *Farm Implement News*, December 5, 1927. Retail price maintenance constraints by the federal and state governments are discussed in Ewald T. Grether, *Price Control under Fair Trade Legislation* (New York: Oxford University Press, 1929), 14–17; A. D. Neale, *The Antitrust Laws of the United States of America* (Cambridge, UK: Cambridge University Press, 1970), 272–99. For the "list and discount" suggestion of the National Federation of Implement Dealers Association, see *Farm Implement News*, November 5, 1925; for the vote of the association, see ibid., January 20, 1927. The Frank Silloway memorandum on "retail prices" is dated September 24, 1928, branch bulletin 653. His bulletin 616 of August 13, 1926, discusses the legal parameters of the Federal Trade Commission rulings. See *Great Atlantic and Pacific Tea Company v. Cream of Wheat Company*, 224 Fed 556 (1915). For Deere's position in the National Malleable Castings Association case, see Butterworth's remarks in Deere & Company *Minutes*, October 25, 1921; ibid., April 16, 1923. For George Crampton's remarks, see ibid., June 25, 1922; ibid., November 25, 1924. For the C. C. Webber quotation, see ibid., October 25, 1921. See also George Crampton to Frank Silloway, September 1, 1923, DA, Scrapbook.

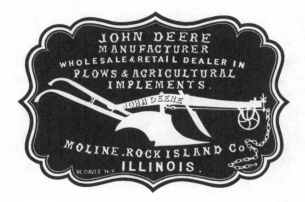

CHAPTER 11

DEPRESSION YEARS

You had a deposit in that bank of about two million dollars when this trouble occurred, and you had it there for some time . . . It is this big Deere & Company deposit that has helped cause this trouble and it has caused our company a great big loss. It should be stopped and I want to tell you and Mr. Wharton that fact and you are the men to stop it and change it and do it just as quickly as you can and don't let this big balance in that bank ever occur again.

C. C. Webber, 1931

If stockbrokers in New York City were indeed jumping out of skyscraper windows, ruined by the "Black Thursday" stock market crash, and if small investors all around the country were in shock after losing their "lifetime earnings," farmers and the agricultural machinery industry serving them were in no such disarray as the new decade began. Though tractor production in 1930 dropped about 12 percent from the peak of the previous year (down to 196,297 from the all-time high in 1929 of 223,081), the first year of the business downturn was not particularly threatening to the seven long-line manufacturers and the major short-line companies. Only Massey-Harris among the seven registered a loss that year, and though the very small and very large short-line companies showed small losses as a group, the middle-range short-liners were themselves profitable.

◂ Kansas farm auction, 1930s. *Grant Heilman*

CHAPTER 11

Frank Silloway had reported to the board on October 29, 1929, that "we believe farmers, generally, are feeling better than they have for some time," and he predicted strong sales for the company in the coming year. Record earnings for the past fiscal year were announced in January 1930, and the board declared not only the regular quarterly cash dividends of $1.50 on the common stock and $1.75 on the preferred, but also an extra stock dividend on the common of 1.5 percent. When the board also affected a 5-for-1 split in both the common and preferred stock, the financial community was dazzled. The annual report of the company was "unusually satisfactory," said the *Chicago Tribune*; the *Wall Street Journal* reported on Deere's "good showing last year in the face of low wheat prices." The *Christian Science Monitor* headlined its article, "Deere & Co. has an unusual earnings record since 1924," and *The Economist* reported, "Deere income and outlook unusual," and followed with a surprisingly strong statement for that conservative journal that "favorable marketing conditions enhance attractiveness of common stock" and a prediction of "continued good results." "Prospects for 1930 are bright," corroborated the brokerage house of Jas. H. Oliphant & Co., which continued: "It has been put in splendid condition and methods of accounting are conservative. . . . The former dividend rate of $6 will doubtless undergo revision upward when payments on new stock begin." Such optimism was not confined to the farm machinery industry, either. Secretary of the Treasury Andrew W. Mellon stated blithely: "I see nothing . . . in the present situation that is either menacing or warrants pessimism."

But such public posturing is notoriously undependable as a predictor of the future. By late in the spring of 1930, Silloway was warning the board of the poor crop in the Northwest and in Canada, the tightness of money in the country areas, and the unfavorable prices that were beginning to show up in farm commodities. Charles Wiman was nudging the factories toward "underrunning rather than overrunning their approved factory schedules." Employment in the company was high, even rising in April to 8,422 employees (though still substantially under the record average in 1929 of 10,820), but by early July Wiman was reporting on the "general slump in business" and some of the factories were temporarily closed. "The factories were told, however," continued Wiman, "that experimental work should go forward without reduction in this account, as it is most important that the design of our goods be kept up to date."

The board mistakenly gauged the situation as temporary. At a round-table discussion at the July 29 meeting, all agreed that "while business is running very much below normal at the present time, prospects seem to be rather bright for a fairly good volume next year." By the end of the fiscal year of 1930, though, the factory pace was very much reduced and employment was down to 4,795 people. When the final results for 1930 were tabulated, total sales had dropped to just over $63 million, down from a record high of $76 million the previous year. There was still a substantial profit for the year—$7.6 million—but this, too, had sagged from the previous year's record of $16 million.

The company was in far better shape to take a downturn than it had been in 1921. Then inventories had been some $36 million, with $14 million owed to banks and $21 million in receivables. At the close of 1930, the inventory was $20 million, there was only $4.9 million in debts owed to the banks, and some $55.9 million in receivables. The surplus account stood in 1930 at $22.9 million, after a conservative board decision to write off the entire goodwill account of $17.9 million; no back dividends were due, and reserve accounts amounting to $33.4 million were built into the financial statement. "Our debts are low and our receivables are high," noted the comptroller. "These comparisons are given to indicate the splendid financial condition of Deere & Company today, and to show our ability to cope with any situation that may arise."

But the situation over the next three years became far more serious than anyone had anticipated, for the country had entered the Great Depression. Industrial output was almost halved at its lowest point, in March 1933; unemployment mounted in the country to about 12.8 million in that year (the official figure, but likely understated)—at least one-quarter of the civilian force was unemployed. The patterns of idleness embraced every calling and condition, every age and locus of residence. The wholesale price index for all commodities other than farm products fell in 1932 to 70.2 (1926 = 100), and farm prices sagged even further, to 48.2 in that year. Wheat, for example, hit a low of $.32 per bushel. Forced sales of farms due to bankruptcies, foreclosures, and tax delinquencies multiplied all over the country and were exceptionally heavy in the Dakotas and Iowa. Nature also took a hand in worsening the farmers' plight. There was a substantial drought in 1930 and then a disastrous one in 1934 that cut output by about one-third. Withered crops allowed the winds to pulverize the exposed land, and all through the Midwest great clouds of soil in the air cut visibility, often to less than 700 feet. Soon the term "dust bowl" struck the public consciousness. Oklahoma was a particularly hard-hit state, and the roads were full of dilapidated vehicles carrying "Okie" families to "somewhere" for a more hopeful life.

With agriculture itself almost prostrate, the machinery industry serving it took the full effects of the Great Depression. At Deere, sales plummeted in 1931 to $27.7 million, sagged again to $8.7 million in 1932, and rose only to $9 million in 1933. All three years were loss years—just over $1 million in 1931, $5.7 million in 1931, and $4.3 million in 1933. Critically important events occurring in these three trying years highlighted some company weaknesses and a number of important company strengths.[1]

CHAPTER 11

FINANCING THE FARMERS

The comptroller's report of 1930 had noted a signal problem that was to haunt the company—those receivables that stood at the end of 1930 at more than $55.9 million. This was a record high total for the company, some 83 percent of sales, compared with 52 percent in 1929, 44 percent in 1928, and a low point of 27 percent in 1919. Indeed, receivables had only risen to 67 percent in the severe year of 1921. The comptroller's report put the proximate reasons succinctly: it was "due to the sales of tractors, combines, etc. on long terms, and poor collections caused by the severe drought, and the drop in prices of farm products."

Back in mid-1929 Frank Silloway had felt that the company was falling behind its competitors, for the latter were taking farmers' paper for direct credit, either through their individual companies or through specially established finance corporations. Silloway recommended that Deere adopt the former practice, and the board enthusiastically agreed. First, the Moline branch was given the privilege experimentally that very month, then the practice was extended to Kansas City and Omaha and later in the year to Minneapolis. Silloway had been optimistic about its potential at that time, predicting: "We would not increase our outstandings thereby . . . it would merely be a different form of doing business."

If the Great Depression had not intruded, Silloway likely would have been right. As it was, the bottom dropped out of farm prices. The farmer now had little or no money to pay off his debts.

The extent of the discrepancy between sales and collections was not fully apparent until the comptroller's report for 1931 was released in January 1932. The total sales for the year of $27.7 million were actually far below the receivables on the books, which now totaled more than $54.6 million, down only $1.3 million from the previous year. The report pinpointed the various debtors:

Farmers' paper taken for direct credit, including Canada	$12,000,000
Dealers' paper for goods sold and secured by collateral, including dealers' current purchases for repairs, etc., which came due from month to month	25,000,000
Dealers' paper for goods on hand, carried over under the terms of our contract	9,000,000
Russia ($1,500,000 since paid)	3,800,000
Argentina and other foreign countries	2,800,000
General company accounts and notes	2,000,000
Total	$54,600,000

Silloway, extremely disconcerted by the shortfall in collections, issued a special branch-house bulletin: "For the first time since 1911, and perhaps ever, our outstandings at the end of the year exceeded our sales for the year. . . . Our problem in 1921 was inventory; our problem during this depression is our outstandings."

How was this to be resolved? Credit and collections have always been two of the most subtle, complex marketing issues facing agriculturally based companies. Farmers typically are debtors—it is in the nature of the equation the agriculturalist faces. Seed, fertilizer, equipment all must be amassed ahead of time, and the exigencies of the ensuing crop season then determine whether the debt can be paid. Now that the era of modern "power farming" had dawned, the time line on these debts had been pushed further out into the future—tractors simply could not be paid off with the proceeds of a single crop season. Implicit in Silloway's decision in 1929 to take direct farmer paper was the need to make a longer-term agreement with the farmer. In the depths of the depression, with the farmer unable to sell his crop at a price that could even remotely amortize the debt, farmers throughout the country were falling in arrears on their promises.

To many managers, the soundest collection policy in a period like this was to put as much pressure as possible on the farmer to put machinery debts high on his agenda—as the ever-blunt M. J. Healey put it in his instructions to his Kansas City branch collectors, "Farmers should pay us before they pay other creditors." Manufacturers were making "wealth-producing farm equipment . . . absolutely necessary for farmers to have" and, therefore, this was "conclusive evidence that we are entitled to the first payment. . . . It would not be fair for the farmers to pay for luxuries purchased with the money produced by our implements before they discharge their obligations to us." Yet such heavy-handedness might well be counterproductive in the longer run. Farmers had the reputation of being responsible debtors, and excessive pressure in a period such as this might produce not heightened payments, but only heightened antagonisms. A delicate touch indeed was needed.

The company's decision—though it was actually hundreds of individual decisions in hundreds of individual farmer contracts—was to carry the farmer as long as necessary to amortize the debt. To be sure, certain farmers had simply given up, abandoning their equipment and moving out of farming altogether. These were lost causes, to be written off as bad debts. The reserve for uncollectable accounts rose from $1.9 million in 1931 to more than $2.4 million in 1932 and just under $4 million in 1933. The uncollectables actually charged against the account were considerably less, though, and the comptroller's report of 1933 pointed out that from 1913 through 1932 only $6.6 million had been written off out of a total sales in that period of $803 million—a .009 ratio of uncollectables to sales. The year-by-year percentages during the Great Depression period were naturally much higher—1932 was just under 7 percent, 1933 more than 8 percent,

1934 at 5.5 percent, and 1935 just over 4 percent. Still, despite four years of extremely difficult conditions, almost 99 percent of all the farmers' debts had been paid over the 1913–1935 period, a record attributable both to the farmers' sense of responsibility and to Deere's conscious decision to stick with the farmer through both good times and bad. To be sure, the company's strong financial position as it entered the Great Depression aided this policy. Its results proved to be an incalculable asset for the company. The farmers' loyalty to Deere products, always strong, was greatly strengthened during the period of the Great Depression. The company's belief in the farmer was reciprocated by a farmer response that gave Deere a priceless asset for the future.[2]

SOVIET DEBTS ARE PAID

The receivables listed in the comptroller's report for 1931 named two other critically important customers whose accounts were due. The first of these was Russia, a customer in the late 1920s for an astounding amount of Deere equipment. After the Russian Revolution, the newly constituted Union of Soviet Socialist Republics embarked upon extensive agricultural machinery purchases, especially from the United States. The most noteworthy was its interest in the early 1920s in the Fordson tractor. Thousands were shipped into the country, and even one of the early postage stamps of the Soviet Union (in 1923) depicted the Fordson. As early as 1923 some Deere Waterloo Boy tractors were also in the Soviet Union, sent there on behalf of American relief efforts. Frank Silloway, ever conscious of commercial possibilities, noted to the board: "It is our plan to send an expert to see that the machines are properly started in the field. . . . Our representative will also report on the general conditions as far as future prospects of trade are concerned." Deere engineers from Waterloo did indeed go, and they sent back some remarkable photographs of the primitive Soviet farming situation (exhibit 11-1).

Deere first obtained several substantial implement orders in the early 1920s; one order alone, in 1924, called for 500 No. 40 plows. This was soon followed by other orders for plows and tractors, the scale of which dumbfounded everyone in Moline. Joseph Stalin's first Five Year Plan had been promulgated in 1928. This rigid, dogmatic document called for the buildup of Russian economic might on a tight, lock-step staging that required particular pieces to fall in place with a hoped-for exactitude. The quantities of output called for by Stalin were stupefying, setting up almost impossible goals. One of the early pieces of this jigsaw puzzle of plans was the heightening of agricultural production. Stalin and his planning coterie had just collectivized much of Soviet farming (in the process killing many thousands of the peasant farmers, the kulaks). The collective farms were presenting a mammoth management problem, and the Soviet experts felt that only through extensive mechanization could the promise of socialist farming be realized. So the country had initiated massive purchases of agricultural machinery—particularly tractors—from abroad. One of the companies to which they turned was Deere.

Another order came for 500 tractor plows; payment terms were 50 percent in cash and the balance over the following two falls. Silloway soon reported, "All accounts have been paid promptly when due. We know that the Russian government is not recognized by the United States and that we are doing business there at our peril. . . . We believe it advisable to maintain a representation of the John Deere Line in that country, even though at some time we might take a loss." By 1929, the Russians were buying Model D tractors in large volume, too; one order alone in late 1929 called for 1,500 machines. Silloway again had to field questions from a skeptical board about the credit worthiness of the Soviet Union: "There are those who strongly maintain that the present Government in Russia is so securely entrenched in power that it will endure; while on the other hand, we receive reports that the present Government is losing, somewhat, its hold on the people." Silloway countered with an essentially entrepreneurial proposal: "If we believe that the present Government will remain in power until November 1, 1930 and pay the money they will owe us at that time we would be justified in accepting this order." After much discussion, the board concurred and the order was implemented into the factory planning process. A few months later, the Soviets raised the total to 2,000 tractors, to be accompanied by additional purchases of disk tillers. By this time, the Soviet orders on the books were well over $2 million. The realization that domestic orders had just about dried up added an incentive to take chances on a buyer whose credit standing was, if not poor, at least unknown. When the Soviets again came back (through their state trading entity, Amtorg) for an additional 2,000 tractors, Silloway once more recommended acceptance and the board, uneasily, went ahead.

The John Deere Model Ds were well suited to the Soviet Union—they were simple and dependable—and soon became revered tractors on the Soviet collective farms and in the machine tractor stations. A Russian children's book was written in 1931 about one of these collective farms, in Yeash district, a place according to its author where "counter-revoluntionary views are known . . . and for that reason [it] was chosen to build a collective farm and a machine-tractor station." The station had eighty-six machines—forty Caterpillar track-laying tractors, thirty-seven Model D John Deeres, and a scattering of other makes. The author waxed lyrical:

> Racing along, runs a two-cylinder tractor of "John Deere" make.
> He shines with spurs and the noise of his motor carries the echo far into the villages. "John Deere" showed himself a punctual good worker

Exhibit 11-1. Top, Waterloo Boy tractors ready for shipment out of Odessa, Russia, for various colonies, and eighty mechanics, also American personnel in charge of the seven squads; middle, Waterloo Boy pulling four implements known as "bookers," used only in Russia; bottom, A view of the vast prairies of Russia, and a part of the eighty Waterloo Boy tractors putting the sod back to productivity after ten to twenty years of lying idle; next page, Five Waterloo Boys after a day's work in Russia, in their stalls for the night. The American in charge, and his automobile; photographs and captions, taken in March and August 1923. *Deere Archives*

on a hard service in machine-tractor station. With obstinateness, given honor to American steel, he carries along with steadiness, without being insured of a good technical attention and professional care. Only after a month and a half of unceased labor, the machine began to demand attention. . . . In Fall the wheat Little Ukraine was planted and gave surprising results. The metallic blue of winter crop was no less surprising than the tireless firmness of American "John Deere."

The author ended his story with a flourish: "The aim of this tale is to show that the tractor 'John Deere' and wheat 'Little Ukraine' are heroes of the success told above."

Fears about the Soviets' punctuality of payment proved to be unfounded. Stalin had pressed ahead with the Five Year Plan despite massive economic dislocations within the country; the tractors, machine tools, and other purchases abroad were being financed in hard currency obtained by selling priceless Russian art on the world market (Andrew Mellon was one of the largest buyers), and even selling quantities of Soviet wheat abroad, despite a famine that was stalking the Ukraine. Yet the Soviets continued to fulfill every contract they made during this period. Indeed, not only during the time of the Great Depression, but through the succeeding years, the Soviets were known as harsh, devious bargainers who always fulfilled a contract, once it had been reduced to writing.

By early 1931 the Soviets had spent $5.6 million for company goods; all the required payments had been made right on time. All through 1931, with massive domestic nonpayment, the Soviets fulfilled the requirements on each note; the last were due in November 1933 and came into the company's coffers on schedule.

The total Soviet sales were unbelievable. In the three years 1929–1931 the Soviets had bought $9.8 million worth of Deere goods. In one year alone, 1930, more than $5.6 million of the company's $63 million in sales had gone just to this one country—almost 10 percent of the total sales of the company. Deere's tractor sales accounted for 20 percent of US tractor exports to the Soviet Union in 1930 and more than 12 percent for the 1927–33 period (exhibit 11-2).

Exhibit 11-2. *Number of Tractors Sold to the Soviet Union, 1927–1933 (numbers of machines)*

	US Industry as Whole	Deere & Company
1927	5,105	13
1928	4,588	51
1929	11,364	2,232
1930	20,447	4,181
1931	22,909	1,679
1932	32	31
1933	8	
Total	64,453	8,187

Source: Deere & Company, "World Wheel Type Tractor Survey, 1927 thru 1944," (December 27, 1945), Deere Archives, 45040.

The end of the outsized Soviet sales came in 1931 (with $2.7 million); in the following year only some $63,000 was spent, with even more miniscule amounts in the following two years. By 1935, the Soviet Union had stopped purchasing Deere equipment and other equipment as well. It had become more or less self-sufficient in tractors. At the end of 1931, the "Russian business" reserve account still stood at something over $1 million. The financial officers transferred this account to a general "Reserve for Uncollectables"—though the Soviets were paying, others were not! The effects of these latter losses were taken in the year 1931. Thus, although the original annual report of 1931 showed a small profit of $409,000, the recast figures put on the permanent record a loss for the year 1931 of just over $1 million. Thus there were losses in three years—1931, 1932, and 1933.[3]

THE ARGENTINE TRADE TURNS SOUR

In contrast to Russia, Argentina had been a major customer for Deere equipment—and other American farm machinery—since well before the turn of the century. Indeed, in 1894, more agricultural implements from the United States were sold to Argentina than to any other country in the world. Deere sold its equipment in South America through an import house, Agar, Cross & Company of Buenos Aires. Relations between Deere and Agar, Cross were close, and sometime in the early 1890s the Buenos Aires company brought out a full-scale catalog of Deere equipment, complete with an unflattering picture of John Deere in its preface. The first and most prominent implement in the catalog was "El Gilpin." Subsequent pages had a number of other Deere plows, some renamed with Argentine words (the Red Jacket became "Le Garibaldino," the Red River breaker became "Le Tucumano"). So important did the Argentine trade become over the next three decades that Frank Silloway told the board in 1923: "I think that the Argentine particularly will always be our most important foreign market." Argentine agricultural prosperity in the mid- and late-1920s did bring large sales, including substantial numbers of tractors; some 2,194 tractors were sold there in just the one year of 1929.

But the Argentine economy quickly took a nose dive after the stock market crash in 1929, just after a very large shipment of Deere tractors was made to Agar, Cross. Sales in Argentina plummeted and Agar, Cross asked for extensions on the Deere notes due early in 1930; these were paid later in the year but a remaining note for more than $1.9 million that was due in early 1931 had to be extended to 1932. Then, in April 1932, the board learned the bad news that "they were unable to pay this amount." Though Agar, Cross did manage to meet the accumulated interest, another extension on the principal had to be made. An Agar, Cross representative came to Moline in that month and reported dolefully that the Argentine company had an inventory worth more than $13 million, with additional receivables amounting to some $6 million. The company's inventory included more than $2 million worth of Deere products, most of them Model D tractors.

When the economy continued to limp badly over the following two years, Agar, Cross extracted an agreement from Deere to take back 900 of the Model D tractors (on the basis of invoice price, with Agar, Cross assuming the freight both ways, but with that figure then credited to their account). Deere also had to take some Argentine treasury notes, to be paid over a fifteen-year period, thus further dragging out the debt. Between the tractors and the notes, about $1 million ended up being credited to the Agar, Cross account, leaving a balance due of some $750,000. Deere shipments to the Argentine did pick up a bit after 1935, though the modest sales through the remainder of the decade never remotely approached the substantial amount of business done before the crash.

Meanwhile, an import house in South Africa—Dunell Edden & Company of Johannesburg—had also been unable to meet its payments to Deere. The amount of business there was not large and the amount overdue not significant, compared with the Russian or Argentine accounts. In late 1930 the obligation stood at some $600,000, including goods on

Exhibit 11-3. John Deere-Agar, Cross tractor in Argentina, ca. 1930s. *Deere Archives*

consignment. Dunell Edden reported that it was near bankruptcy and, rather than default altogether, proposed that Deere itself take stock in the company to the amount of some £70,000 South African. An agreement was struck, and Deere acquired one-half interest in the business. The comptroller valued the purchase at $339,500, the stock being paid for in total by the goods that were already in South Africa. Later the firm expanded into manufacturing—in the process putting Deere into the controversial role of being one of the "American companies doing business in South Africa."[4]

JOHN DEERE OF CANADA

Practically no company in the country was expanding or opening plants in 1931 except Deere, at its Welland plant in Canada. The reasons stemmed from unique features of Canadian–American interactions.

The first recorded shipment of John Deere goods to Canada arrived in Winnipeg, Manitoba, in April 1878; probably John Deere plows or harrows had found their way across the border even earlier than that on some grubstaker's wagon. Deere equipment was sold in Manitoba first by a partnership, Wesbrook & Fairchild, and after 1887 with F. A. Fairchild & Company, which acted as the John Deere agency for Canada until 1907. Deere soon opened new branches in Regina and Saskatoon, Saskatchewan, and Calgary and Edmonton, Alberta, as well as keeping the original branch in Winnipeg. Though there have been other sales branches in Canada over the intervening years, these five have remained the key points, the western provinces being the strongest for the company.

In 1907 Deere purchased the partnership and established its own corporation, the John Deere Plow Company, Ltd. Its first resident manager, H. W. Hutchinson, was a Canadian, and over the years Deere has put most of the management responsibility in Canada in the hands of the residents of that country. Sales to these branches were steady—and significant. In the 1910s the amount sold in Canada ranged from 8 percent to 10 percent of total Deere sales; the percentages in the 1920s ranged a bit higher, reaching almost 15 percent in 1928. These branches were treated the same as a domestic branch and were not considered as "export." Yet Canada is across the border from the United States, and thus the important factor of import duties comes into play.

Canada first exacted a tariff on imports of farm equipment far back in 1847, a 10 percent levy. This figure drifted upward over the next several decades, reaching 25 percent in 1879 and 35 percent in 1883 (the highest level ever recorded). These tariffs were instituted primarily to protect Canadian manufacturers, but with a collateral hope by the Canadian government that they would induce American manufacturers to open Canadian production facilities. Not everyone welcomed these tariffs; Canadian farmers particularly opposed them, as they wanted less expensive machinery and American technological superiority. The farmers' views soon held sway, and the forty years from 1890 to 1930 saw a reversal of the upward movement in the implement tariffs. By 1907 import duties on tillage equipment fell below 20 percent, and the rate on harvesting equipment soon followed. By 1924 the rate for harvesting machinery was down to 6 percent; for tillage machinery it was 7.5 percent and for plows 10 percent. No Canadian manufacturer produced small tractors, so these were let in duty-free. Large tractors, however, were manufactured in Canada and were protected by a duty set at 17.5 percent.

Prior to the turn of the century, the company had never seriously considered opening a factory in Canada. In the several years surrounding the formation of the Deere modern company in 1911, possibilities for manufacturing in Canada had complicated the picture. The question of what to do about Dain's Welland, Ontario, manufacturing operation loomed. Dain, as an independent company but with close ties to Deere, had purchased the Welland property in 1908 to manufacture mowers and rakes for the Canadian market. When Cyrus McCormick and William Butterworth engaged in their convoluted negotiations in 1910 toward combining operations, one of the McCormick suggestions had been that they act in concert in manufacturing in Canada. The deal fell through, but Dain formally came into the Deere organization in that same year and Deere inherited a Canadian manufacturing plant, not particularly by choice. At this very time, one of the company's "great debates" was taking place—whether to build a harvester to take on International Harvester. There was some discussion about manufacturing the harvester in Canada for import into the United States, but this seemed unrealistic from a United States tariff standpoint, so a decision was taken to open a new harvester plant in East Moline. Meanwhile, production of the Dain line of mowers, hay loaders, side rakes, and other hay tools was continued in the Welland plant, and John Deere spreaders and Van Brunt grain drills were also added to the line. The sum of these activities argued for a plant expansion there, and this was done in 1913.

By the early 1920s, however, the Canadian tariff for most of this equipment was low, and it made sense to serve the Canadian trade, which was primarily for the western provinces, out of the Minneapolis branch. In late 1923 Leon Clausen, in one of his last acts with the company before moving to J. I. Case, recommended the closing of the Welland plant and the moving of the drill and cultivator lines back to Horicon, Wisconsin; the spreader lines back to East Moline; and the hay tools back to Dain's Ottumwa, Iowa, plant. A sales branch had been in operation in Welland at this time; it, too, was to be closed. Though all the machine tools and even some of the office fixtures of the Welland plant were either sold outright or transferred back to other factories in the United States, the company fortunately made a decision to "mothball" the plant building rather than to dispose of it. A few employees were left as a skeleton force to ferret out delinquent accounts from past sales and to look after the plant.

This arrangement continued for six years and might have gone on indefinitely had not a new conservative government gained power in Canada in 1930. The latter pledged to "test completely the effectiveness of the protective tariff as an offensive weapon to fight the deepest depression in the world's history" and promptly put its campaign rhetoric into reality with a whopping increase in the tariff. Both implement and large-tractor rates were raised immediately to 25 percent (smaller tractors were still left on the free list). Imports of agricultural machinery into Canada, not unexpectedly, dropped precipitously. If Deere wanted to stay competitive in Canadian sales, it would probably have to manufacture in Canada. Should Deere again do so?

Charles Stone took the advocate's role as the board began to debate the notion. He had been manager at Welland briefly after World War I, and so he was uniquely qualified to speak on the situation. After apprising the board of the details of the tariff situation, he warned them that the Canadian government probably would not accept just an assembly plant—to "ship in goods ready to assemble and paint." To qualify as a "manufacturer," it would be necessary to fabricate a significant portion of the parts of the machines. Stone felt that the company should go ahead, advocating that "if the plant is reopened, it should be with the definite idea that it will be operated for all future time, good times or bad times, high tariff or low tariff, or no tariff. Operations would be increased or decreased as conditions might require, but probably never completely discontinued, so that Deere & Company would cease to be a Canadian manufacturer." The board could not reach consensus, so Charles Wiman himself went to Ottawa, Ontario, for a personal conference with Prime Minister R. B. Bennett to, as Wiman put it, "impress the Prime Minister with our desire to comply strictly with the requirements of the Customs Department," but also to extract some concessions for bringing in machinery and machine tools from the United States. The prime minister agreed, and the decision to move ahead as a Canadian manufacturer was made.

The factory was recommissioned in 1931 and equipment installed for the manufacture of disk tillers, grain drills, grain binders, disc harrows, and gang plows. C. R. Carlson Jr. went to Welland as manager from Deere & Webber in Minneapolis, and there exhibited the energy and enthusiasm that characterized his later tenure as vice president of sales for the entire company. As the Great Depression waned, the Canadian business began to grow again and over the succeeding years the Canadian segment of the company continued to be an important contributor to overall company revenues.[5]

EMPLOYEE TRAUMAS

The period 1930–1933—the depth of the Great Depression—wreaked its havoc on the Moline area and the company, just as it did on thousands of other communities and hundreds of thousands of other companies all through the land. The doleful message, "No Orders," made mass layoffs the rule everywhere, throwing breadwinners onto their own resources, painfully little in most cases. When Deere's sales fell from $63 million in 1930 to $8.7 million in 1932, everyone knew that those miniscule purchases would be made mostly out of existing inventory, with little if any new work being required. In mid-1931 massive layoffs had to be made—total factory employment fell to 1,270 from 4,800 the previous year—and even those working shared a reduced-hour schedule. Small wage cuts were made for those factory employees still working; by mid-1932 the average hourly wage at the Plow Works had fallen four cents, from $.57 to $.53. The pattern was essentially the same in the other factories. More substantial cuts were made in salaried employees' wages, with a 5 to 10 percent reduction being made in 1931 (those making more than $2,500 taking the larger cut) and a further 10 percent late in the year. Then, in mid-1932, a more stringent measure was applied, with those earning more than $3,000 taking a full 25 percent cut, with scaled down amounts for those under this figure.

The greatest challenge was to try to bring the unemployed members of the company through the times without disintegration of their families' living standards. The company's abiding concern for its employees, demonstrated over many years of reciprocal regard, was to be tested under the most trying conditions.

Deere's employee benefit plans aided measurably in this period, particularly the longstanding Thrift Plan. The plan, instituted in the early 1920s, provided two vehicles for employee savings—a conventional savings account, drawing 5 percent from the company, and a mechanism for purchasing savings certificates in fixed denominations, this later to pay 6 percent. By July 1930, more than $1.3 million of the certificates had been purchased and these now functioned as relief measures, George Crampton reporting in that month: "From all quarters we hear of praise and gratification on the part of employees for the accumulated savings to tide over the period of unemployment." Some company benefits had to be put in suspension during these years; for example, vacations with pay were discontinued in 1931 and not reinstituted until 1936. On the other hand, the company continued group insurance for unemployed members of the force and soon was engaging in a wide range of "make work" activities for employees and providing outright relief for others. By 1932, the city of Moline and the state of Illinois began to assume more of the relief burden. Those employees who were renting company houses simply could not pay their rents in many months, and the company made two reductions in the rentals totaling some 17 percent. Temporary reductions were also made in the pension plan, having the effect of reducing the minimum pension from thirty to twenty-five dollars per month, with similar reductions for those retiring with larger amounts.

Deere management clearly suffered less in economic terms, but it shared fully the worries and the disconcerting, unpalatable decisions that had to be made. William Butterworth, while still keeping in touch with the situation as the company's chairman, was several steps removed from the firing line of these decisions. He had been reelected national president of the United States Chamber of Commerce in 1930 and was resident in Washington many days of this period. He often was called for consultation by President Hoover, and he frequently participated in national conferences related to the economic situation. This rarefied perspective seemed to filter the raw facts for Butterworth, who wrote in his diary in July 1932, at the depths of the crisis: "There is very little business, but we are much better off than we were 10 years ago." Inevitably, the burden of day-by-day Deere decision-making fell squarely on Charles Wiman. This was a searing experience for Wiman, one that left permanent marks on the way he viewed the business—and himself.

One incident early in the Great Depression dramatized to Wiman more than any other the wrenching nature of being the final authority, the de facto chief officer. That was the $1,200,000 defalcation at the Peoples Savings Bank in Moline. Peoples had had seventy-four years of uneventful history up to this untimely moment, and all through this period the bank was linked inextricably with Deere & Company. First founded by John Gould, John Deere's old partner, in 1857 as a private bank, it became the First National Bank of Moline in 1863, and John Deere himself was president for one year in 1866. In 1891, a separate bank, the Peoples Savings Bank of Moline, was organized by Charles Deere and others, with Deere himself as president; this entity merged with the First National Bank of Moline in

1905. On Charles Deere's death, William Butterworth had become president, remaining so through the time of the defalcation. George Crampton, the Deere secretary, was also chairman of the board of the bank. Thus, over the years, Deere's name had been synonymous with the Peoples Savings Bank; in 1907, for example, Charles Deere's photograph was featured in the bank's advertising, and the bank was heavily patronized by Deere employees all through its three-quarters of a century of existence.

The swindle was done cleverly, by the cashier and two other inside accomplices. When the full facts became known, the bank examiners found that most of the money had been squandered on real estate investments, none of which would bring the bank much in restitution. The bank's board of directors was faced with not just a scandal, but, given the particular times—in the depths of a depression—the maintenance of the economic lifeline of the town of Moline and the area around it. Butterworth was in Washington and was not available except by telephone, and so it was left to Wiman, Crampton, T. F. Wharton, and Frank Silloway to make an overnight decision about the posture of Deere & Company, in order to forestall closure of the bank the next day by the state bank examiner. The company had maintained an account at the bank of more than $2 million, far beyond its operating necessities; it was in this account that the shortage had been brought about. In essence, the capital, surplus, and undivided profits of the bank were almost entirely wiped out.

Wiman and the other three made the decision that night that Deere would give the bank its check for $1,290,000, for, as Wiman told the Deere board the next morning in a special meeting, "If we do not do this the bank closes. . . . As I view it, there are approximately $7,000,000.00 of savings deposits in this bank, largely made by the wage earners of our factories, and the effects upon them of the closing of the bank, and the resulting consequences to this Company, are beyond calculation." The directors and shareholders of the bank had been contacted and all had agreed to a voluntary assessment of 100 percent upon their own stock, this money to be used as partial repayment to Deere & Company. Nevertheless, it was clear that the company was going to end up with close to $1 million of notes that might not be paid back for many years, if ever.

The embezzlement became public knowledge the next day and that same day a joint announcement was made to the public by the bank through George Crampton as chairman and Deere & Company through Wiman as president assuring the public that the amount of the defalcation had been fully restored by Deere and by the larger shareholders. A run on the bank was prevented, and though several other banks in the Moline-Rock Island-Davenport area closed in later months due to the depression, the Peoples Savings Bank was able to keep its doors open. The "larger shareholders" were, of course, made up mostly of Deere family interests, and when the state auditor ordered that additional funds were needed for cash reserves,

CHAPTER 11

William Butterworth, Charles Wiman, and Dwight Wiman together exchanged their own notes aggregating some $650,000 for notes held by the bank for a like amount. Some were later paid back, but when the final chapter was closed, many years later, the families had taken severe losses.

Other Deere board members from outlying locations rallied to Wiman's side after these decisions were made. Willard Hosford wrote from Omaha about the "favorable notoriety that this action would bring," and S. H. Velie Jr. penned a sympathetic note from the Kansas City branch. "It came to you like a thunderbolt out of a clear sky and you arose to the emergency and saved the bank and its clientele in a manner that will always be a source of comfort and gratification to yourself and to your kind."

Not so C. C. Webber—he was neither solicitous nor kind to Wiman, writing a stinging letter that rambled a good deal but left no doubt about where he stood. "I have not paid any attention to it of late years, but you had a deposit in that bank of about two million dollars when this trouble occurred, and you had it there for some time. These gentlemen tell me that the thought in having Deere & Company keep this big deposit in this bank was so that it would be as big as its neighbor. That is all nonsense, the size of the bank has nothing to do with its solidity or its worth to the community. Never mind the size, run a good bank and a sound one. It is this big Deere & Company deposit that has helped cause this trouble and it has caused our company a great big loss. It should be stopped and I want to tell you and Mr. Wharton that fact and you are the men to stop it and change it and do it just as quickly as you can and don't let this big balance in that bank ever occur again. . . . It is up to you two gentlemen to realize and change this situation and change it right away. . . . Lord knows, stop it now and correct it, because it is to be the interest of the stockholders of Deere & Company that it be stopped. It has caused them loss enough already."

Wiman wrote a conciliatory, apologetic note back: "I agree thoroughly with you about the carrying of bank balances. It has bothered me in the past and I should have appreciated knowing your attitude in the matter. We want to work out of this bank business just as quickly as possible." Wiman vented his own personal frustration to Webber: "My predilection is for Deere & Company, and not for banking. . . . You have no idea how much trouble this whole affair has caused us." Webber, in his return letter, said he knew that the matter had been distressing but, nevertheless: "I very much hope that you gentlemen will act forthwith to correct the situation as to the size of our bank balance, and that never again will our company be subject to the criticism that can rightly, it seems to me, be laid to it now for carrying a balance in this bank so large as not to be in the interest of Deere & Company."

Webber's acerbic views were dissected by the Moline members of the board over the next few days. Frank Silloway commented: "I think there should be a resolution adopted by this board—perhaps in the next five years—that no director of Deere & Company can be a director in the bank.

Within the next five or ten years we want to get out of the banking business and not father it." The board did not follow Silloway's or Webber's advice, though, for in January 1933 a new bank, the Moline National Bank, was organized, with more than 90 percent of the stock held by Deere & Company. The Moline National was put together out of the ashes of the Peoples, which went under at this time. The latter's depositors received only 40 percent of their holdings at first, but by 1941 all the rest had been paid back, due directly to Deere's abiding concern that none of the Peoples' depositors should be left with a loss. The goodwill generated by this was immense, for the company was seen, correctly, as the savior of the "little people." Hosford was right about the public relations effect of the company's forthright action. (Deere, incidentally, kept its shares in the Moline National until 1958, when it sold them to outside interests.)

Charles Wiman truly had a baptism of fire in this sad case; most of the responsibility had fallen on his shoulders, though William Butterworth did return quickly to Moline to help. The weight of the decisions troubled Wiman so much during this period that he became, at least temporarily, an insomniac, pacing the halls of Overlook, perhaps walking in some of the footsteps that his grandfather, Charles Deere, must have made in the trying times of the proposed combinations and plow trusts at the turn of the century.

Wiman now turned more conservative, harking back to some degree to the same caution preached to him by his uncle, William Butterworth. A poignant exchange between Wiman and Wharton, his now elderly treasurer, illustrates the tension Wiman felt in this period. In July 1931 the Chicago *Journal of Commerce* printed a small article, recounting "a bit of Oriental wisdom . . . a story from old Bagdad" in which a wise man replied to a youngster's question as to how to get value received by saying: "A thing that is bought or sold has no value unless it contains that which cannot be bought or sold. Look for the Priceless Ingredient. . . . The Priceless Ingredient of every product in the market place is the honor and integrity of him who makes it. Consider his name before you buy." Wharton happened to read this little homily, and clipped it for Wiman, attaching his own views: "This applies to John Deere. It is up to us to keep together this organization that puts the quality in John Deere goods, and provides this Priceless Ingredient." Wiman's reply was probably a bit of a surprise to Wharton: "That's true—but the 'Priceless Ingredient' must not cost too much or no one can afford to buy it."

The Webber criticisms of company financial acumen were probably deserved—there had been an overwhelming feeling of pride and proprietorship in the bank, and prudent financial wisdom was undoubtedly submerged in keeping the very large balances in the bank. The ugly incident taught the company's executives a painful lesson; Charles Wiman knew he had made a mistake, and he came out of this situation a financially wiser person.

CHAPTER 11

Wiman's experience with the relief programs of the company had also given him a keen sense of the differences, good and bad, among the Deere employees. He spoke of this at length in a remarkably frank statement to the board in October 1932: "Out of the mass of experience coming from this depression, one thing particularly has been brought forcefully to our attention. When the time comes to increase the force in the various plants of Deere & Company, I am convinced more attention must be given to the character and home life of the people we employ. More and more, employers are assuming responsibility for their employees and society is fixing the responsibility upon them. Social legislation of various sorts is more sure to increase throughout the country. Old age pensions and unemployment insurance are examples. It seems to me in the future it will not be enough to know that a man is a satisfactory workman, we shall have to know something about how he manages his home and his affairs so that he will not become a dependent on the industry immediately after any downward swing in production. In other words, we should maintain about the same record of a family that is now maintained by a good welfare agency. Industry can well afford to help those employees who are helping themselves, but should not be saddled with responsibility for the other type. We can only protect ourselves by more selective employment."

For many years after, Wiman would recount the traumas of the Great Depression and the need for caution and care in the company in order to anticipate such happenings again. In 1946, when there was a small downturn at the end of World War II, Wiman spent the first half hour of a speech to a group of factory and branch managers recounting the various depressions that the company had faced—first the downturns of 1913 and 1921, then the crash of 1929–1933, and the smaller downturn after 1937. This speech was noteworthy also for its positive, optimistic thread and its punch-line ending: "The sky is the limit." The ability of Wiman to preach caution, thrift, and spartan surroundings for offices and factories while also infusing the organization with a sense of purpose and excitement turned out to be the hallmark of his management style after the Great Depression. His charge-ahead personality of the late 1920s had been tempered and chastened by the kinds of decisions he had been forced to make in the downturn. Yet his ebullience, verve, and drive made him an arresting and exciting person with whom to share the work of an organization. *Fortune* magazine, in a breezy but penetrating article about the company in 1936, said of Wiman: "His rapid-fire distinctions between the virtues and costs of four-cylinder, two-cylinder and diesel tractors contrast abruptly with the sententious appraisals of human nature that seem to comprise the day's work of many older officers of Deere. . . . He is boyish, fast-talking, superstitious. He punctuates his remarks with such period slang as '*Siest du?*' and reminds you of Scott Fitzgerald's post-War Middle West." Max Sklovsky, the Deere experimental engineer and contemporary of Theo Brown, in writing about Wiman in 1946, again emphasized these qualities: "He possesses a combination of realism and vision.... He is a game, at times ruthless, man, but his delightful personality overcomes some of his deficiencies. At times he indicates an alarming frankness and encourages that in others. He ingratiates himself readily. He is no softy at heart but his driving is not always steady."[6]

INNOVATING IN A DEPRESSION

"With everything so quiet it is very hard to get any enthusiasm in work. It is depressing to learn so much about hard times, cutting expenses, etc. An experimental man needs enthusiasm to do good work. And I can see how this depression affects original thinking." Theo Brown, writing downheartedly in his diary in 1931, captured the essence of the dilemma facing a researcher in those darkest days of the Great Depression. How can one feel excited about new ideas when everything around him cries, "Caution! Cut back! Don't do!" Susceptible to depression in this period, Brown confided to his diary his frustration and sense of a loss of purpose.

Several longstanding company experimental endeavors were indeed severely curtailed, including the soil culture department and several experimental farms in the Moline area. The former had been the brainchild of Dr. W. E. Taylor, who had started it for Deere in 1910 and had developed a set of company publications on soil culture that were advanced concepts in the field. Taylor, a prolific speaker, built a strong following among soil scientists around the country. By 1931, he was ready to retire, and it seemed appropriate, given the economic situation, to close the department. Work in soil science continued in the advertising department, though at a slower pace in the 1930s.

But Charles Wiman had no such doubts about basic product research. Of all Wiman's management decisions during the Great Depression, the one that was probably most long lasting in effect was his decision to aggressively continue new product development through the worst days of the downturn. Already the Model D tractor was highly popular with the trade—the company had sold more than 100,000 units by 1930 and would keep the model in the product line twenty-three more years after this date. The GP and its companion, GP-Wide Track, had not done as well, though more than 35,000 would eventually be made before the latter was terminated in 1933, the former in 1935. Still, competition from the other companies constrained one from resting on laurels with one good model and two somewhat limping models. Also, it was becoming increasingly apparent that farmers around the country were avidly looking for many improvements—adjustable tread widths in the rear wheels, less side draft in the tillage instruments, perhaps even smaller versions of existing tractors.

By 1931, Wiman had put Brown and the other engineers to work on new tractor ideas, and a few months later, in April 1932, he sent Brown on a special trip to Dain, Van Brunt, Syracuse, and the Wagon Works to "pep them up somewhat on experimental work." Out of Wiman's constant press for experimental and engineering effort, and his exhortations for fresh ideas, came two new tractors, the Model A (16.22 drawbar horsepower, 23.5 belt-pulley horsepower) and the Model B (9.28 drawbar horsepower, 14.25 belt horsepower). Tested in the Salt River Valley in Arizona in 1933, the Model A was brought into production in the following year; one year later the Model B was introduced.

The two tractors were strikingly successful; both would stay in the product line until 1952, with more than 293,000 of the As sold by that date, more than 309,000 of the Bs. The two tractors rank first and second in popularity over the entire tractor history of the company.

The Model A was the solution to Deere's need for a two-plow tractor. Its adjustable wheel tread answered the farmer's need for moving his wheels outward or inward from the then-typical 42-inch standard row. (There had to be this much room for the farm horse to continue to walk on solid ground for each pass through the field.) The implement hitch for the A, as well as the power shaft, was located on the center line of the tractor, substantially reducing any side draft. An industry first, a hydraulic-power lift system, increased both the efficiency and speed of operation, as well as providing a "cushion" drop of equipment. There were other important features, for example, a one-piece transmission case that allowed high under-axle clearance.

The simple, powerful two-cylinder engine successfully burned distillate, fuel oil, furnace oil, and similar low-cost fuels, as well as kerosene or gasoline. Thus the engine could be started with gas, then switched over to the lower-grade fuels. These cheaper substitutes burned more effectively in the big cylinders of the two-cylinder engines than when used in the conventional four- and six-cylinder engines. This was a period of hard times, and an economical design with economical fuel was particularly important. Wayne Worthington, in his definitive study of the tractor, evaluated Deere's decision: "Deere & Company had always taken a contrary position with respect to fuels, and as others turned to the use of Regular gasoline (70 octane) they continued to promote the economies made possible by the use of available low cost distillates. A continuing program of combustion research was followed, which resulted in increasing the compression ratio of their distillate burning engines by some 40%. The resulting fuel economy broke existing records when tested at Nebraska. This served to educate tractor users to the importance of fuel economy even though the price of tractor distillate delivered to the farms was in the order of 8 to 8-1/2 cents per gallon."

The two new tractors were just the right sizes—the Model B was described as "two-thirds the size of the A," and company literature stressed that the A gave the pulling capacity of a six-horse team, the daily work output of from eight to ten horses, with the B having the pulling capacity of a four-horse team and the daily output of six to eight horses. The B was made particularly as a one-plow tractor. It had all the features of the A—the adjustable wheel tread, the clearance, the excellent vision, and so on (and both were available with pneumatic rear tires as alternatives to the regular metal wheels). The B had a hydraulic-power lift and, within a year of their introduction, both were also available with single front wheels (the AN and BN versions). By 1936, there were eleven different versions of the company's three basic tractors—the A, B, and D. Both the A and B were also made with standard tread, there were orchard tractors available (with covered fenders for the rear wheels and no protruding stacks at the top of the tractor), and there were versions of the A and B that had adjustable front-wheel treads, as well as rear-wheel adjustments. As the success of the two new tractors sank in, company literature, such as the catalog of 1936, exultantly extolled the many new features and reaffirmed in aggressive advertising prose the efficacy of the two-cylinder tractor (exhibit 11-4).

Wise buyers look for simplicity in buying new farming equipment. Each farmer's experience with modern farm equipment tells him that the machine with the fewest working parts is the machine that gives the least trouble, and takes the least power to run.

Exhibit 11-4. "What—Only Two Cylinders!" *Deere & Company general catalog, 1936*

Probably no single stage in the entire history of the company's product development was any more important than this one. The appearance of the two new tractors in the depth of the Great Depression was a testimony both to the company's optimism about the future and the farmers' desires for and willingness to buy simple, trouble-free farm tractors. The abiding regard for and love of the two new models—the Model A and the Model B—continued all through the 1930s and '40s up to their final models in 1952 and left a residual of acclaim that made the models collectors' items for tractor buffs down to today.

It is likely that the company would not have taken the steps to initiate these two new models had it not been for Charles Wiman. His longstanding love of both innovation and the engineering to back it up stemmed from his own academic training and his continuing enthusiasm. No armchair philosopher, he was always out in the field observing, and this was seen by everyone to enhance product innovation in the company at a time when it was desperately needed. The expenditures made in that uneasy period were substantial, particularly in light of the cuts in every other expense item in the company, both in factory and branch. It was a gamble by Wiman that was based upon faith and sound judgment, and its payoffs for the company in succeeding years were very great indeed.[7]

THE DEATH OF WILLIAM BUTTERWORTH

William Butterworth's tenure as president of the United States Chamber of Commerce extended through two reelections and embraced three critical years in the life of the national economy—1929–1931. After 1931, he continued as a member of the senior council of the Chamber, maintaining his Washington apartment. He graced the inner council of advisors to President Hoover. Despite his Republican politics, he also was called on not infrequently by Franklin Delano Roosevelt. A member of the board of directors of a number of national business organizations, he continued to be active in conferences and important private gatherings of national business figures.

On Wednesday, May 27, 1936, he was in Chicago to have his picture taken for an upcoming article in *Fortune* magazine, going from there to a board meeting at the Illinois Bell Telephone Company. On the next day, he traveled to New York City for a meeting of the National Industrial Conference Board, leaving that afternoon for Absecon, New Jersey, where meetings were to be held at the "Yama" conference center. He took a walk that afternoon with General Charles G. Dawes and David Sarnoff, president of the Radio Corporation of America. There was golf that day, though Butterworth did not take part. At the conference meeting that evening he was animated and made "quite a speech" on Sunday (according to his friend Silas Strahn). After lunch that day (May 31), he went for a walk with another friend and, as they were laughing and talking, he fell backward and was gone, victim of an acute coronary occlusion.

This was the ninety-ninth year for the firm; plans were already underway for the one hundredth anniversary, scheduled to be celebrated a few months later. Few American companies could boast at that time of 100 years of existence. Even fewer still could point to an unbroken family management—and probably no other company could point to 100 years of existence under just three chief executive officers. William Butterworth was mourned for his leadership, though we have documented a number of weaknesses along with his strengths. He was mourned universally for his personal qualities—he was benign and honest, thoughtful and humane. His deep sense of regard for and care about the employees of Deere & Company made the firm a special, unique Deere "family." It was not just that it was a family-owned corporation—the country had a number of these, not infrequently with less-than-effective managerial climates. The quintessence of the Deere story in those first 100 years was that employer and employee were truly close-knit, with progressive policies from the employer and responsible productivity from the employee. Rooted in those first 100 years were cardinal values in being a "Deere employee" (from top management down to the newly hired) that continue to give the company lineaments of human relations and sound personnel interactions that have been the envy of American industry down to today. William Butterworth had a very influential role in this.

MANAGEMENT (AND FAMILY) TRANSITION

Deere & Company was not affected materially by William Butterworth's death, for reasons documented earlier. Top management would miss his counsel, for he had been intimately involved in the company since 1892. In particular, Charles Wiman would mourn the loss of his uncle not only as a revered father figure (William and Katherine Butterworth had had a profound role in raising Charles and his brother, Dwight), but also as an always trusted chairman of the board. Though Charles Wiman was straightforwardly the chief executive officer of the company during the first half of the 1930s, he had drawn upon Butterworth's legal expertise and his wisdom in making the many critical decisions forced upon him during that trying period. Still, the company continued on with scarcely a ripple at William Butterworth's death.

Nor was there any significant change in the pattern of family ownership in the company. At Butterworth's death, he held just over 28,000 shares of common stock and an additional 21,000 of preferred (the total shares outstanding were 1,001,454 for the common and 1,543,000 for the preferred). His shares passed to his heirs in the family but did not much change the basic structure of the family's holdings. These were the owners of the blocs of more than 10,000 shares of the common and the preferred stock in February 1936 (out of a total of 2,401 holders of common, 4,936 of preferred):

	Common	Preferred
William Butterworth	28,636	21,265
Katherine Butterworth	68,172	66,185
Three Charles Deere trusts	259,392	243,205
Charles Deere Wiman	14,654	12,150
Dwight Deere Wiman	10,553	41,000
Pattie Southall Wiman (trust)		14,650
C. C. Webber family holdings	25,598	37,740
Clarence Nason family holdings	23,157	29,330
John Good family holdings	18,255	
Willard Hosford family holdings	12,446	36,735
Charles K. Velie		10,337
George Mixter		12,000
Three brokerage-firm nominee holdings of common stock for unnamed clients	68,594	
Two other brokerage-firm nominee holdings of preferred stock for unnamed clients		30,106

Several facts about these figures are important to note. First, both common and preferred stockholders had (equal) voting power for election of the board of directors (the preferred shareholders had been given this right in 1930). Second, the three Charles Deere trusts, so dominant in comparison to all other holdings, were administered (and their shares voted) by the two Wimans and Burton Peek after William Butterworth's death; the proceeds of the trusts went in the main to the Butterworth and Wiman families. Third, though Charles Deere's descendents no longer held an absolute majority of stock (nor did even the entire combination of John Deere descendants and near relatives), the total of the Butterworth-Wiman holdings alone was enough for practical operating control.

Only at infrequent times were the shareholder records put through the laborious process of following each and every family holding (of record and through nominees) to ascertain exactly how much of the company was owned by each of the five direct lineal descendants of John Deere. This research was carefully done in 1948; though it was some dozen years after William Butterworth's death, the patterns were essentially the same as then (remembering that total shares had increased by intervening stock splits):

January 15, 1948

	Common	Preferred
Jeanette Deere Chapman (Hosford et al.)	60,675	42,983
Ellen S. Deere Webber (Webber, Keator, Mixter, Murphy, et al.)	57,256	40,460
Charles H. Deere (Butterworth, Wiman, et al.)	1,056,500	353,644
Emma C. Deere Velie (Velie et al.)	15,487	17,856
Alice Deere Cady (Cady et al.)		2,085

Total family, both classes   1,646,946

Outstanding, both classes   4,547,362

Percent family holdings   36.22%

(total number of shareholders: 13,294 of common, 5,994 of preferred)

It was palpably clear that at the death of William Butterworth, Charles Wiman was the dominant figure in both company and family, influenced undoubtedly by Butterworth's gracious and respected widow, Katherine, and by the loyal, longstanding friend and lawyer, Burton Peek. Dwight Wiman was too busy with his burgeoning career as a Broadway producer to have much time for company interests.

Charles Wiman continued in the 1930s a pattern of living that had some outward similarities to that of the life style of William Butterworth in this same period, as Butterworth, having turned his business interests and avocations toward a national focus, spent considerable time both in Chicago and in Washington. In the national capital, Butterworth not only had a comfortable apartment on Dupont Circle, in the same building with Secretary of the Treasury Andrew Mellon, but also a farm near Washington, facing a lake and furnished with antiques to fit the architecture of the house. The presidency of the Chamber of Commerce was just one of many other widespread involvements—he was many years a director of the Playground and Recreation Association of America, a member-at-large of the National Council of the Boy Scouts of America, and so on. William Butterworth finished his productive life as a distinguished participant in public service.

Charles Wiman, on the other hand, was not nearly as involved in activities outside of his presidency of Deere; as *Fortune* put it in the article of 1936, "Charley Wiman sticks to the business of changing the presidency of Deere from the influential sinecure he inherited to a high-tension job of work." Wiman did share one Butterworth pattern, and for Wiman it became a lifetime procedure. Wiman enjoyed Chicago immensely and kept an apartment and office there for the remainder of his life. He also had come to love the family places in Arizona and California, particularly at Santa Barbara, and spent much time there. Charles Wiman became a yachtsman par excellence, maintaining a substantial boat on Lake Michigan and a renowned yacht in California, the *Patolita*, which he raced to Honolulu, Hawaii, in 1949 to a third-place finish in that internationally watched event. As a superficial generalization, one might conclude that Charles Wiman was treating his responsibilities to the business a bit lightly, dabbling in dilettantism. The *Fortune* editors were rather too errant in implying this in their article: "After trying for eight years to think of him as the rightful squire, middle-class Moliners are still a little puzzled by Charley Wiman. They understand that if his job weren't there he probably would not live in Moline, and they class him with his kid brother Dwight, who is a New York theatrical producer and never comes home except for directors' meeting."

Available evidence would surely corroborate *Fortune* in its judgment that Charles Wiman would probably not have chosen to live in Moline if the company had not been there. It is also an established fact that Wiman at least once proposed to the board that the headquarters of the company be moved to Chicago (the board demurred).

But *Fortune* was far off course in implying that Wiman was treating the business a bit cavalierly. Wiman had Deere & Company deep down inside of him, a fierce, aggressive pride about its business and its products. Probably much more than his Uncle Butterworth, he looked upon the company as a producer of a critically needed physical good—agricultural machinery. Butterworth would have used more explicitly financial terms in defining his feeling for the company. The company was an overwhelming, dominating influence on Charles Wiman—and he was an overwhelming, dominating influence on it. It is true that he was able to manage the company with frequent visits away, sending back decisions and advice from afar. There never was any doubt, though, that Charles Wiman was the head man at Deere & Company.[8]

RECOVERY FOR THE FARMER AND HIS SUPPLIERS

Franklin Delano Roosevelt's New Deal legislation had an enormous effect on agriculture. Set in motion in the early and mid-1980s was a system of agricultural support built upon a controversial concept—that the country could resort to planned scarcity to raise agricultural prices and, eventually, the agriculturalists' incomes. The key piece of New Deal farm legislation was the Agricultural Adjustment Act, passed just over three months after Roosevelt took the oath of office in 1933. The act was built on the bedrock concept of subsidy—in return for the farmers voluntarily curtailing acreage and/or livestock, they were to be granted direct cash payments. These benefits were to be paid for by levying excise taxes on primary processors, and there were to be collateral marketing agreements among producers, processors, and distributors for the purposes of raising or maintaining prices. The head of the Agricultural Adjustment Administration in its first months was George Peek. What a way he had come toward collective government action from his days as vice president of sales at Deere & Company!

The first Agricultural Adjustment Act was declared unconstitutional in 1936 because of its processing tax and production control features; it was replaced by another in 1938. Other pieces of New Deal agricultural legislation provided for the use of surpluses for relief measures, for soil conservation, and for rural resettlement and rural credit. In the process, the confidence of the agriculturalists of the country was restored.

The farm machinery companies now began to receive the fruits of the farmers' renewed optimism stemming from the New Deal legislative initiatives. Their sales had totaled over $585 million in 1929; this figure had fallen to $259 million in 1931, and the Department of Commerce did not even develop statistics for 1932 and 1933. The eight long-line companies had accounted for $437 million of the total in 1929; in 1932 their total was just $85 million. By 1936 they had climbed back to $373 million. In the process, though, considerable shifting in market share had occurred (exhibit 11-5).

The erosion of International Harvester's early dominance had become substantial; Deere had taken a chunk out of its archrival's market share. But the real surprise was Allis-Chalmers. The latter's growth was testimony once again to the fact that gaining additional market share is not an insuperable task when attempted by firms of sufficient size and resources to ensure adequate manufacturing efficiency and marketing reach. The Allis row-crop tractor became a strong competitor to International Harvester's Farmall almost as soon as it was introduced in 1930. Allis also brought out in 1934 a small-sized 2,800-pound harvester that cut a 5-1/2-foot swath; it was called the All-Crop. A one-man harvester, it was designed especially for family farms of less than 100 acres and was an instant commercial success.

Exhibit 11-5. *Proportion of Sales of Farm Equipment Made by Each Long-Line Company as a Percentage of Total Long-Line Farm Equipment Sales, 1929 and 1936*

	1929	1936
International Harvester	51.8%	44.7%
Deere	20.6	23.3
Allis-Chalmers	2.9	11.3
Case	7.6	7.8
Oliver	7.6	6.5
Minneapolis-Moline	4.1	3.7
Massey-Harris	3.9	1.8
B. F. Avery[a]	1.7	1.0
	100.0	100.0

[a] *The FTC considered B. F. Avery a long-line company, despite the fact that it did not make tractors (though it did have a marketing link for its tractor equipment with Allis-Chalmers).*
Source: *Federal Trade Commission,* Report on the Agricultural Implement and Machinery Industry, *75th Cong., 3rd Session, House Document 702 (1938), 626.*

The significant edge of International Harvester and Deere in their selling and administrative costs continued unabated; it was as strong in 1936 as in the earlier periods reported by the Federal Trade Commission. Warren Shearer, in his well-researched study of the industry, concluded: "Again, as in the 1916-18 period, the major advantage of Deere and IH lay in their lower costs of distribution."[9]

LABOR RELATIONS IN THE 1930s

After the National Industrial Recovery Act came into being in 1933, the dedication of the federal government to the encouragement of collective bargaining quickly brought a resurgence of labor unions (which had been declining in membership since the early 1920s). When the NIRA was declared unconstitutional in 1935, Congress immediately passed an even stronger piece of legislation, the Wagner Act, which the unions called their "Magna Charta" for collective bargaining. Organizing activities mushroomed all over the country, and the labor movement itself went through profound change, the Committee for Industrial Organization being founded in that same year as an industrial-union alternative to the American Federation of Labor's predominant craft unionism. The CIO established the Steel Workers' Organizing Committee (SWOC) and in March 1937 won a contract with the Carnegie-Illinois Steel Company, the chief subsidiary of the United States Steel Corporation. There was violence at the Republic Steel Company in 1937, as the SWOC tried also to organize "Little Steel"; the Memorial Day massacre there resulted in deaths on the picket lines, and the smaller steel companies stayed nonunion until 1941.

The CIO's greatest triumph was at the General Motors Corporation; a "sitdown" strike—in effect, a seizure of plants by the employees—first brought more picket-line tension, but soon the capitulation of the company, after intervention in Michigan by Governor Frank Murphy. Chrysler also signed an agreement in 1937. Ford held out until 1940, signing then with the union involved, the United Automobile Workers (UAW).

In the early NIRA days there were organizing efforts in the Moline–Rock Island area. George Wilson, who handled labor relations for the company, reported to the board in April 1933: "The American Federation is very active both socially and politically, locally and nationally. The communist group, which covers a variety of organizations which have in mind the One Big Union idea, is growing in strength." Left-wing unions had seemed to preoccupy executives in the company for a number of years. Wilson continued: "Our local situation is not particularly bad, although there has been a grouping of organized labor with the various organizations representing Communism, Socialism and the so-called unemployed. It was this Joint Committee that started the hunger march in Springfield. . . . I was waited on by a Committee at home one night last week; rather a tame lot. I think I got more satisfaction out of it than the Committee."

Nothing further happened in Moline that year, but in the summer of 1934 the American Federation of Labor again attempted to organize in that city. The company still had the individual contract as its basic employment policy, and Charles Wiman himself had reinstituted the signing of such contracts earlier that year. The manager at Waterloo, A. H. Head, put out a two-page notice to employees, signed not only by him, but "prepared in collaboration with and jointly subscribed by C. D. Wiman, President Deere & Co." Head alleged that there had been much "misunderstanding and misrepresentation of the facts" connected with the NIRA code—that the code "fully and intentionally protects the individual employee in his right to make his own deal." Head continued: "We have operated an open shop for many years, and in compliance with the Act we shall continue to do so. The mutual welfare of our employees and ourselves is best served by this policy." Through this period of the NIRA, the company vigorously opposed unionization, even contributing money toward employer organizations specifically "to combat unfair propaganda."

CHAPTER 11

When the Wagner Act was enacted in 1935, the company once again interpreted the legislation to embrace the legitimacy of the individual contract. Burton Peek reported to the board in 1937 that Senator Wagner had written the company before the legislation bearing his name had passed that "there is nothing in the bill which prevents employers from dealing individually with their employees, when that has been the basis of satisfactory and peaceful conditions in the past and both the employers and employees desire to continue that arrangement." Peek vowed: "Our existing contracts are valid and in full force and effect, notwithstanding this Act. Until they expire, it is not seen how the Company could engage in collective bargaining." The company once more opted to issue bulletin board notices, to apprise the employees: "You have the right (but are not required) to bargain collectively. . . . You should know that you have the right to deal as individuals with your employer when you desire. Neither the Wagner Act nor the decision of the Supreme Court denies such right."

Apparently the union movements in both Waterloo and Moline–Rock Island felt they were not strong enough to challenge the company, for the individual contract held sway until 1939. At that time, the National Labor Relations Board (NLRB) ruled on a complaint by the Steel Workers' Organizing Committee (SWOC) against the company, alleging "certain unfair labor practices" at the Killefer Manufacturing Corporation. This small Los Angeles maker of road-grading equipment, subsoil plows, and disc harrows had been purchased by Deere in 1937; most of its sales were in California, with Caterpillar Tractor Company handling its marketing through "Cat" dealers. By this date, Caterpillar and Deere had close links in California and the Killefer acquisition was readily fitted to Deere's and Caterpillar's California sales efforts.

When Deere took over Killefer, it introduced the practice of the individual contract. The company also inherited an incipient union effort from the SWOC, and the latter soon alleged that the individual contract was an unfair labor practice under the Wagner Act. The NLRB centered its interest on a particular clause in the written individual contract that asked the employee to agree "not to join in any concerted movement during the life of this contract, for a change in wages, hours or other conditions of this contract," and ruled against the company. Deere was required to "cease and desist" from giving effect to this particular clause. The following year a similar charge was made by the Farm Equipment Workers Organizing Committee (another CIO organizing union) against Deere's Ottumwa and Waterloo factories; again the NLRB ruled against the clause. Though the stipulation did not strike down the individual contract per se, it was the beginning of the end for the practice. Neither union, incidentally, gained a foothold in any factory of the company at that time.

While the company labor policy during the turbulent 1930s was openly anti-union, the general tenor of employee relations was considered by most outsiders to be amicable and effective. The editors of *Fortune,* in their article of 1936, played a bit loose with the facts, but they were right on target about the essence of Deere's labor relations: "Moliners also sense, without understanding, a difference in Deere's labor policies since Charley Wiman took over. Mr. Butterworth, (whose life was threatened more than once by grudge-nursing employees) always liked to find hard working young fellows with the right stuff in them and train them up from the ranks. They do not hear so much about that from Charley Wiman, who used to work in the shops himself, has inaugurated the 40-hour week and vacations with pay, and has had no labor trouble throughout Deere's worst depression." This does some injustice to Butterworth, who had relied heavily on Leon Clausen for his labor policy and the particular use of the individual contract. When Clausen had become too heavy-handed, Butterworth had been instrumental in having Clausen leave the company. Nevertheless, the judgment that the anti-union philosophy of the 1930s was a product of the 1910s and '20s is an accurate assessment.[10]

HENRY DREYFUSS STYLES THE TRACTOR

The farmer loves his tractor, provided, of course, it works well! This machine that had become so important was to him indeed "a thing of beauty." But beauty is in the eyes of the beholder, goes the old adage; the farmer, a practical man, did not worry much about the aesthetics of any of his working life, and his regard for the tractor was rooted in its performance, not its appearance. A few tractor manufacturers had attempted to incorporate aesthetic design features—the Peoria Tractor Company had built a "streamline" version of its tractor in 1920, and one of the Twin City tractors of the Minneapolis Steel and Machinery Company looked a good deal like an automobile racer. But most manufacturers, including all of the major producers, developed strictly utilitarian vehicles never destined to win prizes from outsiders for attractiveness.

Some of Deere's engineers at the Waterloo Tractor Works had the dream, though, of making their tractors more pleasing to the eye, more aesthetically satisfying. By 1937 the notion of good industrial design had been accepted widely among companies, and some well-known designers had achieved prominence in the field. One of these was Henry Dreyfuss, who, with his associates, had formed and nurtured a respected design house based in New York. The engineers pressed Charles Stone, who now headed all of manufacturing, to authorize them to contact Dreyfuss. Stone gruffly allowed that he did not think much of the idea, "but the boys up here want it, so go ahead." One engineer, Elmer McCormick, was delegated to go to New York, unannounced, to engage Dreyfuss. The story, probably apocryphal but often retold, was that McCormick, going to the city in the late fall with a fur coat and a straw hat, so impressed Dreyfuss with the potential for redesigning a tractor, an instrument that had hardly been touched by industrial design, that the renowned designer got on the train that very night and traveled straight to Waterloo to talk with the group.

Dreyfuss first was given the assignment to look at the aesthetics of the Model A and Model B. A wooden mockup of a streamlined Model B was

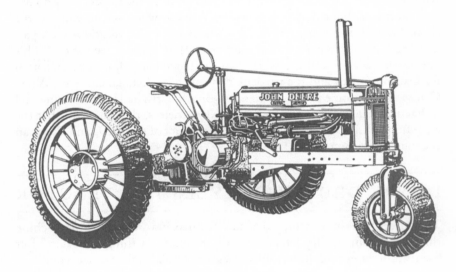

Exhibit 11-6. Model B tractor, before and after restyling by Henry Dreyfuss. *Deere Archives*

constructed by the Dreyfuss team, and in November 1937 the model was unveiled to a startled Waterloo group. It incorporated many new design features relating to the grill and louvres, the shroud over the steering shaft, and design of the radiator cowling. Out of this first glimpse of a more "beautiful" Model B soon came a number of design features that improved the physical appearance of both the tractors. By the 1938 model year, the Waterloo engineers and the Dreyfuss design specialists had combined to redesign both the Model A and the Model B in a most striking way. The radiator cowling had been narrowed for better visibility, the louvres had been redesigned as sweeping horizontal slits (the earlier version had been just a standard radiator front), the dash area had been changed both for more ready readability and for good design.

The changes were not only cosmetic, for the basic safety and workability of the tractor had been significantly improved as well. Beyond this, the Waterloo engineers were delighted to find in the Dreyfuss group a view of the manufacturing process itself that frequently gave the engineers new ideas about easier and less expensive ways of manufacturing. For example, the Dreyfuss group also took a look at some of the Deere harrows and suggested that having one of the frames bent from a single piece of tubing, instead of welding four tubes together, was not only more aesthetically pleasing with the rounded corners, but eliminated in the process four distinct welds.

The farmer was not always easy to persuade on these aesthetic changes. One example in this period before World War II will illustrate. The company a number of years back had begun manufacturing the wheels for its plow with spokes of oval design, this having been determined by the engineers to be the strongest structural shape. Soon complaints began coming back from Argentine farmers about the oval wheel spokes, not because of any wheel failures but because the Argentine agriculturalists thought the spokes looked too "light" in appearance. They were used to the heavy round spokes of the European models and felt the American version might not be as sturdy. In truth, the opposite was the case. So design engineers like Henry Dreyfuss had to be able to incorporate both design aesthetics and practicality in order to relate to the eventual customer's reactions.

From this first experience of working with the Model A and Model B, Henry Dreyfuss and his associates formed a lasting association with Deere & Company that has continued to the present, a relationship that has had a major influence on the appearance, safety, and usability of Deere products.[11]

CHAPTER 11

DEERE AND CATERPILLAR JOIN FORCES

Though agriculturalists were overwhelmingly the clientele for Deere tractors, there had been some industrial use of the Model D from about 1926, when Model DI—the John Deere Industrial Tractor—was announced. It was straightforwardly a Model D, with only two required changes. First, hard rubber tires were put on both front and rear steel wheels (pneumatic tires were not used in the industry until 1931; Deere first made them available in 1934). Second, high-speed sprockets and chains were to be exchanged for the standard ones, allowing a road speed of four miles per hour if the sprockets were changed from thirty-eight to twenty-eight teeth, and a full five miles per hour if a twenty-two-tooth sprocket were used. Additional wheel weights could also be put on, as well as an extension wheel in the rear if additional support and traction were required.

The industrial uses of the Model DI were not exploited very well by the company through the remainder of the 1920s, and it was not until 1935 that a particularly fortuitous association brought Deere's efforts to a more substantial level. This was the link with Caterpillar.

In the early 1930s, the International Harvester Company had added a crawler-type tractor to its line and at the same time the Allis-Chalmers Manufacturing Company, which had previously made only crawler-type tractors, began making wheel-type tractors. Thus, both companies now offered both types of tractors through their dealers. This was particularly important in one key state, California. Many California agriculturalists farmed on a mammoth scale, and for their big tasks, particularly under certain soil conditions, they demanded track-type machines. At the same time, however, on every farm there still remained a number of utility and smaller cropping efforts that were more readily done by wheel-type tractors. Thus a California implement dealer was not handling a "full line" unless he had both types of vehicles. This lack was painfully evident to both Deere and Caterpillar, neither of whom produced both types of tractors.

The impetus for an accommodation came from Caterpillar in January 1935. Frank Silloway reported unequivocally to the Deere board: "We do not desire to manufacture track-type tractors," but, inasmuch as the Caterpillar industrial dealers wanted a wheel-type tractor, "through this Caterpillar arrangement we will have an opportunity to get into this industrial business in an important way."

One of the most attractive features to Deere in the new arrangement was the opportunity to join Caterpillar abroad. "Cat" already had a significant dealer network in a number of foreign countries; often agriculturalists abroad had the same dual need for track-laying tractors for larger jobs, wheel tractors for smaller ones. In truth, a number of these foreign outlets—for example, in some of the Central and South American countries—were really too small to fully support a separate Caterpillar or separate Deere dealer network. Joined together, the two companies could have a larger exposure, both of products and of dealer size, which made for better parts service and other backup needs, as well as more initial sales.

There was one uneasy complication to this seemingly amicable relationship—the farmer quite often had a choice as to which tractor would be more effective, and a bit of persuasion could tilt him one way or the other. Would Caterpillar ever persuade him toward a wheel-type tractor, worried the Deere marketing people (an attitude probably shared just as strongly in the opposite direction by the "Cat" men). Frank Silloway decided to bring this issue into the open and wrote the branches: "We told them that we were not thin-skinned in any way, and were willing that they urge these dealers,

Exhibit 11-7. Top, No. 35 combine, with Caterpillar tractor, 1941; bottom, John Deere DI industrial tractor, pulling Caterpillar Trailer Patrol grader, 1935. *Deere Archives*

in the course of their usual sales activity, to push the sale of the track-type tractor. . . . It is our idea that, after all, the farmer is going to be the final court and jury."

Caterpillar also had doubts about the two-cylinder mindset of Deere. They sent forth an equally frank bulletin to their own sales people, first describing the Deere organization and then facing the issue of the two cylinders. "It is most interesting to note," the bulletin's editors began, "that John Deere, the founder of the steel plow industry and Charles H. and Benjamin Holt, founders of the track-type tractor industry, were all natives of Vermont." This homage out of the way, they continued: "The builders of John Deere tractors have been particularly consistent in regard to the design of their engines . . . two-cylinder design is now exclusively a John Deere feature. In selling the John Deere tractor, 'Caterpillar' salesmen may be met with the sales argument that all other tractor manufacturers building two-cylinder tractors have discontinued this feature." But it would be "false reasoning" to assume that the two-cylinder engine was obsolete, inasmuch as "Caterpillar Tractor, recognized the world over as builders of the finest line of engines in existence," included in their line not only two-cylinder but one-cylinder, three-cylinder, four-cylinder, six-cylinder, and eight-cylinder engines. The latest diesel that Caterpillar had just brought out was itself a two-cylinder engine.

Would there be excessive vibration in the John Deere tractor? "Vibration in an engine is not a product of the numbers of cylinders," the bulletin answered, "both the crankshaft and the flywheel of this John Deere tractor are dynamically balanced." The two-cylinder design would give greater simplicity, excellent fuel economy, the ability to burn a number of different fuels, compactness, excellent accessibility, light weight, and better distribution of weight, and many of the above features were exclusive ones. "Inasmuch as the 'Caterpillar' salesman has been trained to sell exclusive features . . . he should welcome these exclusive John Deere features as sales opportunities."

Within a few weeks new literature had been prepared featuring together Deere's Model D and Caterpillar's Trailer Patrol, a pull-type road scraper with an eight-foot blade. Maneuverability of the combination was stressed, as was the ease of operation of the Model D and the rugged construction of the scraper.

Now the Caterpillar-Deere link took a surprising new direction. In 1926 Caterpillar had offered to sell Deere its harvester business—the Western Harvester Company. The offer had been made when Deere was first developing its own combine and shortly after Caterpillar itself had been put together out of the combination of the Holt and Best organizations. The asking price then was more than $1.2 million, too steep a price for Deere. Over the intervening years, Caterpillar had found its niches to be track-laying tractors and road machinery. Its combine business was confined largely to the West coast, the old Holt and Best territories, the total business not being substantial enough to expand efforts into the Midwest. Indeed, it was beginning to dawn on Caterpillar that the combine business was actually getting in the way of their aggressive moves into the industrial construction and road-maintenance areas.

This time the deal was much more favorable to Deere. Caterpillar had been manufacturing two conventional combines and a well-received hillside combine. Deere did not have the latter machine in its line, and this had hurt in the West. Caterpillar's proposal was to discontinue their two models of conventional, level-land combines and to turn over the hillside combine to Deere to manufacture. All the repair parts common to the three machines then would be furnished by Deere; Caterpillar would continue to stock those parts for the other two combines that were not being made by Deere. Deere would also take over Caterpillar's existing inventory of completely fabricated machines at Caterpillar cost. Deere would be given a royalty-free, nonexclusive license to the use of all of the Caterpillar combine patents; all of Caterpillar's existing templets, dies, jigs, fixtures, samples, patterns, special flasks, special cutting tools, and equipment were to be delivered to Deere without charge. All of Caterpillar's existing quantity of combine advertising would be turned over to Deere without charge. Caterpillar's tracings, blueprints, bills of materials, production records, cost records, specifications, and stock records would also be given to Deere without charge. Caterpillar was willing to supply Deere with a Caterpillar engine for the combine, though it was not a requirement of the agreement. In sum, Caterpillar had given Deere outright the complete rights to make the hillside combine, and it had backed this with all the necessary engineering and manufacturing expertise Caterpillar had assembled over the years. The arrangement was efficacious for Caterpillar, too, for it extracted the company in a clean way from a product line in which it was no longer interested, yet leaving existing users to be serviced by a responsible organization. Deere gained immensely, for it now had a hillside combine to round out its own combine line—and at essentially no cost.

Now the Deere-Caterpillar combined sales effort needed to be put into high gear, for not only would Caterpillar dealers be selling Deere wheeled tractors and Deere dealers selling some Caterpillar equipment (track-laying tractors, road scrapers, etc.), but now Caterpillar dealers would be phasing out their combines, Deere dealers adding the hillside combine.

There was a price to pay, though, and it came in the effects on Deere and Caterpillar dealers. Frank Silloway again: "If the dealer does not want the agency for the Caterpillar tractor, he need not take it." There were a number of places where track-laying tractors would sell, however, and "we believe that in such a town the John Deere man will be interested in having the dealership . . . and if he is not interested now, he certainly will be when the local dealers of the I.H.C., Allis-Chalmers and Cletrac are selling track-type tractors right out of his town." Silloway wanted to "branch out and make a real bid for this industrial tractor business," so that "the sale of John Deere tractors will not be so completely dependent upon the farmer trade." Caterpillar, in turn, exhorted its dealers that there were a number of places where wheel-type tractors would be needed for "light tasks." Their bulletin continued: "In a lot of cases this work has been done by a Fordson tractor, and in just a lot more cases horses are being used. . . . [While] this possibility can't be classed as a new market, it is just a market for which we have nothing to offer."

For the Deere and Caterpillar dealers over most of the country, this was not to be a major new endeavor—one that would require changes in their branch—but just a modest new piece of business that would link two good companies together. There was one place in the country, though, where the wedding had major effects—California.

CALIFORNIA AND THE CATERPILLAR LINK

For years, California's agriculture had been amazingly varied and astoundingly extensive in scope. Individual farms often tended to be very large—by 1935, some 43 percent of California crop land was farmed by only 7 percent of its farmers, in spreads upwards of five hundred acres each (with a number of them many thousands of acres). The average value of a California farm at this time was $15,466, the highest of any state in the country (Nebraska, at $11,696, was next; the country's average was $4,823). Deere had operated two branch houses over the years on the West Coast, one at Portland, Oregon, the other in San Francisco, California. Neither had been outstanding, nor, in turn, had either been accorded much help from Moline (a constant complaint over the years from the West Coast).

In the California branch, the company had been doing business essentially the same as it had in the rest of the country—a quite substantial number of smaller, individual John Deere dealers. In the San Francisco territory at the time of the Caterpillar deal, there were approximately 250 dealers in areas of the state that were to be jointly served by Deere and Caterpillar. Caterpillar's pattern was much different—its dealers tended to be very much larger, much more financially secure, and therefore considerably more independent in relations with the home office, and to serve a wider area. This was a logical pattern, because Caterpillar dealt mostly with large-scale farmers, those who were likely to use the track-laying tractors.

Thus the fit between Deere and Caterpillar dealers, while excellent in the rest of the country, was just not right for California. This fact being self-evident, Deere's San Francisco branch terminated relationships with most of the 250 dealers (who generally then took on competitor brands), and then began selling Deere equipment through the Caterpillar dealers, who totaled only twenty-six at the time of the changeover. By 1940, when Frank Silloway made a major report to the Deere board about California, the John Deere-Caterpillar dealers numbered only eighteen, who had between them thirty-four other branches, for a total of fifty-two towns out of which the combined dealerships operated. In other parts of California, other than the areas of the John Deere-Caterpillar link, there were some ninety-four regular John Deere dealers by 1940, on the same basis as elsewhere in the United States.

The changeover in California to the larger John Deere–Caterpillar dealers brought severe transition problems. The San Francisco branch had to agree with its former dealers to take back stocks of Deere goods and give them credit substantially equal to the cost of such stocks. Unfortunately, much of this stock was shopworn and obsolete and could only be disposed of after reconditioning and at bargain prices. The need for understanding Caterpillar-Deere service and repair concerns required the San Francisco house to invest in a much larger repair department. Further, as the Caterpillar dealers came to the relationship with greater size and financial strength, they were able to buy in carload lots, and Deere had to grant more substantial freight concessions and cash discounts.

So the new John Deere–Caterpillar dealer in California really was an anomaly in the company. The dealerships had a more arm's-length relationship with Deere, and Caterpillar pride and aggressiveness often made the dealer seem to Deere branch-house personnel to be overly caustic and demanding. Benjamin Keator, the branch manager in San Francisco from the time of the Caterpillar link (he had been brought down from the Portland branch, which he had run since 1925), reminisced at the time of his retirement in 1953 about the Caterpillar alliance. Keator concluded that the relationship had been very beneficial to Deere, but also felt that he was "all the time playing second fiddle to the Caterpillar organization, which I can assure you was no fun and would have 'gotten the goat' of most Deere men who had any loyalty in their veins." The feeling grew that Deere was allowing Caterpillar to call the tune in California, and many Moliners felt that Deere should have a larger number of dealers in the joint territory. Frank Silloway was manifestly upset by what he saw as a growing schism, and, in a blunt report in 1941, he cited the figures relating to Deere's market share in California (which had risen all through this period). He stated: "You will, therefore, see from all of the above that we do not have much ground for complaint on our John Deere–Caterpillar setup in California. . . . There is too much of an impression in California at the present time that we do not like the Caterpillar dealers and are tolerating them only on account of the John Deere–Caterpillar setup. They feel that our side of the business transaction is influenced by some grudge we bear. Of course, that is not true." Silloway knew there really was a problem and concluded: "We must change our entire attitude and show these dealers that we do appreciate what they are doing for themselves and for us; that we are their friends and are cooperatively with them for a bigger and better business for John Deere in California."

The most effective times for the joint dealerships were those years just before World War II—from 1935 to, roughly, 1941. In this period, the companies had mutual interests in building strong dealerships and interactive sales. The war itself was a dividing line. After Pearl Harbor, Caterpillar sent its equipment all over the world. The Army Corps of Engineers and the Navy's renowned Sea Bees built landing strips, ports, and bridges all over the European and Far Eastern theaters. The Caterpillar tractor (the second word soon not even used, "Caterpillar" becoming almost a generic name) was the chosen instrument for a great amount of this work. The International Harvester crawler was also widely used by the armed forces. By the end of the war, Caterpillar had left its equipment all over the world, with a justly deserved reputation for quality and durability. It had the beginnings of a mammoth, worldwide business, and it made the most of it. Concurrently, many of the domestic joint dealerships were expanding enormously in the sale of new machines and in repairs and parts, with most of this expansion on the Caterpillar side of the aisle. The war had left a residue of need for agricultural equipment, too, but not on the scale of Caterpillar's more industrially based opportunities. In many of the domestic joint dealerships, and widely among the foreign operations, the Caterpillar faction of management now tended to dominate—and probably rightly so.

This raised plenty of hackles all through Deere and seemed to threaten the relationship. R. B. Lourie addressed a joint meeting of branch-house managers and Caterpillar representatives in early 1947 and decided to put the matter frankly to them. He pointed out that though top management in both companies knew and understood the arrangements, some of the people down in each of the organizations were really not completely "sold" on the joint link. (Left unsaid was that a few were even sabotaging it.) As some of the people Lourie was speaking to were these people, Lourie pleaded, "I want to suggest that all of us let by-gones be by-gones and start fresh as of now to give this Deere-Caterpillar program a real chance to work to our mutual advantage." Lourie had to tiptoe with his own Deere people, and he vowed, "I want to make it very clear right here that we are not interested in cancelling any good, loyal John Deere dealer in order to make room for a Caterpillar distributor or his branch store." But Lourie also wanted to gain a quid pro quo, so he proposed that Deere would consider new dealerships for Caterpillar people if Caterpillar people were, in turn, willing to do so for Deere. The tenor of the meeting was positive, but the undercurrent of combativeness in Lourie's remarks telegraphed the chary feelings of both parties.

It was only a matter of time, though—the relationship was star-crossed. A number of Caterpillar dealers now began to question the validity of diffusing their efforts by also attempting to sell Deere equipment, now not a major part of their business. Finally, more and more dealers were inquiring about the possibility of terminating the joint relationship.

From Deere's side, a severance was not nearly so attractive. To be sure, everyone knew that there was not a full synergy between Deere's and Caterpillar's business, but Deere was not yet ready to develop a competitor track-laying tractor to compete head-on with Caterpillar. (At a later point Deere did decide to move into industrial equipment in a major way and to build a range of machinery in the smaller sizes directly competitive with those of Caterpillar.) Thus the question of the relationship between Deere and Caterpillar was not an inconsequential one for either party. Both knew that there was always the possibility of each moving into the other's product lines.

Over a number of years beginning in the early 1950s and extending into the mid-1960s, the joint dealerships began to split apart (except at certain locations abroad). This left Deere with similar enigmas in regard to dealer choice, except precisely in reverse. Now Deere needed to rebuild a dealer organization, particularly in California, and go through a laborious, sensitive "divorce," moving away from a joint relationship to begin again completely on its own.[12]

PRODUCT DEVELOPMENT IN THE LATE 1930s

Despite some drift in both the economy and the agricultural machinery industry in 1937 and '38, the late 1930s saw aggressive product development in the industry. The small tractor had been welcomed by the farmer as in no other period before, and there were an estimated 1.6 million tractors in use in the country in 1939—almost double that reported in 1930. Perhaps as many as half of these were general purpose. The United States Department of Agriculture, in a special report in 1940 chronicling the onrush of technology on the farm, estimated that upwards of 60 percent of those farms large enough for motor equipment were using tractors by 1940; they flatly predicted that the number of tractors would increase by 500,000 by 1950. The USDA's authors also reported that what they called the "baby combine," one that cut a swath of five to six feet and could be operated by one man, comprised some 80 percent of the combines sold in 1939.

Deere was nicely positioned in the tractor market, particularly after adding the Model A and Model B in 1934–1935. A larger version, the Model G, with a 20.7 drawbar horsepower, was marketed beginning in 1937 and a smaller version, the Model H, with a 9.68 drawbar horsepower, joined the line in 1939. The company also added a small utility tractor, the Model L, with a 7.01 drawbar horsepower. The last was built at the John Deere Wagon Works in Moline.

Exhibit 11-8. Left, No. 55 combine working a field near Findlay, IL, ca. 1950; right, No. 55-H sidehill combine in winter wheat, Wenatchee, WA, 1954. *Deere Archives*

The company was not as strong in the combine market. The hillside combine from Caterpillar had been very effective and the other larger Deere combines were selling well. Mushrooming sales of smaller combines, however, made it necessary to bring out Deere's own versions, and here the company stubbed its collective toe. A six-foot version, the No. 6, was developed in 1936 and tested over the following two years. It was to be Deere's answer to Allis-Chalmers' All Crop. Deere engineers compared the two machines side by side and Theo Brown dolefully reported, "Saw a lot of Allis-Chalmers, all doing well and found #6 John Deere a real lemon." Brown continued a few days later in his diary: "Deere's position in the small combine field is very poor. We must do something soon. . . . It is a tremendous problem." M. J. Healey, the aggressive Kansas City branch manager, was particularly vocal: "It was not a good machine. . . . We are not anywhere up to competition on the small combine." Healey wanted Brown to intrude into the picture but the latter was reluctant, given the continuing decentralization of authority for product development. "I told him my position was that when anything like that was wanted, that I could not butt in but should be asked and wanted." Charles Wiman, sensing the critical nature of the problem with the small combine, stepped into the situation and promptly assigned Brown to the Harvester Works, once again demonstrating his ability to affect product development from the outside, yet to leave prime responsibility in the hands of the individual organization.

As the demands for smaller combines continued to increase, the company built two new machines, the Nos. 11A and 12A, and put them in the product line in 1939. These were left-hand cut machines, built this way for an interesting reason. During the 1930s there had been a major switch from grain binders to combines. The company branches in the areas of the country

where windrowing was the practice felt that if the farmer could not use his binder at all, he would swamp the branch with trade-ins for combines. If a left-hand cut combine were developed, the binders could be converted into windrowers. These two machines were better engineered than the No. 6, and the No. 12A proved to be particularly popular in its thirteen-year life, 116,000 units being built and shipped in that period. There were some problems with these two machines at the start, though, their lightweight construction not holding up in all field conditions, with some of them being returned to the factory for rebuilding.

Thus both the farmers and the farm machinery industry came through the 1930s stronger than any could have imagined at the depths of the Great Depression in 1932. Though the smaller long-line competitors lost market share and relative profitability, the industry seemed characterized at the end of the Depression by more vigorous competition, partly because of the erosion of International Harvester's dominance but particularly because of the aggressive marketing of Allis-Chalmers. Deere, too, was now well positioned for the challenge that lay ahead with the onrushing war.

Endnotes

1. Production statistics on tractors are reported annually by the Bureau of Census; a useful summary of the figures for the 1920s and '30s is in *Implement and Tractor*, May 15, 1948, 60–61. For the quotations on Deere results, see *Chicago Tribune*, January 30, 1930; *Wall Street Journal*, January 30, 1930; *Christian Science Monitor*, May 10, 1930; Jas H. Oliphant & Co. release, February 1, 1930. The Andrew Mellon quotation is from *Commercial and Financial Chronicle*, January 4, 1930. The Deere stock dividend and stock split is noted in Deere & Company *Minutes*, January 28, 1930.
2. For F. Silloway's remarks, see branch bulletin 714, February 9, 1932. See M. J. Healey, "Rules and Instructions for the Guidance of our Collectors," Kansas City branch, March 1, 1935, DA, 28397.
3. The shipment of 1923 on behalf of American relief is noted in Deere & Company *Minutes*, January 30, 1923. For the quotations on Russian credit worthiness, see ibid., October 30, 1928; ibid., July 29, 1930; ibid., August 4, 1930; ibid., January 20 and 27, 1931. For the text

CHAPTER 11

of the children's story, written by Boris Kushner and dated January 20, 1931, see DA, 1125. For additional discussion of Russia's purchases of machinery in the 1929–1932 period, see Wayne G. Broehl Jr., *Precision Valley: The Machine Tool Companies of Springfield, Vt.* (New York: Prentice-Hall, 1959), 128–35.

4. For Frank Silloway's remarks on Argentina, see Deere & Company *Minutes*, July 31, 1923; for the Agar, Cross representative's remarks, see ibid., April 26, 1932. The agreement to take back the Model D tractors is first mentioned in ibid., April 24, 1934 (where the number was to be only 500); for the final agreement to take back 900, see ibid., July 5 and 31, 1934. For Deere's purchase of one-half interest in Dunnell Edden, see ibid., October 28, 1930; the comptroller's report of 1932 elaborates on the details.

5. The early history of Deere in Canada is documented in C. A. Carter, "John Deere of Canada" (ca. January 1952), DA, 2300. For Wesbrook & Fairchild, see DA, 19357. For early branch-house history, see DA, 3570. For the Charles Stone memorandum on returning to Canada for manufacturing, see Deere & Company *Minutes*, January 27, 1931; for Charles Wiman's visit to the Canadian prime minister, see ibid., March 31, 1931. For United States–Canadian tariff relations for agricultural machinery, see William G. Phillips, *The Agricultural Implement Industry in Canada: A Study of Competition*, Canadian Studies in Economics, no. 7 (Toronto: University of Toronto Press, 1956), 40–45, 54–61, 68–76.

6. Wage and salary cuts are documented in Deere & Company *Minutes*, April 21, 1931; ibid., July 28, 1931; ibid., October 27, 1931; ibid., April 26, 1932; and ibid., July 28, 1932. For the quotation on the Thrift Plan, see ibid., July 29, 1930; for the decision on suspension of vacations with pay, see ibid., March 31, 1932; for reduction of rents, see ibid., June 21, 1931. The Peoples Bank defalcation is described in ibid., March 14 and 15, 1931; ibid., April 21, 1931; ibid., September 25, 1931; ibid., October 27, 1931. See C. C. Webber to Charles Wiman, March 28 and 31, 1931; C. Wiman to C. C. Webber, March 30, 1931; see also the S. H. Velie Jr. letter, March 21, 1931, the Willard Hosford letter, March 20, 1931, DA, 1105. For the early history of the bank, see Newton Bateman, ed., *Historical Encyclopedia of Illinois and History of Rock Island County*, vol. 1 (Chicago: Munsell Publishing Company, 1914), 749; for the final liquidation in 1941, see DA, 1107, 1108. See clipping from *Chicago Journal of Commerce*, July 16, 1931; for the Wharton and Wiman comments, see DA, 19389. For Wiman's quotation on employees, see Deere & Company *Minutes*, October 25, 1932. See "Deere Plow," *Fortune*, August 1936, 73; Max Sklovsky to Charles Wiman, October 22, 1946, DA, Overlook file and attached manuscript, "Excerpts from 'Creators of Deere & Company.'" Charles Wiman's speech to the factory and branch-house managers was at Dubuque, Iowa, on October 11, 1946.

7. See Theo Brown, "Theo Brown Diaries," October 13, 1931; ibid., April 29, 1932. The termination of the soil culture department and Dr. W. E. Taylor's retirement are discussed in William Butterworth to Maurice Block, September 19, 1931, DA. For the decision to terminate the department, see Deere & Company *Minutes*, October 27, 1931. The closing of the experimental farm is discussed in T. F. Wharton to W. Butterworth, June 25, 1931, DA. For Model A and Model B tractor development, see Deere & Company *Minutes*, October 25, 1932; ibid., April 25, 1933; ibid., January 30, 1934. Wayne Worthington, *50 Years of Agricultural Tractor Development*.

8. *Fortune*, August 1936, 155, 159.

9. For index figures, see *Historical Statistics of the United States, Colonial Times to 1970, Bicentennial Edition*, part 1 (Washington, DC: 1975), 200. For the development of New Deal farm policy, see Edwin G. Nourse, Joseph S. Davis, and John D. Black, *Three Years of the Agricultural Adjustment Administration* (Washington, DC: The Brookings Institution, 1934). George Peek's role in the Agricultural Adjustment Administration is described in G. Peek (with Samuel Crowther), *Why Quit Our Own* (New York: D. Van Nostrand Company, Inc., 1936). Shearer, "Competition through Merger," 276.

10. See George Wilson's report to the board in Deere & Company *Minutes*, April 25, 1933; for A. E. Head's notice to foremen and employees, see ibid., July 31, 1934; for the contribution to an unnamed fund by "a number of prominent businessmen in the Midwest States," see ibid., October 5, 1934. For the purchase of the Killefer Manufacturing Corporation, see ibid., April 27, 1937; ibid., May 25, 1937. Burton Peek's testimony in the National Labor Relations Board case involving Killefer and the Steel Workers Organizing Committee is dated May 10, 1938, DA, 10401; the NLRB cease and desist order is dated August 10, 1939, the bulletin board notice on it September 5, 1939, DA, 11482. *Fortune*, August 1936, 155.

ENDNOTES

11. The early design work on the Model A and Model B tractors by Henry Dreyfuss is documented in DA, 2330; see, especially, progress report 2, November 24, 1937. For the story relating to Argentine attitudes on wheel spokes, see H. M. Railsback to Henry Dreyfuss, December 2, 1940, DA, 19765.
12. For Frank Silloway's report on Caterpillar, see Deere & Company *Minutes*, January 29, 1935; see also branch bulletins 773, 774, and 781. Caterpillar's remarks on the Deere two-cylinder engine were first made in a sales meeting, November 1935, and then reprinted in branch bulletin 794. The memorandum of agreement between Caterpillar and Deere is dated March 5, 1935, DA, 19123. For changes in the California branch-house and dealer structure after the Caterpillar cooperative selling arrangement, see "Statement of Facts—Organization and Business of John Deere Plow Company of Moline," ca. 1943, files of Killefer—State of California case, DA. For R. B. Lourie's speech to Deere branch-house and Caterpillar sales representatives on September 10, 1947, in Davenport, Iowa, see DA, 44141. For Frank Silloway's report on California of January 28, 1941, see DA, 19268. See "Brief Outline of B.C. Keator's Service," June 26, 1953, DA, 25589.

CHAPTER 12

WORLD WAR II AND THE POSTWAR DECADE

I feel this Board must have the vision, sanely—at the same time boldly—to expand production facilities. . . . Surely as we are sitting here, if we fail to provide the necessary facilities, someone else will. I know we don't want this someone else to be Ford, General Motors, Curtiss-Wright, Graham-Paige, or the XYZ Company of New Mexico, nor do we want this opportunity to be cashed in by IHC, Allis-Chalmers, Case, Oliver, Minneapolis-Moline or any other company not now in the business. We must have courage.

Charles Wiman, 1944

Germany invaded Poland on September 1, 1939; Great Britain and France declared war on Germany and World War II had begun. Though the United States Congress passed a Neutrality Act in that year, by May of the following year extensive aid was being sent to Britain and massive national defense preparedness measures for the United States were underway. When the Office of Production Management was formed in December 1940, the country was already launched on a buildup of military production; a year later, the attack on Pearl Harbor drew the United States into armed conflict.

Incredibly complex questions of priorities faced the country's leaders, and the farm machinery manufacturers were squarely in the middle of some of

◄ Patriotic appeal to farmers, World War II. *Deere Archives*

CHAPTER 12

the most important of them. How were competing needs for food production and war materials to be reconciled? How much new farm equipment and how much replacement and repair equipment would be needed to maintain the necessary level of food production? How much of the farm equipment industry should be shifted into war production and how much of it should continue to make farm machinery? What amount of civilian production was to be allowed?

Who should be able to buy? What of the implementation of these steps—who should work on production (as against going to war), how should they be compensated for their work, what prices should be charged for the products sold both to government and to civilian users, how should critically short supplies of key materials be allocated? Finally, how should the industry reconvert to civilian production after the war?

Just before the United States entered the war, the Office of Production Management issued a "limitation order," holding civilian production of farm equipment for the industry to 80 percent of 1940's production of new machines, 150 percent of 1940's repair and maintenance parts, and 96 percent of 1940's exports as the allowable production for 1942. Vehement protests from both the industry and the farmers soon persuaded the government to relax the restrictions somewhat, but at the same time the industry was chided for not converting enough resources to war production. There was even an implied threat by the Department of Justice that it was considering instituting grand jury proceedings "in an endeavor to obtain an indictment against corporations engaged in the farm machinery industry." In essence, the industry was accused of taking the opportunity to build its inventories for civilian needs, and indeed the farmer did appear to be well taken care of in 1942.

By mid-1942 severe cutbacks in civilian production were demanded by the government. An important limitation order, issued in October of that year, introduced the concept of "concentration" of farm machinery with small- and medium-sized producers, thereby "freeing" the larger manufacturers for mostly war production. This classification scheme had the effect of tilting the farm machinery business in 1942 and '43 toward smaller firms. The manufacture of tractors necessarily had to be left with the major long-line companies, but the amount of civilian business done by International Harvester and Deere in this period was considerably less than it had been before the war. The wartime period gave the smaller long-line companies—Minneapolis-Moline, Oliver, and Massey-Harris—the chance to strengthen their finances and win a much larger market share than they had enjoyed before the outbreak of hostilities. Many farmers were forced by shortages of equipment to purchase new makes; in the process some brand loyalties, built up over many years, were eroded.

Ford-Ferguson, which had been able to put its small, lightweight tractor into production quickly in 1939 and '40, now began to make sharp gains in market share. When the limitation order of 1942 was being debated, the Ford-Ferguson people made a stunning proposal to the War Production Board. Inasmuch as their tractors were lightweight, they argued, why not "exchange the old standard wheel heavy weight, pre-depression made tractors for lightweight tractors of modern design and manufacture." There were almost half a million farm tractors of more than ten years of age, they noted, and if these were turned in and replaced by the lighter Ford tractor, a savings in weight could be made of approximately one ton per machine. By scrapping the old timers, they promised, more than one million tons of steel might be recaptured, enough they said to construct forty-two average battleships. (Underlying this startlingly high figure was an additional assumption that the light tractor would be substituted for the 1942 quota of heavier weight tractors, which they argued then would not need to be built.) Ford-Ferguson would, they proposed, make all the tractors required for the country for this year—100,000 tractors more than Ford was already producing.

Predictably, the rest of the industry was dumbfounded at the idea. In one stroke, Ford wanted to grab the entire market for tractors! The other companies vigorously disputed the claim that a small tractor was inherently better than a large tractor, or that any one tractor would be adequate for all farming conditions. Concentrating repair service in one company also would be devastating for the repair of all the other makes. After much lobbying in both industry and government, the notion was finally shelved.

The industry as a whole now was sensitized to gambits by fellow manufacturers. They did not have long to wait. In early 1944 the government, through the War Food Administration, exhorted farmers in the American wheat belt to increase production by some one billion bushels. Yet large numbers of farmers had been drafted into the armed services, so labor in farm areas was in critically short supply. Massey-Harris sensed a major commercial opportunity. They had just brought out a small, self-propelled combine and now proposed to the War Production Board that they be permitted to produce 500 machines in addition to their regular quota for 1944. These machines would be sold to operators who would agree to harvest, under Massey-Harris supervision, a minimum of 2,000 acres each. Many hundreds of tractors would be released for other work and, in the process, self-propelled combines would supplant a like number of conventional combines. This time the War Production Board agreed to the scheme, and in March 1944 the "Massey-Harris Brigade" was sent south from Canada to begin its well-publicized move up through the harvest areas of the Plains states. It was a signal public relations coup; *Fortune* magazine reported: "The self-propelled carried out an elegant light-armored blitz of the U.S. wheat belt." It was not only great public relations, it carried tremendous commercial implications. Massey-Harris made a technicolor motion picture, *Wonder Harvest*, and by the time the operation was finished Massey had projected its name into the minds of farmers throughout the American farm belt.

The industry continued to protest the concept of concentration and the severity of the limitation orders, and by mid-1943 somewhat more liberal limitation orders were issued. By 1944, as the tide of the war turned toward the Allies, the limitations on civilian production were progressively terminated. By the end of the year the industry was back to virtually unrestricted civilian production.[1]

DEERE'S WORLD WAR II EFFORTS

By mid-1940 Deere's pace had quickened; by December of that year it was making war materiel in the Welland plant, to be delivered to the Canadian government. By March 1941, the company was prime contractor for the fabrication and production of transmissions and final-drive units for the M3 medium tank. The work was done in the Waterloo plant, but under the aegis of a separate company called the Iowa Transmission Company.

After this first large defense contract, the company joined the rest of American industry over the following months of 1941 to gear up for greater defense production. In March 1942 a major new assignment came; Deere was to be a subcontractor to the Cleveland Tractor Company in the manufacture of MG-1 military tractors for the United States armed forces. Heavy tractors (weighing approximately seven tons) with track-type traction, they were intended primarily for use at airports in servicing and pulling aircraft. The Cleveland Tractor Company was one of the important manufacturers of track-laying tractors, under the trade name Cletrac. For some time, Wiman had seemed concerned about how Deere would keep its organization together in the face of war demands (in several previous board meetings he had articulated his own personal worries about this), and now he apparently saw a solution in the Cleveland Tractor link. His idea (as espoused in the board meeting of April 28, 1942) was a merger of the two companies! The notion was debated vigorously in the meeting amid much opposition.

Almost immediately, however, the Cleveland Tractor merger proposal was sidetracked. Charles Wiman's career had built him a wide reputation among government administrators; now he was asked to come to Washington as a colonel in the Ordnance Corps of the United States Army, to work directly in the tank and combat vehicle division. Wiman felt that he had no choice. He resigned as president of Deere & Company and left immediately for Washington. Burton Peek was elected president of the company, to hold responsibility while Wiman was gone. This sudden change in top management doomed the Cleveland Tractor merger proposal, since Wiman had been its chief advocate.

Deere also engaged in a wide range of other defense efforts—contracts for ammunition (75-millimeter and 3-inch shells), subcontracts to provide parts to aircraft manufacturers, and the building of mobile laundry units.

Deere assembled the laundry units, which were pulled as semitrailers by large trucks, to accompany combat troops and thus reduce the age-old problem of war, the vermin that cause so many diseases. The Wagon Works of the company also made various cargo units, and just about all of the factories of the company were devoted in a major way to some war production. By the end of the war, more than 4,500 employees had entered military service; even a "John Deere" ordnance battalion had been formed, recruited mainly from employees of factory, branch-house, and dealer organizations, and had served in the European theater of operations.[2]

WIMAN'S WASHINGTON YEARS

Charles Wiman threw himself into the high-tension Washington job in the spring of 1942 with characteristic tenacity and dedication, and later he was awarded the country's Legion of Merit. Then, in January 1944, Donald Nelson, the head of the War Production Board, requested that the Ordnance Department release Wiman so that he could come to the board and assume the directorship of its farm machinery and equipment division.

This job was to be a critically important one, for Nelson and his cohorts were both calibrating the war effort, now coming to a climax in Europe, and planning for the postwar reconversion that was now seen by all as upcoming.

When Wiman arrived at the War Production Board, he found the agency full of tensions and antagonisms concerning reconversion. Throughout industry, manufacturers maneuvered for position and watched over their shoulders the board's policies that would demark civilian from military production. Wiman was on the job no more than three months when he became publicly embroiled in a donnybrook. By this time in the war (early 1944), difficulties on the domestic agricultural front were being aired widely by farmers and their representatives. Farm leaders charged that shortages in farm machinery were threatening "international disaster" on the food front. They accused the War Production Board of overstating figures about totals of civilian production—that the board was "dishing out glamorized reports." A special Senate committee, the so-called Truman Committee, investigated the national defense program and reported that civilian production was indeed behind schedule, but the committee reminded everyone that farmers were still better equipped than they had been a year earlier. Despite this, the farmers around the country remained vexed. In particular, the so-called quota system was the focus of attack. Production quotas for civilian goods were established on the basis of each firm's production in the immediate prewar period. Many farmers felt that some of the larger manufacturers could not reconvert quickly enough to fulfill their quotas, and that therefore the total amount of goods available for purchase was being kept

CHAPTER 12

Exhibit 12-1. Deere & Company in World War II: upper left, Assembling a mobile laundry unit; lower left, Eleven Deere factories had a hand in producing the MG-1 military tractor; upper right, John Deere Model A armored tractor, Aberdeen Proving Ground, 1941; lower right, Assembling a tail wheel for the P47 fighter plane. *Deere Archives*

CHAPTER 12

Exhibit 12-2. Top left, Charles Wiman with No. 52 plow, 1936; bottom left, Charles Wiman laying the cornerstone at the Deere Tractor Company, Waterloo, November 1946; right, The opening of the Atlanta branch house at Chamblee, November 4, 1949; left to right, Burton Peek, C. M. Haasl, Bruce Lourie, W. O. Washburn, Charles Wiman.
Deere Archives

artificially low. The War Food Administration took the side of the farmers and accused the War Production Board of blocking "prompt and aggressive shifting of quotas" to some of the smaller companies that did not have much production in the prewar period. In early May, Edward O'Neal, president of the American Farm Bureau Federation, appeared before a House committee and bluntly stated: "We simply cannot understand why rigid production quotas are reserved for some manufacturers when they are not able to produce the full amount of their quotas, while other manufacturers who have met their quotas and have the labor and materials to produce far more are not permitted to do so."

At this point, Charles Wiman was dragged into the issue by the press. Representative Marion Bennett of Missouri charged publicly that the farm machinery division of the War Production Board was "in complete control of representatives of two companies," and named Wiman and the latter's assistant, Harold Boyle of International Harvester, as the two men manipulating the situation. Representative Fred Crawford of Michigan explicitly blamed the situation on Wiman who, he said, "muscles in and parcels out the business." Crawford introduced a resolution calling for a subcommittee of the House Committee on Agriculture to investigate the situation. Though the House committee did not go further into the matter, Donald Nelson intervened and promptly eliminated most of the quotas that had been in effect.

Wiman was in a very sensitive position as a "dollar-a-year" man, trying to administer programs that affected all of the companies in the industry from which he had come. Inevitably, there were conflicts of interest for him, as there

were for many businessmen serving the government during World War II. In order to have the war effort staffed by the people best able to do the job, the government had turned extensively to industry people. In this case, it was natural for the stronger long-line companies—particularly International Harvester and Deere—to desire the perpetuation of the competitive positions that they had earned before the war started. It would not have seemed fair to them to have weak companies bailed out by war contracts and then given market shares at the beginning of the postwar reconversion that were larger than those in effect at the beginning of the war.

But in an imbroglio like this one, politics intrudes in a major way. Allied against the two major long-line companies, Deere and International Harvester, was a potent group of farmers, abetted by the scare tactics of the War Food Administration. This was an unequal match, and it was foreordained that the quotas would fall, with the smaller companies gaining a larger market share in the process. A few weeks later, in June 1944, Wiman suffered a severe attack of pneumonia, and in mid-July he resigned from the War Production Board. Fate had intervened to take him out of an almost untenable situation, after more than two years in Washington.[3]

UNIONS COME TO DEERE

Wartime labor relations policies of the federal government were designed to keep employee effort at top pace, to prevent or settle work stoppages, and, in general, to keep the country's production on as even a keel as possible. President Franklin Roosevelt summoned labor and business leaders to a conference just after the Pearl Harbor attack; out of this came an agreement to foreswear strikes or lockouts for the duration of the war and for the creation of a board to settle all disputes that might result from the suspension of traditional collective bargaining practices. The tripartite National War Labor Board (with membership from management, labor, and the public) enunciated strict wage controls during the war, and labor increasingly chafed under the narrow cost-of-living base. As compensation, the board promulgated a new concept of union security, the "maintenance of membership" clause, that aided unions in holding their membership ranks during the period.

The first unions to be recognized by Deere & Company in its modern period were on the West Coast. In November 1937 the Garage Employees (AFL) obtained a contract at the Portland, Oregon, branch, and in August 1938 the San Francisco branch began bargaining with the International Longshoremen's and Warehousemen's Union (the company acting through an employer group, the Distributors Association of Northern California). These unions were only in the branches. Despite considerable organizing activity by competing unions in Ottumwa, Waterloo, and Moline, no factory-line production union had yet been able to organize any Deere plant, including the Killefer operation in Los Angeles, the site of the unfair labor practice case involving the individual contract. Beginning in 1939, a skilled-trades union, the Pattern Makers League (AFL), obtained a contract with the company (at the Plow Works, Harvester Works, and Spreader Works and at Union Malleable), the bargaining unit confined to the skilled journeymen pattern makers and their apprentices.

Not until 1941, at Ottumwa, was a production-workers union finally able to organize a Deere plant. Inconclusive organizing efforts had been going on there for a number of years; one of several competing unions, the Farm Equipment Workers Organizing Committee, affiliated with the CIO, won a National Labor Relations Board decision in 1939, striking down the same part of the company's individual contract that had been excised in the Los Angeles Killefer plant. In June 1941, that union contested with the International Association of Machinists (AFL) and a nonaffiliated local group, the Independent Farm Implement Workers, in an NLRB certification election. The Independent won but apparently was not up to the task, and a year later, in a second election a newcomer, the United Automobile, Aircraft and Agricultural Implement Workers (UAW), a CIO union, won handily.

At this time, the CIO was allowing, albeit uneasily, competition among its own national unions in representation elections. One of the most tense and acrimonious of all these intra-CIO battles was that between the Farm Equipment Workers and the UAW. Each union was wracked by internecine battles, both attempting to grapple with the left-wing leanings of factions within their leaderships. Critics from not only outside but within the CIO leadership itself were accusing the leftist groups of being Communist-dominated. In the UAW, there were seesaw battles between the "right" and the "left" all through the war, but by 1946 Walter Reuther had succeeded in driving out the left, leaving the UAW a solidly cohesive, strong postwar union, one of the stalwarts of the CIO. In the Farm Equipment Workers, the left held sway, particularly strengthened after 1941 by surprising victories at several key plants of International Harvester, which had been troubled by particularly bitter labor relations all through this period.

The more propitious climate for employee organization nurtured by the War Labor Board soon led to increased organizing activities at other Deere plants. In October 1942 the UAW began organizing workers at the Harvester Works and Spreader Works of the company, as well as at Union Malleable. Meanwhile, the Farm Equipment Workers was attempting to organize the Planter Works (Deere & Mansur). Most people in management did not believe the unions had much of a chance; indeed, most of the bets among the managers were whether the unions could poll a 5 percent favorable vote. To the great consternation and disappointment of most company officials, the UAW won all three of its contests, and the Farm Equipment Workers also was victorious. The tallies were reasonably close at the Spreader Works and Deere & Mansur, but much wider in the other two plants.

The results were a watershed in company labor relations—the first outside union in the company's Moline-area plants. Given the company's long history of benevolent employee relations policies, which had often bordered on being paternalistic, it was no surprise that the presence of the union saddened many people in management. The plant managers felt betrayed—"Yesterday my men were loyal Deere employees, today they are not." Yet their crestfallen view was not rooted in reality. The employees clearly wanted a union at this moment in Deere's history, despite the widespread feeling of goodwill toward the company, built up over many years, and this had not really been changed by the vote.

The Moline elections were soon followed by similar ones in Waterloo, and again the UAW won (March 1943). At that same time, the Van Brunt workers in the Horicon, Wisconsin, plant voted in an AFL union, a "Federal Labor Union," which was the AFL's answer to the CIO's drive to organize production workers. In May 1943, the company had its first union strikes; the piece-rate workers at the John Deere Spreader Works walked out over the issue of piece-rate price adjustments. A few days later, the employees of Union Malleable went out on strike over piece rates for molders in the plant.

Meanwhile, the company had been notified that the Army-Navy E award was to be given to all seven thousand employees at a mass meeting scheduled for Sunday afternoon, June 20. As the two strikes dragged on nearer and nearer to the date of the award, plans were announced for the celebration. Just two days before the celebration, however, the War Department telegraphed Deere that it was indefinitely postponing the ceremony. Although the War Department did not announce the reasons publicly, the delay was clearly attributable to the strikes, for two provisions required for the coveted award were "avoidance of work stoppage" and "cooperation with the war program." The company put out a bulletin board announcement and press release, bluntly stating these facts and in effect accusing the striking employees of forcing the postponement. The local newspapers sided with the company, one paper headlining its article "Shame! Shame! Shame!" The union retorted that Deere had not followed one other provision, "maintenance of fair labor standards," and that the company's record of accidents, health, and sanitation also were not good. (There was little evidence to support these allegations, though.) The strike was settled shortly thereafter, and the E award belatedly made.

The Farm Equipment Workers registered a further inroad in February 1943 when it won an election at the Plow Works. A month later the company discharged about fifty workers in the plant for allegedly instigating and participating in an unauthorized slowdown. The union cried "lockout," and threatened to press "the government" to "take over" the plant. The issue was settled by the two parties without outside intervention (the employees were rehired), but the episode generated considerable ill will on both sides.

In December 1943 the Farm Equipment Workers gained ground on the UAW by winning the right to bargain for employees at the Moline branch house. A few months later, in May 1944, there was a strike at Deere's Harvester Works that was not even authorized by the leadership of the UAW local. The War Labor Board threatened reprisals, but the Harvester Works employees voted to stay out "until the Company lives up to the contract with the union or the Army takes over the plant." The issue related to vacations. The company had been required to fit its vacation plan to War Labor Board edicts, and it so stated in a bulletin board announcement. The union then implied that the company was attempting to abrogate vacation plans. In effect, the company was caught in the middle between union demands for prewar conditions and War Labor Board requirements for patterns fitting national labor policies. The strike was terminated in June 1944, and later that month the War Labor Board denied the company's more liberal vacation plan, bluntly excoriating the strikers: "This is no time to create the impression that unions violating the no-strike pledge will be rewarded for such conduct."

And so it went during the war—it was a disruptive, acrimonious period for the company and the country, and employee relations were laced with grievances and irritations. Labor relations at Deere during this period were certainly no more troublesome than in most other industrial plants, but compared with the earlier days of industrial peace and direct relations with individual employees, the collective approach of World War II was a change of style that made most company officials uneasy and unhappy. George Wilson, who handled the company's labor relations in the early part of the war, probably summed up the feeling of most of top management when he told the board in October 1944: "The control of labor matters is rapidly passing out of the hands of the employer. There is practically no matter left that cannot be brought before some governmental tribunal for settlement. Bonuses, vacations, merit rating, retrenchment in employment, wages, incentive plans, job evaluation, seniority and even technological improvement are little by little being established as matters for collective bargaining and, therefore, matters on which governmental agencies or arbitrators can make final decisions. Managements' rights as formerly recognized have practically disappeared. Largely your labor costs and your choice of personnel are passing out of your control." The company had existed for 105 years under a philosophy of individual relations with individual employees. Now there was to be a new era, that of collective bargaining.[4]

POSTWAR ANATOMY OF THE COMPANY

Wartime management of companies entails highly centralized decision making. Defense requirements, rather than those of the market, dictate production targets. The price maker is the government; at the same time, the government is also the dominant customer. Tight control from the top—rigid rules from a nation's production "czar"—breeds a similar pattern down into individual organizations. Thus Deere & Company found itself becoming a more centrally managed organization during World War II.

The leadership of the company changed significantly over the war years. In October 1943, T. F. Wharton, long-time comptroller and wartime treasurer of the company, passed away. Coming to Deere at the time of the formation of the modern company in 1911, Wharton had gone on the board in 1918 to become an influential counselor to the company's chief executive officers, first William Butterworth and then Charles Wiman. Lloyd Kennedy became secretary and treasurer of the company, to succeed Wharton as the chief financial officer.

A second death was an even more serious loss to the company. In October 1944, Charles C. Webber died at the age of eighty-five, after an unparalleled sixty-seven years of service with Deere & Company. He had been on the board of directors from early in 1886; thus he had been involved in the general management of the company for more than fifty-eight years. His perspective was always broad, reflecting not only the requirements of the Minneapolis branch but of the entire company as well. His judgment on personnel and on top management organization and interaction was unmatched in the company's history. His blunt frankness was softened by a thoughtfulness and patience that won respect from the rest of top management. It seems almost incredible, given today's retirement ages, that an individual could be fully productive for so many years. From the moment he began work to the last day of his active business life, he was a contributing, integral member of the organization that he saw grow from a small set of loosely knit factories and branches into an integrated company taking an important role in the war. His peers rightly had designated him "the dean of the American implement industry."

In that same month, F. H. Clausen also died. He had been with the Van Brunt organization from 1899 and its president from 1920. In 1923, he became a member of the parent Deere board, and one of its vice presidents in 1941. Clausen was more insular then Webber, and his contributions to the parent were different. Still, a company cannot lose a trusted and capable executive like Clausen without suffering a management gap.

With these losses, the locus of power in top management shifted. Burton Peek clearly viewed himself as a "caretaker" right from the start; everyone sensed that he did not wish to hold the reins of the presidency any longer than was necessary. He was assisted by three other senior members who carried the rank of vice president and who gradually assumed increasing responsibilities: Charles Stone, the Harvester Works manager and Wiman confidant; Frank Silloway, longstanding head of sales, a board member since 1914, a vice president from 1919; and Lawrence "Pat" Murphy, on the board since 1937 but elevated during the war to a vice presidency. (Murphy was a descendant of the Ellen Webber side of the family, and thus was related to C. C. Webber.)

Murphy was the dynamic leader of a major part of the line organization; he and Stone shared the manufacturing responsibility in the company. Murphy was the dominant representative of his generation for the Webber side of the family. His position was strengthened when Edmund "Budge" Cook joined the company as legal counsel in 1943; Cook was related by marriage to both Pat Murphy and C. C. Webber. As the war wore on, key policy making gravitated toward this smaller group of Silloway, Murphy, Kennedy, Cook, and Stone (with Peek in the chair as a more passive chief executive than Wiman).

In June 1944, while Charles Wiman was in the midst of his final battle at the War Production Board, Charles Stone was shifted out of his direct line role in manufacturing and put in charge of product development. Murphy, in turn, was given responsibility for all of the factories, as well as the engineering department, the testing and research laboratories, and the personnel and industrial relations department. Lloyd Kennedy, who had worked closely with Peek and Wiman in the financial affairs of both the company and the family, was elevated to vice president, responsible for the overall financial policy of the company.

This far-reaching decision was debated over two separate board meetings; at the second, Burton Peek elaborated on its rationale: "War contracts have interfered with the policy of decentralization, which has been an outstanding characteristic of our operations . . . and have created a trend toward centralization. This trend must be reversed. . . . The purpose is to leave with and restore to the Factories and Branches the maximum feasible operating latitude as in the past."

These announcements of June 1944 had a great effect on the whole organization of the company; for years afterward, the bulletins elaborating these changes were remembered as the singular manifesto that spelled out again the company's basic commitment to decentralization and that restored the power to the branches and the factories. If the war had indeed centralized many functions, now this centralization was to be reversed.

But was it? On the one hand, Frank Silloway had had a thoroughgoing dedication to giving the branch managers "their heads" and had presided over the continuing branch-house decentralization into the hands of such strong leaders as C. C. Webber and Michael Healey. Though Webber and Healey were now gone, there were other new branch-house managers, displaying the same independence—W. D. Hosford in Omaha, R. B. Lourie in Moline, and others. By instinct and longstanding operational style, Silloway was committed to decentralization, and he had developed a relationship with the branch managers over the years that had provided bonds of understanding and communication that gave the system a special Silloway mark. Still, Silloway was in his late sixties, just a few years away from retirement.

CHAPTER 12

Murphy, on the other hand, was in his mid-forties and had been a general company vice president only since 1942. Murphy was quite cognizant of factory prerogatives—he had headed the Plow Works before coming to the general company post. But the consolidation in June 1944 of all general company manufacturing under his responsibility was unprecedented—never before had there been a vice president with this breadth of responsibility for manufacturing. Peek's announcement seemed inconsistent in promoting, on the one hand, the "return" to decentralization, yet at the same time centralizing all manufacturing under one strong vice president, Pat Murphy. The move gave mixed signals to the factories. Murphy needed to be cautious about contradicting his factory managers' decrees (his relationship with "Duke" Rowland, the Waterloo Tractor Works manager, was a particularly prickly one). Yet Murphy's instinct for detail and desire to involve himself in all decision making centralized a substantial amount of authority and power in his hands. Stone's long involvement in the line had ended; as a staff vice president concerned with product development, Stone was now to be more advisory, less directly managerial.

Peek's announcement left no doubt about the sincerity of the board in returning to past ways of managing the company. Still, the messages were not clear—there were mixed meanings in the organizational changes, complex and subtle new roles to work out in new, postwar managerial interactions.

Budge Cook, who had become a director in 1943 and general counsel in 1944, was given the assignment of articulating some of the subtleties in the new moves. In a major speech to the factory managers in July 1944, he exhorted the managers not to leave "your problems" to the general company departments. Cook had been delegated by Murphy to make the speech; surprisingly his remarks seemed to speak directly to Murphy. "Let us suppose that a General Company man is an individual who likes to dominate or suppose there is no one to whom he can refer the situation in which you and he fail to agree; or suppose that he knows that you will be resentful if he does refer the matter to higher authority. If any of these situations persist, the General Company man will strive the harder to force edicts upon you and your people." Cook continued, "I am clear in my own belief that a General Company Department should not be permitted to force that move onto the Tractor Works against its opposition." Cook had moved subtly from a general statement about nuances of policy to a specific example, a carefully chosen one, the Tractor Works. Clearly, stronger, old line factories—the Tractor Works, the Harvester Works, and so forth—were to be maintained as almost autonomous units, subject only to general communication and report, after the fact, with corporate headquarters. This was not Murphy's view, for he was a demanding executive who wanted extensive communication with and involvement in his operating department entities. On the other hand, strong factory executives, like Rowland and Harold White of the Planter Works, were not likely to take direction very readily.

Thus, as Pat Murphy, Lloyd Kennedy, and Budge Cook emerged as the next generation of management, subtle differences of opinion had surfaced among them. Kennedy was foremost a representative of the family, only early in the process of becoming the chief financial officer of the company itself. Murphy was a young member of the next generation representing part of the family, a logical person to step into a major role in line management. Cook, though also related by marriage to the Webber side of the family, was a professional; his lawyer's instincts combined with his remarkable philosophical understanding of organization structure to give the company a sophisticated spokesman for the restructuring process, independent of any particular faction within the organization.[5]

CHARLES WIMAN'S "POSTWAR RECONVERSION"

This interplay of responsibility among the three men had to be fitted to an important organization enigma—would Charles Wiman return? Wiman had left abruptly for war as a strong, even dominant chief executive officer of Deere & Company. Wiman had found his assignment with the US Army rewarding; during his illness he had confided to his friend, Edward R. Stettinius Jr. (just named secretary of state) that he would like to have gone back to ordnance "but there was no chance of my again getting active service. . . . I could not pass the physicals." His next assignment, with the War Production Board, had made him once more privy to the evolving status of the industry. But clearly he had lost touch with the day-by-day activities of Deere, having attended only a segment of one board meeting since leaving in mid-1942 (where, incidentally, he had startled everyone by proposing that the company make airplanes and deep-freeze apparatus after the war). Apparently the challenge of Deere was deep inside him, though, just as it had been all through his years prior to World War II. After more than a month of serious illness and an extended convalescence in the fall of 1944, Wiman made his decision to return to active management in the company.

Wiman's reentry was tension-filled. The major executive change of the previous June had thrust Murphy into a pivotal role in the company—in charge of all manufacturing and engineering. Cook was a new face to Wiman—Cook's entry into the company had come after Wiman had left for Washington. Lloyd Kennedy's elevation to a vice presidency was a change; was he now more independent than in his previous role as the family's financial counsel?

Charles Wiman had practiced a personal management philosophy of not giving any one individual in the company such a strong base of authority. Wiman seemed to have an abiding faith in a constellation of strong people, with the inevitable tensions among them adding to the effectiveness of the whole. As Wiman returned and surveyed his top management relationships, he seemed uncertain as to Murphy's and Cook's places within the group. Indeed, there is substantial evidence that the nagging thought had crossed his mind that there was a "barracks revolt" against him. Deere & Company seldom had been subject to rank factionalism, erupting into power plays by one group against another (the only serious instance being the Butterworth-Velie acrimony in the 1910s). On the other hand, what would have been the leadership succession had Wiman not returned? One of the options would likely have been the choice of Murphy as president.

Once back on the job, Wiman quickly moved to reassert his leadership in the company. He was elected to the presidency in a formal board meeting of October 31, 1944, and at this meeting made a speech that was vintage Wiman. He praised the monumental job done by the entire organization during the war, then noted explicitly the restructured organization stemming from the board decisions of the previous June. He singled out the Murphy assignment as "a real constructive organization change" but left no doubt that Murphy only had been put in charge of all the factories "from the *production-management* standpoint."

"I turn now to the question of my re-entering Deere Company," Wiman continued, "and, as we say in the Army, taking the oath of office, or as we say in industry, getting on the payroll. This Board made me the Company's chief executive officer in 1936. I acted in that capacity for 6 years. *If* you want me to come back, my return should be on the same basis."

Wiman then stated his view of the chief executive officer's role with no equivocation: "Mr. Peek and I *privately* (and publicly as far as that is concerned) have agreed that there should be but one chief executive officer. There is nothing in the world that is *worse* than a divided command." There were now to be "local board meetings" held at least once a month during the interim between regular board meetings, but Wiman made it clear that the president would always act as the presiding officer.

Wiman also addressed the new role of Cook. After remarking that Burton Peek had acted as general counsel of the company since the formation of the modern company in 1911, Wiman explicitly suggested that when Peek decided to select his successor, it be Budge Cook. This was a curious tactic, for seldom does a chief executive officer desire to tie his hands on a future appointment. In this case, Wiman seemed to want to acknowledge the contributions made by Cook, at the same time delineating a narrower role for him within the constraints of the legal counsel's portfolio.

Throughout, Wiman exhibited his unique brand of enthusiasm and optimism; he concluded with a ringing call to arms: "I feel this Board must have the vision, sanely—at the same time boldly—to *expand* production facilities.... Surely as we are sitting here, if we fail to provide the necessary facilities, *someone else will*. I know we don't want this someone else to be Ford, General Motors, Curtiss-Wright, Graham-Paige, or the XYZ Company of

New Mexico, nor do we want this opportunity to be cashed in by IHC, Allis-Chalmers, Case, Oliver, Minneapolis-Moline or any other company not now in the business. We must have *courage,* after due consideration and analysis of the problems to *spend* our money wisely . . . to expand this position and to acquire an even *greater* percentage of the total implement and tractor business than we have had in times past."

The entire speech was a vivid demonstration of Wiman's strong-minded personality and power to persuade. Its effect on the board at that time was great—remembrances are unanimous in characterizing this particular board meeting as the critical one in the postwar reconversion of Deere & Company. The return of Wiman once again reaffirmed the dominance of the Charles Deere side of family ownership and brought back in the process a chief executive officer of uncommon abilities, one who could dramatically energize the management group as no other person within the company at that time could have done. For the health of Deere & Company, it was a crucially important positive step.

Wiman had been back on the job less than three weeks when he made an important change in certain top management assignments. L. A. "Duke" Rowland had headed the Waterloo tractor factory since 1936 and had been elected a director in the company in 1942. Now Wiman decided to give him an even broader assignment, as general manager of all tractor production (that is, Waterloo and the Moline Tractor Works). The bulletin announcing the change came from Murphy and stated additionally that Rowland "is hereby appointed assistant to this writer." Rowland was a doughty, precise man, born in England, a feisty and independent person who would insist on "running his own show"—the word "assistant" inevitably would have limited meaning as applied to him. Wiman's message was clear—Rowland was going to "head," literally, all tractor production, Murphy to have implements and to continue responsibility for the engineering department, the laboratories, industrial relations, and other staff groups. Murphy's tenure as head of all manufacturing had lasted less than six months. The demarcation was made even more sharply in 1947, when Rowland was elected a vice president, co-equal with Murphy.[6]

DEFINING DECENTRALIZATION, ONCE AGAIN

Though Charles Wiman had laid at rest the issue of executive succession, there was still confusion about where the decentralization policy stood. All the board agreed that too much centralization had been forced by wartime conditions, but just what relation was now to exist between general company vice presidents and constituent factory and branch managers was not well understood.

CHAPTER 12

Wiman had brought many new ways back with him from Washington. He changed the layout of his office, eliminated the prewar pattern of having a secretarial pool next to him, shifted his former private secretary, Maurice Block, to another assignment out of the president's office. Wiman also brought a new person with him when he returned, and the effect of this soon led to more questions about the relationship between central office and field operations.

Wiman had had an assistant at the War Production Board, Harold L. Boyle, in whom Wiman had vested considerable trust. Boyle had been an International Harvester executive for a number of years; when Wiman returned to Deere, he persuaded Boyle to join the company. The titular post for Boyle was purchasing; in reality, Boyle acted more as a personal assistant to Wiman in the first few months of the latter's return to Deere. Boyle seemed for a while to be taking over the role held for a number of years by Charles Stone, that of personal confidant for Wiman. In May 1945 Boyle was elected vice president; the previous month, Boyle had also been elected to the board (along with C. R. Carlson Jr., a hard-driving executive who had headed the Minneapolis branch after C. C. Webber's death).

Wiman now used Boyle's appointment as purchasing vice president to announce a surprising further decentralization of purchasing functions. Citing again Burton Peek's statement in 1944 that "the trend toward centralization . . . must be reversed," Wiman made it clear that while Boyle would have a major role in company policy making, he was not to be given the central responsibility for the effecting of purchasing decisions. In key respects, Wiman's move here was incongruous—establishing a major officer for purchasing, yet at the same time decentralizing purchasing into the hands of individual factory and branch purchasing officers.

This apparent inconsistency soon triggered much concern. Wiman sensed the unease and apparently felt that the time had come to take on the divisive issue of decentralization in a more organized way. His vehicle was to choose a small group of directors—Frank Silloway, Charles Stone, Pat Murphy, Lloyd Kennedy, Harold Boyle, and Budge Cook—to meet over a number of months and return at the end of the year with firm recommendations that would, he hoped, lay at rest once and for all the tensions concerning lines of authority.

But there were surprisingly deep divisions among even this committee about the philosophy of organization at Deere. Cook and Boyle were particularly in disagreement. Cook espoused a militantly decentralized point of view, Boyle a single-minded concern to enhance the power and authority of the presidency. Cook believed that the various officers should be the "eyes and ears of the Directors," in effect bypassing the president if more direct communication was desired. Boyle vehemently resisted this notion, holding that officers should report to the board "only when requested to do so by the President." Cook stated bluntly that "with rare exceptions of an emergency nature, the president's power to give orders to key people at lower levels will not be exercised." Boyle argued: "No matter what we write in our memo the President will run his job as he sees fit and each of us individually will have to gear our activities to accomplish the end results for which he is striving. He is our boss and he takes full responsibility with the Board of Directors for his activities and ours." Cook had a view of decision making that put complete responsibility in the hands of individuals out in the field (factory managers, branch managers, etc.) and held that officers were "not chargeable with preventing errors . . . not chargeable with the duty to take corrective measures . . . not subject to criticism for lack of action . . . where the subordinate is authorized to finalize decisions." This seemed to Boyle to be excessive autonomy and he retorted: "Important and even isolated mistakes with a key man's field of action should never pass unchallenged. . . . I cannot imagine a ball team making errors and the coach not criticizing the players in an effective and constructive manner. The officers are the coaches on the Deere & Company team and their staffs are the assistant coaches."

The committee continued to meet over a four-month period and produced a remarkable set of minutes. Though the discussions were not acrimonious, so strong was the disagreement on principles that Wiman backed off and decided not to make the committee's report public within the company.

Why this sudden timidity on the part of Wiman, not one of his characteristics in the past? The final minutes of the committee had been frank in stating that there was "unrest, uncertainty and a degree of frustration among key people in the organization. . . . We are in danger of losing or seriously impairing the high degree of morale, loyalty, and efficiency that existed in the past." Wiman knew that if he made definitive judgments at this point, he would bruise some egos. Further, he privately worried that if an attempt was made to codify in writing all of the organization structure this process might not give proper weight to the traditions, habits, and practices that had grown up through the years. It seemed best to Wiman to beat a strategic retreat.

But Wiman was astute enough to know that the fuzzy lines of organization structure would continue to trouble top management until they were more openly resolved. He bided his time, and in October 1947 unilaterally brought the issue of organization structure back to the table for consideration by the officers. This was a period of relative calm—it was at the end of a product and fiscal year that had been quite successful (sales were up to $194 million from the previous year's $131 million, net income had risen to $18 million from the previous year's $12 million). Reconversion problems were largely over; the upcoming year promised to be a good one. It seemed a propitious moment to face up to a controversial issue.

This time Wiman put together the outline of the structure by himself, writing out an organization chart and functional responsibilities in his own hand at home. Once ready, he drew Cook and Charles Stone into his confidence. The three then developed a set of three organization charts. The first, a conventional "line of authority" chart, pictured the organization as it existed at that moment. A second chart showed the same organization, but developed a set of dotted-line relationships among various departments to show advisory and informal linkages.

Stone then suggested an innovative new way of viewing the organization. He likened the lines of communication to lines of travel within a city, observing that the residence of any individual is connected with every other building in the city and with every other building in the United States by a series of highways. These highways were not exclusive, but were for general public use. Stone then suggested the analogy of the streets to the Deere organization—that every key individual in the organization could promptly get onto a highway that would take him to any other part of the organization. Out of this notion, he and Cook developed the third organizational chart, a pictorial effort to show the "highways" of Deere & Company, complete with traffic lights at intersections!

A few simple traffic rules were established—streets and highways were to be used expressly for the purposes of obtaining information and making agreements; orders, directives, and demands were only to issue through the line of authority. Still, information or agreements should be obtained at the nearest practical points, and executives were to go no further or higher than necessary. As each individual used these channels of communication, he was to inform his immediate supervisor of all significant facts obtained and agreements made. Finally, all executives in the entire organization were to apply these concepts, not only in their dealings with the general company, but within their own organizations as such.

With this material in his hand, Wiman made a major decision, to call all branch-house managers and factory managers to Moline to discuss the subject of organization structure at a general meeting, at which the three charts would be used as the points of departure.

The meeting was held on December 12, 1947, amid pregnant anticipation among the executives present. Wiman's opening remarks set the stage: "I wish at this time to pay special tribute to Messrs. C. H. Deere, William Butterworth and C. C. Webber, all of whom carried the policy of decentralization through to the very 'nth' degree. . . . The question is: 'Are we ourselves thoroughly sold on decentralization so that we can practice decentralization in a very efficient way?'" Wiman then discussed the three charts, putting particular focus on the "street chart" (Wiman's words). He spelled out the suggested traffic rules of Cook and Stone and emphasized that traffic was on a two-way street—that the general company officers also were to be free to come into the factories and branches to bring their counsel, advice, and guidance.

The day-long discussion turned up widely different views. On the one hand, Burton Peek spoke vehemently for central control: "I am not one of those who think that there is no necessity for some degree of centralized management. Quite to the contrary. Centralized management must map out and state overall Company policies to the end that so far as that is humanly possible there be no overstepping of bounds and no interference one with another." Pat Murphy took an opposite stance and rejected out of hand any "interference" by central engineering: "There is no Company policy that makes it mandatory to obtain the approval of these departments on any given project. It is the responsibility of these departments to so conduct themselves that the factories and the branches will have confidence in their judgment, and consequently seek their advice." (Murphy must have worried that he had gone too far, for he later requested Wiman's permission to revise his remarks for the formal minutes; he substituted a more innocuous statement that the outside "staff of experts" were to "work with the factories and to keep us informed." The record is mute as to whether this more diplomatic concept of staff was Murphy's own.)

So deep was the division that Wiman ended by asking the entire group to go back to their respective organizations, there to take up to ninety days to give intensive study to all the matters that came forward in the meeting. When each person felt comfortable with his own views, then he was to forward to Wiman suggestions for revision of the charts, together with a statement of what he was doing "toward making fully effective and insuring the continuance of decentralization . . . under your jurisdiction." Wiman also asked each to decide whether it was feasible to publish an organization chart and a general-company bulletin, explicitly stating the organization structure.

The ensuing statements that came to Wiman over the next three months provided remarkable testimony to the interest and concern that the members of top management felt about the principles of Deere organization. Wiman collated all of the comments and, as promised, issued a general-company bulletin to both factory and branch in May 1948, under the title "John Deere Decentralization Policy." It was clear in the bulletin that Wiman still felt it unpolitic to state definitively the organized policy: "While this is being issued in Blue Bulletin form, this method is being used solely as a means of getting the collective opinion and my thoughts into your hands, and I think it would be better if you think of it as a letter rather than a Blue Bulletin outlining a policy."

As to the charts, there was "little doubt as to your general preference for the street (or highway) type of chart, although the line of authority chart did not lack proponents, and at least two replies indicated desire for all three charts." Nevertheless, Wiman stated: "No centralized directive could resolve all your circumstances, nor meet your varying problems on furthering this or any other policy as effectively *as your own initiative and decision.*" Again, Wiman had backed off. He concluded: "Your replies leave no doubt as to your intent to maintain and improve John Deere decentralization. . . . They show that you propose to be aggressive leaders, rather than passive followers. They prove that each of you fully plans to apply decentralization in the way he considers best under the particular circumstances in his factory, branch house or department. . . . I am satisfied that a repetition every few years of last December's meeting will keep our tradition alive and effective. . . . Decentralization formally by manual or bulletin at this time would detract from the tradition."

Although this six-month saga did not result in a definitive statement at the end, the sum of all of the efforts was to infuse everyone with the excitement and enthusiasm of continuing to work together as the "Deere family." The concept that there needed to be certain centralized control and coordination, backed by enough authority to make this process stick, was implanted. Not resolved, however, were the many individual rough spots in communication, or the excesses of overly arrogant individual action and the unthinking protection of fiefdoms. After all had been said, the strong factory and branch managers had maintained their personal control more or less untouched.[7]

FINANCING EXPANSION

Deere had not borrowed any money for long-range purposes since it retired the issue in 1921 of $10 million of 7.5 percent "Gold Notes," fully paid off ahead of time in 1925. The board had depended on a conservative combination of short-term bank financing and the plowing of as much of the profits back into the business as was prudent. The effect of the policy of caution, though, was that the company for years had been making only as much equipment as it "could afford," not as much as it could actually sell.

The company's balance sheet on October 31, 1944—the end of that fiscal year—was one that any good conservative executive would be proud of; here in outline is how it stood in that year's annual report:

Current Assets		Current Liabilities	$40.6
Cash	$78.6	Reserves (benefits, price declines, war losses, contingencies)	36.6
Government securities	24.4		
Canadian bonds	3.7		
Notes and accounts receivable	18.8	Capital and surplus preferred	31.0
Inventories	38.5	Common	30.0
Property and equipment	21.6	Earned surplus	53.6
Other assets and deferred charges	6.2		$191.8
	$191.8		

Charles Wiman had come back from the war enthusiastic enough about postwar prospects to consider a more aggressive policy. Manufacturing capacity for tractors seemed to demand the highest priority. The board already was on record for "producing a low-cost model tractor" and wanted this to be done at a location other than the Tri-City or Waterloo areas (preferably, also on a water route).

A thorough study finally centered on Dubuque, Iowa; the Ordnance Corps of the United States Army wanted Deere to continue manufacturing 75-millimeter shells and offered to share the building of the new factory, part of which could be used also for tractor manufacturing. In January 1945 the board unanimously approved the purchase of land and the immediate construction of the plant. The total cost of the project was estimated to be $9.4 million.

Wiman's postwar ebullience seemed to give him a new perspective about the past financial conservatism of the company; though he did believe that "management had followed the proper course in these years," nevertheless, he now felt that "the freedom of action of management had been restricted" at that time. So, pushed by Wiman's unequivocal commitment to expansion, the board decided to issue new debt instruments. The needs were estimated to be a startling $30 million. A package was put together with Deere's lead Chicago bank, Continental Illinois National Bank and Trust Company, whereby Continental and a group of other banks would subscribe to $10.5 million of serial notes (1.5 percent and 1.75 percent) and the remainder of $19.5 million in twenty-year 2.75 percent debentures that would be offered to the public. The issue sold with no snags, and the company had established the underpinnings of its ambitious postwar financing.

The new debt did raise some eyebrows—many people had taken as immutable Deere's very conservative financial policies. Max Sklovsky, who had retired from his post as the company's chief engineer, wrote Wiman "as a stockholder," questioning the need for the issue and, in the process, gave Wiman an unusually barbed personal criticism, implying too much dominance by him: "It can be assumed normally that the collective wisdom of a Board of Directors transcends that of an individual, yet it is well known that a Board is led by one or two and the rest following saying 'Amen.' It does appear that the results aimed at could be reached in a more simplified manner than that of building factories for complete goods." Wiman seemed stung by the queries of Sklovsky and others, and he drafted a letter outlining the expected uses of the funds. The letter seemed apologetic, ending: "The Directors took into account the obvious truth that they could not read the future with certainty, and recognized that a decision either way could be wrong."

The expansion of plant and equipment moved rapidly over the next two years. The Dubuque plant was completed as scheduled, with housing in such short supply that the company also built some 111 houses, which were first rented and then later sold to employees. By mid-1947 the factory was producing the company's new Model M tractor.

This machine had been in the making since early in World War II (Charles Wiman having insisted that product development continue apace all during the war). The Model M was a general-purpose utility tractor, rated at 14.39 horsepower at the drawbar. It incorporated an innovative hydraulic unit—called the Touch-O-Matic—for attaching and controlling integral equipment. When the tricycle version, the MT, came out in 1949, it had the same hydraulic concept, but with a dual system for independent adjustments of tools working on either side of the tractor.

Dubuque at this point also took on the manufacture of another tractor—the Lindeman crawler tractor—that was the forerunner of a major division of the company. Since 1940, Deere had been supplying to a Yakima, Washington, company, Lindeman Manufacturing, Inc., the Deere Model BO tractor chassis (this was the orchard version of the B, with fenders so that protruding fruit-laden branches would not catch in the wheels). Lindeman attached its own crawler tracks to the Deere chassis, and the machine was marketed as a "John Deere-Lindeman." There was substantial demand for this type of traction among the Pacific Northwest orchard owners; crawler traction was also popular in logging areas—in New England's snow and in the swamps of the Southern timber regions.

In May 1945 Deere purchased Lindeman. When the deal was consummated, nearly two years later, the Lindeman was phased out and the Deere Model MC was made in Dubuque but still using tracks made in the newly

Exhibit 12-3. Deere-Lindeman tractor working in an orchard, ca. 1941. *Deere Archives*

renamed Yakima Works. The Yakima plant also made some of the Deere tillage and land-leveling equipment. By 1953 it became apparent that the plant's freight costs for shipping back to the Midwest were too high and the Yakima Works was closed. In this year, Dubuque brought out a new larger crawler, the Model 40C, with 15.11 drawbar horsepower; Dubuque began manufacturing the track, too.

Charles Wiman's view of the world, so different after he returned from World War II, continued to expand. In February 1947 he embarked on a three-month trip to South America to visit Peru, Chile, Argentina, and Uruguay. The trip made a profound impression on him, for he sensed great business potential there. Indeed, he cabled back instructions to segregate a significant amount of Deere products for shipment abroad. This was at the height of the shortage of products in the United States, with farmers clamoring for Deere machines and often expressing irritation over the long waits.

Budge Cook, apparently one of the most expansion-minded members of top management, wrote Wiman while the latter was in Chile: "My impression is that many of our people are becoming imbued with a sense of failure. . . . Our factory production is at a record peak and yet we are working in an atmosphere of failure . . . failure to supply an unprecedented demand. . . . We must create that atmosphere of success, for otherwise many of our key people will simply wear out and let down under the strain."

Wiman appeared upset by this and wrote back immediately, "Sense of failure? No, indeed! We have and are spending all we can." As to the unnamed people whom Cook felt were "wearing out," Wiman responded: "I suppose you've certain individuals, or individual in mind—OK, let's give special treatment . . . but we also don't want a Pollyanna 'what a good boy am I' feeling to creep in." Still, from this exchange, Wiman became convinced that additional capacity had to be acquired. When Pat Murphy proposed to increase the Plow Works production by purchase of a new plant, Wiman immediately wrote Murphy from South America, giving his approval for the purchase of a vacant plant in Des Moines, Iowa (earlier offers of plants at Tulsa, Oklahoma, and Longview, Texas, had been rejected). This was the former Des Moines Ordnance Plant, at Ankeny, Iowa, and it was to be bought from the War Assets Administration for $4.15 million. The deal was completed in June 1947; by the end of the year the plant was in production, making the company's cotton picker as well as some cultivating tools.

Significant additions were also made to existing operations in Moline, Waterloo, and Ottumwa; by October 1947, at the end of that fiscal year, the depreciated value of the property and equipment of the company had soared from $21.6 million in 1944 to more than $52 million. This breathtaking acquisition of bricks and mortar, so exciting when it was in the formative stage, now began to make Wiman quite uneasy. The financial ratios had shifted during this three-year period. The ratio of cash and securities to current liabilities had declined a bit; the notes and accounts receivable had risen significantly, given the quickened pace of sales out in the field; and, of course, the property and equipment account had made the fixed assets side of the ledger much larger. Though there was certainly nothing in the ratios to trouble a financial expert, Wiman nevertheless warned the board in January 1948: "Pressures from customers and Government have caused us to place our emphasis very heavily upon maximum production which, in turn, has involved steps that in other days would have been looked upon as wasteful if not extravagant." Plans were already on the drawing board for $30.9 million of capital expenditures; Wiman now proposed that the sum be slashed by more than $24 million, and that there be a tightening of spending all through the organization. He proposed a sharp cutback on further expenditures for facilities and a tightening of the belt throughout the organization.

Later that year, as Wiman reported on the results of fiscal 1948, he was able to say that the company was "closing the finest year" it had ever had. Sales had tripled—rising from $94 million the previous year to $283 million; profits for 1948 were $27.6 million and, after $10.4 million of this was paid in dividends on the common and preferred stock, more than $17 million was retained. The working capital of the firm rose to $147 million from $92 million in 1941 on the eve of the war. The pullback from the earlier expansion plans had put the company in a conservative position. Once again, though, Wiman was ambivalent about his turn toward conservatism, and he ended his statement to the board: "At the same time we must be aware of the business truism that if we stop using money for improvement of our factories and our products, we will cease to progress. We have before us the ceaseless choice between progress on the one hand and financial stability on the other hand.... A major mistake in either direction could be serious to an extreme degree—could be fatal." It seems that Wiman could never truly feel he was off the knife's edge.[8]

POSTWAR LABOR CONFLICTS

As V-J Day approached in the late summer of 1945, industrial relations around the country were in a state of turmoil, as both labor and management prepared to reassert their powers after the tight controls of the War Labor Board were lifted. The period 1945–1946 turned out to be strife-ridden. The historic 113-day strike at the General Motors Corporation that began in November 1945 was soon followed by strikes by 100,000 telephone workers, 640,000 steel workers, 75,000 retail clerks, and others. In May 1946 the greatest strike in the history of American railroads took place; it lasted just three days before President Harry Truman went before the Congress to seek the authority to seize the properties and make the walkout illegal. A coal strike at the same time persisted for almost two months; in this case, the government took over the mines and promulgated a settlement. Certainly the two years after World War II vied with the late 1930s as a period of sustained acrimony and overt strike action.

It was not surprising, therefore, that Deere & Company had trouble, too. After the initial signing of contracts with the two CIO unions—the United Automobile Workers and the Farm Equipment Workers—in the early part of World War II, there had been long and inconclusive negotiations for their renewal in 1944. Locals signed extensions, worked without contracts at certain times, then renewed previous extensions. By 1945, with the prospect that negotiations would be freed of governmental restrictions, the unions' aggressiveness grew apace. By late August 1945, employees at the company's Harvester Works and Spreader Works had walked off the job in what seemed, if not a wildcat strike, at least a walkout authorized by a very small segment of the union members (these were UAW locals). The issue primarily centered on wages (the company had offered a 10 percent increase, the unions were demanding 30 percent); there also were issues concerning piece rates and related issues. By early September the tie-up had spread to the Plow Works, where the Farm Equipment Workers local had taken the employees out; again the vote for the strike authorization had been carried through by a very small segment of the union. A week or so later the strike had spread to Union Malleable (UAW) and to the Ottumwa plant (also UAW).

Both the company and the unions had additional agendas beyond the economic issues ostensibly involved in the strikes. Deere officials strongly felt that bargaining should proceed only on a plant-by-plant basis, in keeping with the company's longstanding policy of decentralization. The United Automobile Workers took the opposite position and from the start had pressed the company for multi-plant bargaining, covering all of their jurisdictions as one. The situation was complicated by the fact that another CIO union was involved—the Farm Equipment Workers. At one point early in the strikes of 1945, both unions approached the company about joint negotiations involving all the plants. This suggestion had a short life; not only was the company implacably against such an idea, but the two unions were so competitive that joint action seemed unworkable.

As the strike continued into October, feelings began to run higher, and the two unions decided to band together under the aegis of the Quad-City Industrial Union Council (the overall CIO unit in the area) to jointly promote some mass picketing and demonstrations. Two places were targets. As expected, the first was the Deere main office; the second choice was a shocker—the homes of Burton Peek and Charles Wiman (who lived across the street from each other). Both demonstrations were orderly, though Burton Peek was lustily booed as he drove away from his home shortly after the pickets arrived. The demonstration outside the homes of the two executives seemed to backfire on the unions, for many of the oldtimers were outraged by it, one employee being quoted as saying: "I have always been a union man, and I am still a union man, but I told them bastards that when they picketed a good old s__ of a b____ like Burt Peek, they were going too far."

By early November an ad hoc committee of employees' wives met with the Moline mayor, hoping to persuade him to intervene in the situation, but he did not. The union's strength was waning, though, and a few days later the Harvester Works and Spreader Works strikes were over, settled essentially on the same basis as the initial offer of the company (a 10 percent wage hike being the central feature). The Farm Equipment Workers local at the Plow Works prolonged its strike for about a month after the UAW plants had settled, finally agreeing to similar terms, as did the Ottumwa and Waterloo locals.

It had been an inconclusive endeavor for all concerned. The negotiations themselves had not served the process of collective bargaining very well, given the primitive stage of the unions' organizational structures, the competition between the two CIO unions, and the company's difficulties in working out its own role (particularly on the issue of centralized bargaining).

In the industry, meanwhile, industrial relations had become even more acrimonious at several other companies. At International Harvester, the Farm Equipment Workers had taken its locals out on strike in early 1946. Simultaneously, there were prolonged strikes by the UAW at Allis-Chalmers and J. I. Case. In all three of these strikes the major issue was the withdrawal by the companies of wartime union security provisions. (During the war, the War Labor Board had promulgated a maintenance-of-membership clause in these companies' contracts.)

Some very large issues were also being fought out within the UAW itself, where a pitched battle between "right" and "left" factions was coming to a climax. There had been widespread accusations that the leadership of the UAW was Communist-leaning; the left had been particularly in evidence in the violence-ridden 329-day strike at the West Allis, Wisconsin, plant of Allis-Chalmers. There the local leader was alleged to be following the Communist Party line throughout the strike, seemingly prolonging the conflict beyond the time when a settlement had seemed assured. Walter Reuther, the leader of the right-wing faction in the union, had had to go to Wisconsin to bring the situation to a close. Reuther had just won the presidency a few months earlier in the turbulent convention of 1946, one of the most furious gatherings of labor leaders in the modern period of the labor movement. His margin then had been a narrow 104 votes and the membership promptly turned around and gave a majority of the executive board to his opponents (with R. J. Thomas, the former president, reelected as vice president).

In the spring of 1947, the faction led by Thomas proposed that the Farm Equipment Workers be brought into the UAW; inasmuch as the union

Exhibit 12-4. Top, Deere employees picketing in front of the Charles Wiman residence during the strike of October 1945; bottom, Strikers also visit the Burton Peek residence on that day.

was also strongly left leaning, Thomas felt that the votes in the national convention from the small Farm Equipment Workers group would be enough to enable the Thomas faction to recapture control. In Moline, the local Farm Equipment Workers leaders made desperate attempts to persuade the UAW locals to vote favorably on the proposal, but the motion was rejected summarily by members at both the Harvester Works and Union Malleable. Reuther, holding up the West Allis debacle as a warning, engineered the defeat of the merger proposal by a more than two-to-one vote. This breakthrough now allowed Reuther to wrest the leadership from Thomas. The convention of 1947 was much less turbulent than the one the year before, Reuther winning a clearcut victory for the presidency and a majority on the executive board.

Farm machinery negotiations with the UAW in 1947 followed the union's negotiations with the automobile industry, and when the latter bargaining set a pattern, Deere and the UAW agreed to the same terms. They provided for an $.11-an-hour increase and liberalization of vacation pay and night-work bonuses. In separate negotiations, the Farm Equipment Workers quietly concluded the same agreement with the company. Two months later, another quarter was heard from—the International Association of Machinists, who represented the skilled machinists in the company. The walkout lasted three days before a settlement was reached that was essentially the same as that accepted by the UAW.

The lines were now drawn for a full-scale local confrontation between the Farm Equipment Workers and the UAW. The Farm Equipment Workers had also controlled the local at Dubuque, and in the negotiations of 1947 they had rejected the proposed pattern contract and subsequently engaged in mass picketing. The company had petitioned for an injunction to prevent the picketing, and only after some minor picket-line violence was a contract finally consummated there in late June. Apparently there was widespread disenchantment among Deere employees with that union's tactics, and the UAW sensed an opportunity to defeat its rival. In a National Labor Relations Board election in July 1948, the UAW won bargaining rights handily (the Farm Equipment Workers at the last minute even asking that its name be withdrawn from the ballot).

This set the stage for a major battle between the two unions in the Moline area. The UAW sent national organizers in to entice local members of the Farm Equipment Workers at the Deere plants as well as at the International Harvester plant in East Moline. The situation was ripe for violence, and it did not take long in coming. In early February 1949 a bloody battle of fists took place outside International Harvester's plant. The fight began when Farm Equipment Workers members left the day shift and found UAW men lined up outside the gate, passing out handbills. The Farm Equipment Workers then organized a phalanx of its own members behind the gates and, when the gates were suddenly opened, they rushed the UAW men. The latter threw their handbills in the air, and hand-to-hand fighting broke out. The UAW men were forced back along the street to where their cars were parked; by the time the twenty-minute battle had subsided, some twenty-two men were injured, most of them from the UAW faction. "The East Moline battle was a surprise to local authorities," one of the newspapers piously reported; it seems hard to believe, though, that anyone would not have foreseen the potential for violence.

Over the next weeks, the UAW pushed leaflets at Deere's Plow Works, and the Farm Equipment Workers retaliated in kind at the Harvester Works. Bargaining on contracts at the various plants was also going on at the same time, and the negotiations became confused in the face of the internecine union squabble outside the gates of the plants. The Farm Equipment Workers soon began to have internal troubles within its Plow Works local and ten officers were ousted in late March. Finally, an NLRB election was set at the Plow Works, pitting the Farm Equipment Workers against the attacking UAW. When the election was held in late April 1947, it turned out to be a major surprise—by more than a four-to-one margin, the Farm Equipment Workers handily retained bargaining rights.

It was a real setback to the UAW. During the jurisdictional struggle between the unions there were also difficulties on the bargaining front. The UAW called strikes at Des Moines and at one of the locals in Moline, and it threatened to strike at other plants as the contracts expired in May. Within a few weeks, though, after negotiations in the face of these strike threats, an agreement was reached that provided for a wage increase of $.05. But the UAW's demand for a $100-per-month pension for workers at age sixty was not included in the agreement.

As the isolation of the left-wing faction within the UAW was occurring, similar battles between right and left had been taking place within a number of other CIO unions. The CIO itself faced a serious dilemma, for an independent report by the Research Institute of America had listed some eighteen unions, including the very large United Electrical, Radio and Machine Workers of America, as Communist controlled. Several of these unions finally did oust their Communist officers over the period 1945–1948. In a landmark decision of the convention of 1949, ten of its affiliated unions were expelled as having "consistently directed their policies and activities toward the achievement of the program or purposes of the Communist Party."

The huge United Electrical Workers was one of those unions; more important to the agricultural machinery industry, the Farm Equipment Workers was another. The latter was a small splinter union by this time, and because of its precarious position it decided to merge with the United Electrical Workers. Thus the Deere & Company contracts at the Plow Works and the Planter Works shifted to the new independent combined union, no longer in the CIO. Now the lines were drawn for an even more vigorous, probably bitter competition at the local level between the UAW and the newly constituted United Electrical Workers/Farm Equipment Workers.

By the time of the negotiations of 1949, Deere had become the first contract renewal for the UAW, so Deere had become the "target" as the union tried to establish an industry pattern. It was the year the UAW sought to bring pension programs into its contracts, hoping in the process to gain joint administration of the plans. Deere already had a very liberal retirement plan and remained unwilling to include the plan in the contract. The final settlement did provide for a reopening after one year for discussion of pensions and insurance and for wage adjustments.

Later, after the Deere agreement of 1949 had been signed, the UAW did obtain pension agreements from the Ford Motor Company, from Chrysler (after a one-hundred-day strike), and finally from General Motors. The last agreement, with the largest company involved with the UAW, was a landmark one. It was a five-year contract, incorporating two types of automatic wage increases during its term. First, an "annual improvement factor" of $.04 an hour per year was to be given to reflect increased productivity. (The contract terminology put it straightforwardly: "A continuing improvement in the standard of living of employees depends upon technological progress, better tools, methods, processes and equipment, and a cooperative attitude . . . to produce more with the same amount of human effort.") In addition, a built-in wage escalator was provided to offset increases in the cost of living. The contract also included a pension and insurance agreement, to be jointly administered by company and union.

As Deere began its "reopener" contract negotiations in June 1950, company officials decided to strive for a similar five-year agreement incorporating the cost-of-living escalator and the annual improvement factor adopted at General Motors. In addition, the company was willing to upgrade its pension program to at least the level of the General Motors plan (but not to have it jointly administered). The company decided its key bargaining demand was to be retention of a tightly worded no-strike clause.

The union, in turn, sought the centralization of bargaining for all seven plants under the UAW contract and the elimination of the no-strike clause that had been in the contract of 1949. In addition, they wanted a central pension board for all plants, to be jointly administered by the union and the company. Pensions by their nature were company-wide in their implications, and Deere industrial relations officials now were forced to modify somewhat their internal organization for bargaining. Local negotiations had been conducted concurrently and at a central location in Moline since 1945; the same pattern was now used again, but the industrial relations director now bargained on the pensions with a company-wide UAW group.

Early in the negotiations, the company decided to "go public" with its proposals, sending a general letter to all employees and releasing statements to the newspapers detailing the company's proposals. By this time, the

UAW was already out on strike against International Harvester, and those negotiations complicated the situation at Deere. The UAW rejected Deere's offer (its hand apparently having been forced by the public announcement by the company) and a strike was called for September 1. All seven plants with UAW locals went out on that date, and Deere almost had its first company-wide strike. It soon settled into a harsh battle, with much bitterness on both sides.

As positions hardened, both parties took to the press to win employee and public support. While the Moline area employees seemed to accept the strike as necessary, the company sensed that employees at Waterloo, Des Moines, and Dubuque were less enthusiastic. There had been allegations at Des Moines right from the start that the strike vote had been fraudulent, that the ballot boxes had been depleted of an overwhelming "No" vote and stuffed with a requisite number of fraudulent "Yes" votes. These charges were never definitively proved or disproved, but the press felt it had enough corroboration to print extensive details on them after the strike was over.

In early November the company made a critical decision—to encourage a back-to-work movement at the Waterloo and Des Moines operations. It was a miscalculation. In Des Moines the situation was the most tense. The newspapers had played up the embryonic movement widely through the press, and the union countered with harsh tactics, using flying squadrons of union members (aided by a number of people from outside the area) to terrorize employees thought to be involved in the back-to-work movement. The unidentified groups broke up a fish fry and a skating party, threw bricks through windows of certain employees' homes, and actually assaulted one or two people. At the plant itself, mass picketing was established for the proposed first day's return and there was some scuffling and damage to cars. The picketing subsequently was constrained by local officials, but the situation was so dangerous that few employees seemed willing to risk the picket lines. The events at Waterloo were less violent, but there, too, the back-to-work movement fizzled. Though there was a small increase in returning employees over the succeeding days, in neither plant did the back-to-work movement become substantial.

There was also a violent incident in Moline about the same time. The UAW had struck the Union Malleable plant, and the local negotiations had been quite tense as the strike progressed. In late November, the union added a large number of people to its picket lines and imprisoned two management employees in the factory overnight. No further incidents occurred, but a residue of bitter feelings remained about the imprisonment.

The strike dragged on until December 19. At that point, the union gave up its demands for a master contract and for a central pension board, two of its key goals. In turn, the company altered its position on the no-strike article and agreed to a limited right-to-strike over grievances related to piece work and incentive rates. In handling such grievances, employees had the option of obtaining an outside opinion from independent engineers (a mechanism set up in the contract) or resorting to a strike.

Unfortunately, there were grievances connected with the incentive system, for it had been set up far back in the past as a hodgepodge of individually negotiated rates, with little in the way of comparability among them. Individual factory managers had developed their own incentive systems, and there was an appalling lack of uniformity in the company. Try as they had, the industrial relations group could not persuade factory managers to work out a realistic uniformity, for the managers typically wanted to keep things as they were.

The negotiations of 1950 brought these issues to the forefront again. With the right of the union to have an authorized strike if it disagreed with piece-rate structures now established, it became doubly important to revise the entire incentive structure. Just as soon as the negotiations of 1950 were concluded, Deere began the analytical work necessary to develop some new company-wide form of "standard hour" system. Wisely, they drew the UAW leadership into this process right from the start. As the studies progressed over the period of the five-year contract, the union leadership became as expert as management on the proposed changes in the incentive concepts. Indeed, the union officials probably were more sophisticated and knowledgeable about the system than were many of the people in the factory. This mutual work on the system over the years of this five-year contract proved to be very efficacious as the two parties approached the very important renegotiation of the contract in 1955.

The remaining parts of the contract of 1950 were essentially those proposed by the company at the beginning of negotiations and promulgated in the company's public announcement shortly after the strike started. The contract was for five years and contained the annual improvement factor and cost-of-living escalator adjustments, the pension-plan improvements, and certain hourly-rate adjustments. The strike had been monetarily costly to both parties. The 110-day loss of production and wages was very substantial; the union also expended large sums in strike benefits (and found itself with its own internal squabble after paying extra benefits in Waterloo and Des Moines to counter the back-to-work movement). Arlyn John Melcher, in his definitive study of collective bargaining in the agricultural implement industry, concluded about the strike: "The company finally decided to compromise on the key issue and granted the union the right to strike during the contract period. The log jam was broken and in the next two weeks the negotiations for a new contract were completed. . . . The union was firmly in control of the strike and J. D. accepted this condition with as good grace as possible." The strike had been seen as a test of will by both parties, and each came away from the strike feeling that it had established its own independence and resolve. Each party also came out of the long strike with

ANTITRUST CHARGES AND DEERE'S DEFENSE

On September 10, 1948, in the US District Court for the district of Minnesota, lawyers for the attorney general of the United States filed a complaint against Deere & Company, International Harvester, and J. I. Case, alleging that the three companies had violated the antitrust statutes of the United States. The government charged that the three were selling farm machinery to dealers on the condition that the dealers refuse to handle competing manufacturers' goods. This, it was alleged, unreasonably limited the market outlets for the farm machinery of the three companies' competitors and therefore was in restraint of commerce under the Sherman Act, as well as a violation of the Clayton Act.

The period right after World War II was one of vigorous antitrust activity by the government, comparable to the prewar crackdown by Thurman Arnold. It seemed particularly preoccupied with corporate concentration. Earlier, Wendell Berge, assistant attorney general, had singled out the farm machinery industry for particular censure, stating to a Wisconsin Farmers' Union convention in 1945: "Unfortunately, monopolies do not expect to continue full production for peace. Their purpose is not to achieve the greatest volume of production and employment, but the maximum profits on the upswing and the minimum losses on what then becomes the inevitable downswing.... In farm tractors, for example, four companies control 91% of the production.... In the depression, agricultural machinery payrolls were cut nearly 85% but prices were cut less than 15%." Thus the government again raised the specter of large, dominant companies being able to manipulate and control total production of an industry, holding it down for their own purposes—precisely the argument made in the War Production Board flap when Charles Wiman was head of the farm equipment division in 1944.

Deere already had a stated policy, in writing, against pressuring dealers to trade exclusively with the company (so, too, did International Harvester). A number of people in the company felt that the government case was almost frivolous, destined to be easily defeated by the company—just a "political" suit that had no particular consequent. Budge Cook, on whose shoulders as general counsel would fall the greatest burden of the defense, vehemently disagreed, sensing a fundamental threat to existing marketing patterns in the industry. "This is not a money matter, but involves the future functioning of our sales organization.... The stake ... is our ability and our right to market our products."

CHAPTER 12

The government's prayer for relief to the court asked "that the defendant . . . be perpetually enjoined from refusing to sell farm machinery to any dealer because said dealer purchases or sells farm machinery manufactured or sold by another than a defendant." How was this to be interpreted? At the minimum, this might constrain the ability of the company to exhort against sloppy or ineffective practices by dealers; already Cook was getting reports from branch managers in the company that territory managers were becoming timid in their dealings with some of the weaker dealers because of the government suit. If standards of good dealership could not be required by the company, overall dealership productivity would probably slip. There was an even more devastating threat; if the court were to rule that no company had the right to refuse selling their equipment to any dealer who requested it, this truly would mean a revolution in the methods of distribution. No longer would the privilege of being a "dealer" have much meaning—just about anyone could demand to sell machinery.

Given some 4,000 Deere dealers, it was likely that at least a few territory managers (travelers) dealing with them in the field had resorted to heavy-handed methods, despite the company's written policy. Cook explained why in a memorandum to key members of top management: "Nearly 50% of our travelers have never sold goods in a buyer's market. When they come under heavy pressure for sales, their natural tendency will be to reach for the 'big stick' and to force our products rather than to sell them. Even though we are in a position today to defeat the pending suit, still those new travelers carry the possibility of making a case against us in the first year or two of their struggles in a buyer's market." Violations by company travelers might be much the exception, Cook guessed, yet the company would need to prove just how "exceptional" these instances really were.

First there was need to know just how many of Deere's dealers had been handling competing goods in the years in question (the government investigation was to cover the years 1944–1948). Such information could be obtained from Deere's territory managers, but such statements would likely be considered by the government to be tainted—and by Deere top management, too, for it would be in a field man's personal interest to gloss over any pressure tactics he might have used in those years. The alternative was to take depositions from some 4,000 separate dealers, a task of quite a different magnitude. Further, the job had to be done by outsiders, not by company personnel; the company finally chose Dun & Bradstreet for the task.

A survey of this ilk was ticklish business—dealers were skittish about discussing their own affairs with any outsider; further, they could not be certain just how Deere really felt about the matter. Some dealers might worry that the company was indirectly threatening a termination. It finally took a number of second visits, coupled with assurances that Deere really wanted to have the full story told, before Dun & Bradstreet got a viable sample. The results were clear-cut—more than 75 percent of the dealers handled at least some competing machinery. This reinforced what Deere marketeers strongly suspected but had not been able to document before. Of course, this was "good" news for the legal case, but it must have given pause to the territory managers and travelers about the obvious competitor inroads. "Why are all these people selling others' goods?"

The company files themselves produced some unexpected benefits. Deere had slipped in efficiency of record keeping in the Wiman period (as contrasted to the Butterworth period, when extensive historical data were maintained). So the company was undeservedly lucky now in finding several key letters of the recent past that documented unequivocally the dedication of the company to the policy of never using pressure for full lines. These letters complemented Burton Peek's blunt bulletin to all factory and branch personnel in 1938, written shortly after the publication of the Federal Trade Commission's report, exhorting all company personnel: "Don't use 'the big stick' in getting business. The use of excessive coercion is unbusinesslike. . . . Sometimes the salesman is inclined to use the 'big stick.' This is not our policy."

The most sensitive issue was that of involuntary termination of dealerships, for one of the first items requested by the government investigators was a list of former dealers, many of whom were to then be interviewed by government field personnel. The total of these terminated dealerships was not inconsequential. In the years just prior to the war, the company had mounted a program to strengthen individual dealers, in the process winnowing out the weaker ones, leaving the remaining ones financially stronger and more effective. There were just under 4,000 Deere dealers in the United States at the time of this case, down from approximately 10,000 in the early 1920s. The very concept itself of being a "dealer" was not nearly as well refined in those earlier periods—many then had seen themselves just as order takers, with little responsibility for follow-up service.

This incisive winnowing process led to Deere's "key dealer" program, pinpointing those dealers in each particular location in every territory that could be considered as the most effective. The territory managers then were to put particular emphasis on working with the company's most important outlets. The earlier experience with the joint Deere-Caterpillar dealerships in California had been instructive, pointing up the values of having a few large, well-financed independent dealers, rather than a string of smaller, weaker groups.

There were various reasons for terminations: the weakness of too small a scale of operation, personal ethical behavior of the dealer himself, and so on. For whatever reason, terminations almost always left hard feelings and recriminations. No one wishes to have a right taken away, whether or not it has been earned by good performance. There also can be unfair terminations, and for this reason several states more recently have enacted legislation constraining the right of companies to unilaterally sever relationships with dealers. The company's interviews of discontinued dealers who also had been interviewed by the government turned up a number who were hostile to Deere. To the surprise of company interviewers, however, a substantial number of them were willing to state frankly that their own failings had led to the terminations, and they were willing to put their views on record for the court. Thus, by the time the company had put together its full list of potential company witnesses, it had found a number of former dealers who would testify on behalf of the company.

Deere top management made an astute decision in the process of conducting this extensive in-company research. Rather than leaving all the legwork to junior personnel within the organization, a deliberate effort was made to involve top-level executives in the field work itself. All through the organization, both leading branch personnel and general-company officers involved themselves in the field interviewing and field research, in the process learning more about the company than probably they ever would have had they stayed back in their own offices.

This introspection and self-examination turned up in the process a number of new ideas that could give the company a more intelligent approach to all of its operations, particularly marketing. Cook wrote Charles Wiman after the case was over: "Branch house personnel have become so acclimated to the problems presented by the lawsuit and actually are doing a more intelligent job today in the field of dealer relationships than was true before the suit was filed." In effect, the case became the venue for an undesignated but nevertheless highly effective executive development program for training of future top management.

THE GOVERNMENT TRIES THE J. I. CASE COMPANY

In April 1950, after an enormous expenditure of time and money on preparation of the case (the direct costs alone were estimated to be more than $1 million), the company was notified by the Department of Justice that Deere was not, after all, to be subjected to trial in the foreseeable future. Rather, the government had decided that the J. I. Case Company would be the first target. All eyes turned to Minnesota, where the Case proceedings were scheduled to come to trial in April 1951 before District Court Judge Gunnar H. Nordbye. The trial lasted most of April and May, and by the end of formal testimony there were some 1,940 pages of documentation on Case's policies and practices. In many respects, Case's practices differed from Deere's; for example, Case did not have an explicit written policy similar to that promulgated by Burton Peek in 1938. The name of Deere & Company did come up from time to time in the trial—it was inevitable that when Leon Clausen, the chairman of Case and former head of all of Deere's manufacturing, was called to the stand, he would be required to answer questions comparing Case with Deere. Clausen did this adroitly, with no prejudicial remarks about Deere itself that would hinder in any substantial way the company's defense, were it also to come to trial later. Each company's situation was different; no one of the three wanted to damage the others' defenses, deliberately or inadvertently—the collective stakes were too high. On the other hand, they really were not in the case "together" and were wary of any overt cooperation, not only for fear of being accused of "conspiring" together, but because each felt that there were sound competitive reasons for keeping a distance from the others.

Deere's day in court never came. Judge Nordbye's decision four months later was a shocker. The entire "case against Case" was unequivocally turned away, all pleas by the government turned aside, the bill dismissed. The judge's decision was a detailed one, extending over a dozen pages. The opinion was considered so uncompromisingly clear that it came to be a "landmark decision" in antitrust suits, soon appearing in texts and law books devoted to pivotal law cases.

Nordbye left no doubt that he thought the government's evidence weak and unconvincing. He stated: "Efforts in salesmanship were used to convince the prospective dealer that his future was more promising if he would devote his efforts to the Case line rather than being occupied principally with other activities, whether they be diversions in selling hardware, automobiles or competing farm machinery. But such a policy, in absence of coercion and pressure, merely reflects sound business judgment and reasonable competitive efforts which do not necessarily constitute competitive restrictions." There were examples of "flagrant attempts to coerce and put pressure on a few dealers, but they sprang up here and there disconnectedly and contrary to the announced policy of the company, and without indicating that there was any pattern."

More important than the issue of exclusivity was the allegation of restraint of trade, and here Nordbye was equally definitive: "The burden of proof rests upon the Government to establish that defendant has engaged in practices, agreements or understandings, and acts unreasonably restraining commerce . . . and tending substantially to lessen competition and to create a monopoly. . . . This it has failed to do. . . . Here there is no 'potential clog on competition,' which the Supreme Court referred to in the Standard Oil case, and upon which decision the Government principally relies. . . . The bill must therefore be dismissed."

This was a district court decision and could, of course, be appealed to higher courts. The government chose not to do so. At the request of the Department of Justice, the case against Deere was dismissed in April 1952. The court's dismissal was "without prejudice"; the Department of Justice reserved the right to file the same kind of suit at a later date. But it never did.

What judgments can be made about this case? Inasmuch as the government case collapsed completely after Judge Nordbye's decision, one can legitimately infer that the arguments of the Department of Justice lawyers were weak. Moreover, it seems incongruous that the government pursued the industry at this particular time; the pattern of production limitations imposed on the larger farm machinery companies during World War II (particularly the controversial concept of "concentration") gave the smaller companies a remarkable opportunity to increase their market shares and steady their shaky financial statements. The government had depended heavily on market-share statistics to buttress the case that the three big long-line companies were dominating the industry. Judge Nordbye repeated these same figures in his opinion, prefacing them with the following statement: "The evidence reflects that there is healthy competition among all farm machinery manufacturers. The existence of a vigorous growth in the farm machinery business reflecting sound competition seems evident from the following table [exhibit 12-5]."

Exhibit 12-5. *Market Shares of the Seven Full-Line Agricultural Machinery Manufacturers, 1944–1948 (percent of total industry sales)*

	1944	1945	1946	1947	1948
J. I. Case	3.7	4.4	3.8	5.3	7.0
International Harvester	15.8	17.6	22.3	23.4	22.8
Deere	9.6	9.9	15.1	14.1	15.3
Allis-Chalmers	3.2	4.3	3.5	6.3	6.8
Oliver	2.1	2.9	4.3	3.9	4.2
Minneapolis-Moline	1.7	1.9	3.1	3.3	3.6
Massey-Harris	1.1	1.3	2.9	2.8	3.8
Full-line companies' share of industry sales	37.2	42.3	55.0	59.1	63.5

Source: *United States v. J. I. Case Co.*, 101 Fed. Sup. 856.

As shown by the figures for 1944, the market shares of the seven long-line companies declined during the war years, due to the limitation orders. By 1948 all seven had made strong recoveries; the four smaller of the seven registered proportionally larger gains from 1944 than did the three large companies that initially were parties to the suit. The two smallest, Massey-Harris and Minneapolis-Moline, had not only more than doubled their 1944 market share, but they had also exceeded their positions in 1936 by substantial margins. International Harvester was the least successful in making a comeback, though it continued to lead the industry. Warren Shearer's definitive study of the industry's competitive relationships, completed just before Judge Nordbye's decision, came to the same conclusions. Shearer commented after lengthy analysis, "The industry was far more evenly balanced than it had been in 1938 and . . . the prospects for continued gain by the smaller of the long-line concerns is excellent. . . . If we draw our conclusions from the present market shares obtained by the leading concerns, it would be necessary to conclude that healthy and effective competition actually prevails." Shearer felt that the stronger financial position of the smaller firms in the industry allowed more resources for investment in research and design, and that "it is this technological battle of which implement executives speak when they talk of 'fierce' competition." Shearer argued that this "design" competition was the key to the industry: "Although IH and Deere remain the leaders, their position is being vigorously contested. Relative equality of product will prevent many of the abuses which are likely to creep into an oligopolistic situation, as long as that equality is based on legitimate product differentiation. . . . As long as the industry is developing new products or distinctly better products at the rate they have been developed since 1930 there is no danger that the consumer can be made to suffer from price agreements." Shearer concluded, "On economic grounds alone the farm machinery industry meets with fair success the tests we have established for 'workable' competition."[10]

INNOVATIONS IN ENGINE DESIGN

The insatiable demand for agricultural equipment at the end of World War II momentarily gave the manufacturers a booming "sellers market," in which existing models, however obsolete, were snapped up by avid buyers. But fierce competition soon encouraged innovation and change, and the ten-year period from 1945 to 1955 became noteworthy for product improvement. Deere remained in the forefront in yearly, shorter-term product development, but found itself in an increasing dilemma about one crucial longer-term question: How long should Deere stay with the two-cylinder tractor?

During the 1930s the two-cylinder engine uniquely fulfilled some important demands—for low fuel cost, for simple cooling systems, for straight camshafts with no belts or pulleys, and, in general, for an easily understood, farmer-repairable piece of machinery. Distillates could be used as fuel, the "all fuel" concept at that time giving lower costs. Deere had been successful in selling this concept on its fundamentals—that the two-cylinder engine was as good, or better, on its merits than competitor machines for a wide range of farming conditions. For more than two decades company salesmen had vigorously defended the "Popping Johnnies" against competitor sneers; almost yearly, some trade rival would float rumors that "Deere is going to junk the two-cylinders," hoping to panic buyers away from a model threatened by obsolescence. It became an article of faith among Deere marketing men to fight across the board any attack on the two-cylinder concept. The defense worked very well indeed, for the Models D, A, and B continued to sell in large volume.

After World War II the situation began to change. Many thousands of young farm men had gone off to war to drive tanks, jeeps, weapons carriers, and countless other vehicles, almost all of which were either four- or six-cylinder. When they returned to the farm, often to take over from their weary fathers, would they be willing to stay with their two-cylinder "John Deeres"? Further, some of the most potent substantive arguments for the "two" had lost their credence. Distillate fuels, now relatively higher in price, no longer carried a significant cost advantage. Wartime advances in the catalytic cracking process had made heavy fuels sometimes even more expensive than gasoline; moreover, the heavy fuels required lower compression and, thus, less horsepower. Further, the farmer was demanding larger tractors with higher horsepower, and there were upper limits to the size of two-cylinder engines that could be mounted on a tractor. The horizontal-bore configuration and the diameter needed for such power would soon force so large a size that the tractor could not fit on the row of crops it was to work. The slower speed and lower compression of the distillate-burning engine was an advantage in terms of wear, but a disadvantage in gaining higher horsepower. For this reason, right after World War II, the Waterloo tractors were made available in both a distillate "all fuel" version and a higher-compression gasoline-burning model.

Deere lived under a further cloud—it did not have a diesel engine. Rudolph D. Diesel had perfected his engine back in 1892, and adaptations of his design were successfully operating in Germany in 1897 and were sold in the United States before the turn of the century. In these early years, diesels were not so easily adapted to vehicles, for the difficulties in starting the engine made it unpredictable. The first commercial use of a diesel engine in a crawler tractor did not come until Caterpillar's adaptation in 1931. International Harvester followed with a successful wheel tractor version in 1933, and soon many manufacturers were competing with successful models. The diesel had earlier become very popular in Europe, and by the end of World War II a substantial number of the tractors built in the world were diesel-powered.

Thus the Deere dilemma. On the one hand, farmers continued to press for refinements in existing tractors. They wanted better steering control, an independent power take-off, increased hydraulic power for operating integral equipment (particularly hay balers and forage harvesters), more fuel efficiency, added horsepower for heavier jobs, and, at the same time, more flexible shift-up, throttle-back, fuel-efficient ways of doing lighter work. All of these demanded immediate attention. Competitors were making changes, and if one wanted to stay alive in the marketplace, improvements in existing models had to be constantly provided. This always commanded great amounts of engineering and product planning time.

Yet adaptations to meet competition in the short run always borrowed from longer-term product development. More and more farmers were interested in diesel tractors; if Deere was going to bring out a diesel—say, a two-cylinder version—this would take many months, even years, of hundreds of engineers and designers. If Deere decided to move to a four-cylinder or six-cylinder tractor, or to both, a much longer commitment would be required. Deere manufacturing men had often espoused the rule of thumb that a new model of an older machine generally required about 33 percent new parts; a new line incorporating a major engine change would require new designs for 95 percent of the parts.

The Deere management at both of the tractor factories—Waterloo and Dubuque—now faced hard choices. A turn in one direction might foreclose all other opportunities. Perhaps the two-cylinder concept could no longer be sold alone on the fundamental argument that two cylinders per se were as good or better than four or six. Rather, the particular values of the Deere versions needed to be stressed—they were so well made, so dependable, and so easy to repair and carried such an excellent reputation with the farmer. On this basis it might be possible to sell those same tractors over many future years; the continued success of the Models A, B, and D held considerable promise for this scenario. There would come a point where this could no longer be the case, but had this point been reached?

By the end of the war, the Deere manufacturing and marketing teams came to believe that the Model D was no longer commercially viable. Though it had been upgraded throughout its product life—mounted on rubber in the early 1930s, restyled in 1939, increased in horsepower in 1940—it was nevertheless becoming inadequate for the wheatland farms, which were growing bigger and demanding a standard tractor with even greater power. The goal now became twofold—to increase power, and to generate that power from a diesel. By 1949 the company was ready with its Model R, the first diesel-powered tractor Deere had ever manufactured. It was also the most powerful, rated 34.27 horsepower at the drawbar, 43.32 at the belt. Charles Stone, by this time the vice president for product development, impressed by the enormity of the new machines, said unequivocally, "We see the limit here to the rubber-tired tractor—after this, it has to be a crawler." (Today wheeled tractors of more than 200 horsepower are not uncommon!)

The Model R had a gasoline-fueled auxiliary starting engine, which eased two problems that had plagued earlier diesels. Most diesel tractors had been started by the operator standing on the ground, manually turning a flywheel; the Model R was started by the operator in his seat, pulling a lever. The earlier diesels had been notoriously difficult to start; the Model R auxiliary starting engine enhanced warm-up, allowing a fast start even in cold weather. To the relief of everyone, the new tractor was greeted with enthusiasm by the trade; Deere was back in the running, with a good diesel engine. It was a two-cylinder engine, though—the company had taken the calculated risk that a two-cylinder diesel could compete with the four- and six-cylinder counterparts in other companies.

With the change in the Model D, the question now came of what to do about Models A and B. These enormously popular tractors had been in the line since the mid-1930s; by 1952, there were more than 600,000 of these two Deere models in operation. But, despite their success, they could not continue to compete forever without change. Once again there was the nagging dilemma—the changes could either be minor (cosmetic or small refinements), or major, or, more basically, they could entail scrapping old models and introducing tractors with more cylinders.

By this time there were strong advocates, including Charles Wiman, for phasing out the two-cylinder tractors. Dubuque engineers had already developed an upright two-cylinder engine for the Model M, and they were experimenting with another upright multiple-cylinder engine for self-propelled harvesters.

Duke Rowland, who still headed all of the company's tractor production, was adamantly reluctant to scrap Waterloo's very successful tractors, however. They were selling well in the field and his organization had its manufacturing on so efficient a basis that the factory was the biggest profit-making unit in the system. Over and over, he told the management in Moline, "We would be run over by our competition if we stopped now to change—we couldn't make the shift within nine months of downtime—we'll go broke, we'd be just like Ford when it changed from the Model T to the A." These arguments worried Wiman, for they called up images of the Great Depression that he had faced early in his career; still, though Wiman was genuinely ambivalent on the timing of the change, he nevertheless advocated making it earlier rather than later.

Once again the authority vested in the factories by the policy of "decentralization" intruded to decide the question. Rowland's conservative posture held sway, and the decision was made to upgrade Models A and B, but to retain the two-cylinder motor. In 1952, two new models were introduced—the Model 50, to replace the B, and the Model 60, to replace the A. Both had many innovations. Each had duplex carburetion (a separate carburetor for each cylinder), which allowed more precise fuel metering and helped increase horsepower (the Model 50 was rated at 20.62 drawbar horsepower and 26.32 belt horsepower). Both were row-crop tractors, like their predecessors, and the 60 was also available in a "high-crop" model of particular value to producers of tall, bushy, bedded crops like sugarcane, flowers, and so forth. These two models were also the first of the Deere tractors available with "live" power shafts, which provided continuous power for operating equipment driven by a power take-off (PTO). A continuous running or an independent PTO became a necessity for the postwar versions of hay balers and field forage harvesters, for both machines contemplated stopping the

> **A FOUR-CYLINDER ENGINE—NOT ON YOUR LIFE**
>
> Some of our dealer and tractor owner friends occasionally tell us they have heard that John Deere is coming out with a four-cylinder engine in all John Deere Tractors. Dealers and others who are well informed know that this rumor is false.
>
> We want to assure those who are not so well informed that we are *NOT* coming out with a four-cylinder engine. **THE JOHN DEERE TWO-CYLINDER ENGINE HAS BEEN SO OUTSTANDINGLY SUCCESSFUL THAT THERE IS NO THOUGHT OF A CHANGE.**
>
> JOHN DEERE TRACTOR COMPANY
L. A. Rowland
Vice Pres. and Gen'l Manager

Exhibit 12-6. "A Four-Cylinder Engine—Not on Your Life." *John Deere tractor field service bulletin, February 15, 1937*

forward travel of the tractor from time to time while continuing the PTO for the baler or forage harvester.

An interesting issue surfaced at this time, that of interchangeability. Farmers wanted to be able to use any given power equipment, whatever the make, in whatever combination desired; thus the lack of compatibility of PTOs became an industry-wide problem. Once again, as the industry had done just after World War II with hydraulic control standardization, the Farm Equipment Industry Engineering Advisory Committee set up an industry-wide subcommittee on the PTO to standardize such things as hub and spline dimensions, drawbar hitch and power take-off location, and drawbar vertical load dimensions. Out of this came a set of standards in 1958 that were uniformly adopted by the industry. Each manufacturer was free to develop his own engineering of the internal mechanisms that would lead to the external PTO, but the couplings themselves would all be standard.

The drive for industry standardization led to an unusual contest between two groups at Deere. Once again it involved Waterloo (and Rowland). The company had fallen behind in the technology of the hitch. One of the most innovative ideas in the Ford-Ferguson tractor, when it was introduced just before the war, was its three-point hitch, which incorporated a hydraulic system that was draft-sensitive, thus making the hitch work for both draft and position. Deere's hitches could not be controlled hydraulically for draft, so several company factories began experimenting, each on its own (decentralization again!). The real push and tug was between the Waterloo Tractor Works and the Plow Works, longstanding rivals in the company. Before the advent of the tractor, the Plow Works had been center stage, the main actor—after all, was not the plow the most important, bread-and-butter product? Pride, a sense of importance was still the legacy there. Similarly, the Tractor Works saw itself in the same light—was not the tractor now the real bread-and-butter product, dominating all others in dollar volume, in profit? Pride and sense of importance were in abundance there, too.

An important step toward bridging the gap between the two touchy groups came in 1949 with the establishment of the tractor and implement committee, which had high-level representatives from both operating groups, as well as central staff people from the experimental and product planning groups (and was chaired by one of the latter). Still, rivalry and jealousy between the two operating entities persisted and now surfaced once again as each insisted it had the right to develop the definitive version of the hitch. "After all, the hitch is at the end of our tractor," said Waterloo. "No, it's our implement that is to be positioned," said the Plow Works.

Each finally developed its own version, but neither would give ground. The Plow Works prototype was an adaptation of one the Lindeman brothers had developed at the Yakima Works, and there was considerable sentimental feeling that it should be adapted because of this. But sentiment is more often than not a bad decision-making criterion; the hitch was an "add-on" version, not self-contained, and it had enough other disabilities to be dubbed by its detractors "the Stinger." Even this failed to convince the "plow boys," so top management finally decided to hold a field contest between the two. In the ensuing match, the Waterloo version won, hands down. The Plow Works remained unreconciled, predicting "disaster" if the Waterloo hitch was adopted. Only after a compromise was arranged—that the Plow Works could build 1,000 of its hitches, to be stockpiled until the disaster struck—was the conflict settled. The fact that Pat Murphy formerly had been head of the Plow Works had more than a bit to do with this contingency concession. The Waterloo hitches turned out to work very well and soon were adopted for all the company machinery; the 1,000 stockpiled Plow Works hitches were scrapped.

In 1953, the Model 70 arrived in the product line. Originally the model was available with gasoline, "all-fuel," or LP gasoline options but soon was offered with a diesel option, and thus it became the first Deere diesel row-crop tractor. In its Nebraska tractor test, the Model 70 diesel set a new industry fuel economy record, bettering all previously tested row-crop tractors of all manufacturers. By 1954 Deere engineers had perfected another "industry first," an optional factory-installed power-steering system for all three of the new models. It utilized built-in hydraulics to control the steering mechanism, differing from so-called "add-on" systems that used externally mounted motors on the steering shafts or hydraulic cylinders hooked up to tie rods. Deere's power steering was a breakthrough for the industry, soon widely adopted by other manufacturers. Again, though, the engineering time taken to develop it was at the expense of an upright, multiple-cylinder engine.

All of these models were manufactured at Waterloo, with all of the engineering done there. Meanwhile, the Dubuque engineers were equally busy. In 1953 they brought out a new Model 40 in both a standard tractor, a tricycle, and a crawler, replacing the Dubuque-manufactured Models M, MT, and MC. In effect, the Model 40 series was a smaller counterpart of the Model 50, 60, and 70 tractors built in Waterloo. (The 40 tricycle model was rated at 17.16 drawbar horsepower and 21.45 belt horsepower, the crawler was rated at 15.11 drawbar horsepower and 21.45 belt horsepower.) In 1955, six years after the introduction of the historic Model R diesel, its own replacement arrived, the Model 80 diesel. This was a gargantuan machine, weighing some 7,850 pounds, and it was quite powerful—57.49 horsepower at the belt, 46.32 at the drawbar. It incorporated many of the most popular features on the other numbered models, such as optional power steering and "live" power shaft, and also included a six-speed transmission. The 80 was capable of pulling a twenty-one-foot disc and working up to 126 acres with it in a single day. A large 32.5-gallon fuel tank allowed the operator to stay in the field many hours before making a refueling stop.

The new models—the 40-50-60-70-80 group—were well received, particularly for the numbers of refinements and innovative new ideas they incorporated. But the handwriting was on the wall about the staying power of any two-cylinder tractor, no matter how good. Waterloo knew this as well as anyone. By this time, Duke Rowland had left his position heading tractor production and Maurice Fraher assumed his post. Both Fraher and Gust Olson, the Waterloo Works manager, were less resistant to change. So a decision was taken, sometime in early 1953. (The exact date is no longer determinable, the whole operation being so secret that no records were kept!) A select group of Waterloo engineers and design men were to be pulled off their regular duties and set to work on the task of developing an upright, multiple-cylinder motor. Their charge: "Start with a blank drawing board." An abandoned building that had housed a grocery store was rented (quickly dubbed by the Deere men "the butcher shop," perhaps an unintended double entendre!), and, amid great security against any communication leaks, the process began.

Right away, there was a serious miscalculation. The marketing managers in the branches had been uneasy about the decision to go ahead: "For 30 years, we've been selling an 'exclusive' . . . you'll have to give us another 'exclusive,' else we will look like we've recanted on all of our claims." So in part to placate them, the first efforts of the "Butcher Shop" group concentrated on a V-4/V-6 notion. The V-design engine had worked well on the automobile, but the idea was flawed for tractors—the narrowness required of the row-crop tractor would not allow the requisite juxtaposition of the pistons. After months of work on the V, the notion was abandoned. In the process, valuable time had been lost. The engineers were forced to turn to vertical positioning, despite the remonstrations of some sales leaders that this was too "conventional." Nevertheless, an irrevocable decision had been made—the company would produce four-cylinder engines.

COMBINE AND CULTIVATING PRODUCT DEVELOPMENT

In combines, the new star was the self-propelled combine. The Massey-Harris coup of providing self-propelled combines for the crop season of 1944, the "Harvest Brigade" story recounted earlier in this chapter, alerted all other manufacturers to the need for self-propulsion for their own combines. Massey's version dated back to 1938, when its first models were tested in Argentina. International Harvester also brought out its own successful self-propelled model in 1942 (and made thousands of them during World War II). Deere engineers knew that they had to get into this fray, and by 1947 the company was marketing its No. 55, with a top center–mounted engine and grain tank and using a twelve-foot platform. At first, a canvas platform was utilized, but then the engineers developed a successful auger feed, with retractable fingers at its center; this proved to give a very even feed. Deere's self-propelled combine was an instant success, one of the most widely accepted models of harvesting equipment ever brought out by the company. By the time of its demise, more than twenty-two years later, nearly 84,000 units had been built and sold. The farmers in the windrow areas, particularly in the Northwest, demanded a machine that had the same capacity of the No. 55 but they wanted a pull-type machine because of their large horsepower being idle during the harvest period. This led the company in 1949 to bring out the No. 65, a model similar to the self-propelled. Small combines were also still in demand, but farmers now insisted on more rapid harvesting, so the Deere engineers modified the old No. 12A to accommodate a seven-foot platform, and this was reborn in 1952 as the No. 25. In 1950, the company also brought out its first two-row, self-propelled cotton picker, a model that would pick from both sides of a row.

Deere's popular hillside combine, in the line since the purchase from Caterpillar in 1935, had been a pull-behind machine all through the period of the 1930s and '40s. With the great success of the self-propelled combine in the Midwest, several harvester manufacturers began marketing similar models for use on the slopes of the Northwest farms. A problem arose, though. On a hillside combine, the cutter bar necessarily had to follow the ground slope, while the separator stayed level. With a self-propelled machine, the mechanics of this became more difficult. A particular problem was in the required leveling, which had to be done manually on the early machines. A few agriculturalists in the Northwest began experimenting on their own with self-leveling devices, though most of their compatriots ridiculed the idea. Deere saw in these amateur efforts the nucleus of a viable self-propelled, self-leveling machine and purchased some of the farmers' best ideas. The result was the Model 55H, brought out in 1954. It incorporated an automatic lateral leveling device that could keep the separator body level on side slopes of up to 45 percent—a concept (shared by International Harvester) that revolutionized harvesting in the Pacific Northwest. Not only was the machine more effective—it now could readily be run by one man—but it was a good deal safer.

Probably the most spectacular of Deere's postwar harvesting triumphs was the combine corn head. In 1952, the company brought out its Model No. 45 combine, a spot-welded machine of more robust construction, especially designed for heavy work in the Midwest corn and soybean areas. Deere engineers had been experimenting with the use of the combine in corn since 1950, at first using modified grain heads to harvest the whole plant. In 1954 they developed the No. 10 two-row corn attachment, especially designed for use with the No. 45. With it, the farmer could pick, shell, and clean up to twenty acres of corn per day in a single operation. Field shelling losses had been reduced as much as 75 percent and the requisite storage space had been cut in half; the corn could be cut while it had as much as 30 percent moisture content, earlier than with other harvesting devices. By 1956, two-row and four-row attachments were also designed to be fitted to the No. 55. All sizes of the corn-head attachment incorporated a unique snapping unit that prevented damage to the ear, thus preventing inadvertent shelling during the snapping operation. The combine corn head was an "industry first" of major importance, one of the most successful of Deere's harvesting innovations.

There were other important implement innovations in this period. In 1947, the first front-mounted, "Quik-Tatch" row-crop cultivators were marketed, and the first planters having remote-control hydraulic cylinders were sold. In 1950 a safety trip standard for moldboard plows was developed, again an industry first. The ten years—1945–1955—had been rich ones for product innovation; now to be awaited was the crucial work being done in the "Butcher Shop" on the new "more-cylinder-than two" tractor.[11]

MANUFACTURING ABROAD: THE SCOTLAND PROPOSAL

By 1950, European recovery, buttressed by United States aid under the Marshall Plan, had borne fruit. All through Western Europe the pace of development had hastened agricultural and business recovery, and American business measurably increased its trade with its Western partners. The "Cold War" that began in 1949 did curtail some industrial reconversion, leading to certain restrictions upon civilian production, and this was further complicated by the effects of the Korean War in 1950–1953. But despite these hostilities, the first half-dozen years after World War II provided great opportunities for American business expansion, now on more of a worldwide basis than it had ever been.

Deere & Company had shared in this postwar expansion of world markets only in a selective way that had telegraphed some disturbing implications. There was a sharp rise in Deere foreign sales (all those outside of the United States and Canada), from only 5.7 percent of total company business in 1946 to 14.7 percent in 1949, but while there were some burgeoning markets, there were also some lagging ones. The imbalance shows graphically in Deere's percentages of total United States machinery sales in five key international areas:

Market	Deere Percentage of US Sales			
	1946	1947	1948	1949
South America	11.8%	16.4%	19.6%	18.4%
Mexico	9.3	11.0	18.4	17.9
Africa	13.2	15.5	19.6	16.8
Europe	5.0	7.5	9.4	9.9
Asia	7.6	4.2	14.2	17.6

The anomaly, obviously, was Europe—Deere's percentage in each of the years from 1946 to 1949 was less than 10 percent of the US agricultural machinery going to European countries.

In March 1950 John Good, second-in-command in the export department, was sent on a long trip through Western Europe—his assignment: to discover why Deere was lagging in Europe, in contrast with the rest of the world. The answer was unsettling. European buyers of agricultural machinery were finding it difficult to obtain hard currency (dollars) to pay for imported machinery and were turning to their own domestic manufacturers, as well as those in contiguous countries, with whom they could finance in local currencies. Several North American companies had already anticipated this and had their own manufacturing plants in Europe. International Harvester and Massey-Ferguson were particularly well positioned, serving much of their European market from their own European factories; International Harvester was even expanding its operations in order to make its small tractors in Europe, and Allis-Chalmers had just announced plans to go into England to make its five-foot combine and one of its tractors. It seemed apparent that if Deere wanted to stay alive in European sales, it had to contemplate doing the same.

The notion of manufacturing abroad had not been considered seriously since the early days of the modern company, when the board had summarily rejected George Mixter's quixotic proposal to manufacture in

Russia. Export sales were considered as sideline or incidental sales, to be welcomed, of course, but only if they did not interfere with domestic sales, and if most would be paid for in cash and shipped on an irrevocable letter of credit.

When Charles Wiman had plumped in 1946 for an allocation of machinery for Chile and Argentina in the face of domestic shortages and allocations, the sales department had complained loudly. So when he floated the trial balloon in July 1950 of manufacturing abroad, he was greeted with skepticism by a number of board members. But he was not easily discouraged, and he cajoled his colleagues over the remainder of the year and the early months of 1951 to change their minds. Finally, after considerable research, a decision was made to build a plant at East Kilbride, Scotland. The proposal contemplated construction of a plant of approximately 425,000 square feet, to be financed by the British government and rented back to Deere by the Scottish Industrial Estates.

The debates in the board became quite argumentative, for profound differences emerged not only about the project but about the concept of the corporation itself. Was Deere & Company primarily a domestic producer, concerned mostly about the United States and Canada and only peripherally about the rest of the world? Or was the company to turn its sights outward to the rest of the world, becoming in the process a "multinational" corporation?

The amount of capital would dictate the scale of operations abroad. It would be one thing to open a small factory to produce a limited set of tillage implements. Adding perhaps some spreaders, mowers, or even a combine would be yet another scale. But if one contemplated the production of a tractor, including the engine, then the project would entail a very heavy commitment. Where along this continuum did Deere want to buy in?

The minimum price tag seemed to be about $5 million, enough to provide the working capital, including inventory, for a small combine and a spreader, with a few miscellaneous implements. Charles Stone wanted to hold to this scale: "The worst thing would be to start over there with a 12-A combine and spreader, and then find in two or three years that we were not in business unless we make a tractor. . . . I am not saying you should not make a start, but I think if we make a start you want to be very sure not to obligate ourselves to carry this thing on at a very rapid rate. I watched I.H.C. lose $6 million before they broke even . . . pioneering is a hard life." Duke Rowland, still at that time heading Waterloo tractor production, expressed a manufacturing viewpoint, worrying that the small scale of operations might not give production runs of enough size to reach the break-even point. Yet he was not willing to advocate a larger operation. Rowland also worried about where the right men might be found to staff the new factory—"We are spread pretty thin." Pat Murphy was a pivotal person, for he headed all remaining company manufacturing. Murphy had always been domestically oriented, and he now gave only lukewarm backing to even the more modest plans. Murphy worried on the one hand that the operation would eventually require a tractor and that the price tag for producing one would be upwards of $15 million. Indeed, Murphy pointed out: "If we go into France and England, it might be closer to $50 million for plant layout, receivables and inventory. I think we should go into this with our eyes open." Murphy, like Rowland, also was hypersensitive about effects on domestic manufacturing: "If we ever got to the production shown and we were back to normal times, it would take this production away from United States factories and would reduce production in this country." Bruce Lourie, representing sales, worried about how Deere-manufactured goods would be marketed; European jobbers were for the most part undercapitalized, not willing to extend their areas in the face of new opportunities—lukewarm representatives at best. A branch house or houses would probably be needed, and these would raise sensitive questions about when and how to sever relationships with existing jobbers. New dealers would be difficult to find, given the highly competitive nature of the European markets; the best ones were already committed.

The most enthusiastic proponents were, expectedly, from the export department. They were not reconciled to Scotland, however; as late as ten days before the final board vote, they were lobbying for France—and not just for combines and mowers, but for a "single package—25-30 hp tractor, diesel preferred, and five or six integral tools . . . similar to the Ford package."

One thing was clear: Charles Wiman not only wanted to go abroad, but he wanted to go to Britain. Further, he desired to do so on a scale that would ensure that Deere would become a substantial influence. He had made it plain to the British that Deere was contemplating building not just a combine and mower, but other equipment as well. He wanted to keep Deere's options open so that it would be free to make tractors or anything else at a later date. Wiman seemed to have as much fear of too narrow a line as the others did of too wide: "It is important to keep that freedom of action and not get euchred into a short line." He concluded: "If we do not ever start, where will we be in the potential business for the rest of the world? We may have to put a tractor over there, so instead of $10 million it may be around $22 million, but it would increase our business over there, and that might be on the plus side for the long pull. If we wait too long, the parade may go marching by." Wiman gave the board members a week to contemplate their final decisions.

Given the prevailing sentiment of his colleagues, Wiman finally bowed to the practicalities and decided to bring forward the more modest proposal of an initial expenditure of $5 million to manufacture just the small combines, mowers, and spreaders and a restricted line of tillage tools. When it came to a vote, it passed unanimously, in keeping with the longstanding Deere board tradition of final unanimity, however much there had been prior disagreement or even wrangling. Here, though, Kennedy announced for the record that his vote was a "reluctant yes," and others, particularly Pat Murphy, were equally unenthusiastic.

A small group was to be posted almost immediately to Scotland to begin the planning of the factory, as well as the coordination with the British government. But when the operating departments were approached about potential candidates for these important posts, they seemed very reluctant to give up any of their senior, key personnel. These were manufacturing and engineering units, directly responsible to Pat Murphy, and the latter's insular views about any foreign manufacturing influenced their thinking. As a result, no first-rank senior personnel were assigned to the project, and this proved to be a serious limitation. This same narrow, proprietary stockpiling of subordinate management personnel by factory and branch managers had strongly characterized past personnel policies. Seldom if ever did junior managerial talent have the opportunity to transfer from one factory to another, or from one branch to another. This insular view of executive development was surely a pernicious aspect of the policy of unbridled decentralization.

Many months went by, with extensive efforts by the Deere team on-site in Scotland. Then, just after the new year 1952 had begun, an unexpected set of events intruded, stemming from the change in the British government from Labor to Conservative. At the same time an apparently unrelated set of events took place in the company's domestic operations, which proved to have a profound influence on the effort in Scotland.[12]

DEERE'S FERTILIZER VENTURE

In October 1951, just at the time the project in Scotland was being staffed and planned, Charles Wiman visited officials of the US Department of Agriculture in Washington and came away persuaded that fertilizers were, as he put it to the board, the "most attractive" possibility of all projects he had seen. "The production of fertilizer involves a market with which we are familiar and the same facilities and organization capable of producing nitrates for fertilizer can be used as an entree into the broader aspects of the chemical industry, which is one of the newest and most rapidly expanding industries today."

There were some negatives, however, as the study team picked by Wiman soon reported. "A comparison between that industry and our industry makes it obvious that as presently constituted we have very little to contribute in the way of experience, trained personnel or operating methods. Presumably in such a venture we would bring nothing more than a certain amount of capital and willingness to engage in a different business quite unrelated to our normal activities." Further, the chemical industry was characterized by high investment in facilities and low labor cost, and Deere's distribution organization seemed not particularly well adapted to the distribution of fertilizer—a separate organization probably would be needed, both to handle production and distribution. Finally, fertilizer was not really the best kind of diversification, inasmuch as the same economic conditions that lowered the demand for farm machinery would also lower the demand for fertilizer.

All of these caveats by the study team and members of the board did not deter Wiman. Promising the board that he was "not asking for an appropriation, no money with which to build the plant," he persuaded the board to allow negotiations to proceed, and he even induced them to "look with favor upon issuance of new securities, either in connection with a merger or construction of a new plant."

From their lukewarm endorsement, Wiman proceeded rapidly. First, he appointed Duke Rowland as the titular head of the project, taking him off his assignment as head of tractor production. Certificates of necessity were needed from the government; these were promptly obtained. By January, he reported back to the board that new personnel had been hired for the project and that a joint venture was being discussed with the Standard Oil Company of Indiana.

But there was now real resistance among the board members, enough so that the minutes of the meeting of January 29, 1952, noted the "concern . . .

Exhibit 12-7. Deere directors at the company's Grand River Chemical Company, Pryor, OK, October 1963. *Deere Archives*

expressed that embarking upon the synthetic nitrogen project might jeopardize the financial position and the functioning of our farm machinery and tractor business." Wiman ended the meeting by throwing down the gauntlet: "If this Board as a Board is opposed to entering the business and is opposed to giving thoughtful consideration to future requests for appropriations, then it would be well at this time to give the matter extensive consideration and discussion. If there is no such objection, then it seems to me that the program as outlined above can go forward." Few people would want to be considered as in opposition to "thoughtful consideration." No actual vote was taken at this meeting, but neither were there overt efforts to back Wiman.

This still left unanswered the question of financing such a mammoth endeavor. Already the East Kilbride, Scotland, program was calling for capital upwards of $5 million, probably closer to the $20 million that Wiman had estimated. Now this second project was going to demand an additional $20 million. Wiman proposed a public offering to the board and, in July 1952, the company sold 691,276 additional shares of its common stock (at an excellent price of $30.40) and additionally marketed $50 million of 25-year, 3-1/8 percent debentures. "Moneymen quickly snapped up the issues," *Forbes* reported. The proceeds to the company from these two sales of securities totaled $69 million.

At this point, Wiman came back to the board, demanding a final commitment to go ahead in the chemical venture; an option on land near the Grand River Dam Authority power plant near Pryor, Oklahoma, already had been obtained by Rowland. There was animated discussion around the table about the disadvantages of the project, particularly the lack of compatibility between the farm machinery business and the chemical business. The capital commitments were very high, too, though the success of the public offering now gave the company considerable new leeway. Finally, Wiman put the issue to vote, and for the first time in many years negative votes were cast. Pat Murphy voted against the project; so, too, did George French. (French was immediately under Murphy, in charge of all implement production; he, Elwood "Woody" Curtis, and William Hewitt had been brought on the board the previous year.) As if two negative votes were not enough, Charles Stone asked that he be recorded as "not voting."

The die was cast, though—Wiman had singlehandedly persuaded a majority of the board to turn the company in a major new direction. *Forbes* magazine featured the move in a major story, reporting that "entrepreneur Charley Wiman's test tubes will begin to burp, barring delays, early next year." The *Forbes* reporters were not privy to the internal dissension within the board about the venture; their article oozed optimism: "Deere's soundskulled country boys can see a mite beyond the visible horizon. A foothold in synthetic fertilizers will work double dividends, they calculate. In the 13-state midwestern corn and wheat belt which Deere figures to supply from Pryor, nitro-fertilizer demand in 1955 is expected to quadruple 1950's consumption. Corn planters are learning that nitrates increase per-acre yield from 40 to 100 bushels. By thus aiding the farmer (tentative plans call for selling bulk urea, non-trademarked, to fertilizer makers) to more abundant harvests, demand for machinery will become more abundant too."[13]

THE SCOTLAND PROJECT FALLS THROUGH

Just as the chemical company came into being, the project in Scotland was derailed. A change in the British government from Labor to Conservative had brought new views about development, particularly that from outside the country. British officials, while denying that they were terminating the project, nevertheless postponed government financing. The board had three alternatives: Deere could finance construction itself, continue to maintain the company's organization in Scotland and wait out the delay, or disband the Scottish organization. Charles Wiman, preoccupied with the chemical endeavor, had obviously lost some interest in the Scotland venture. This was all it took. With almost a visible sigh of relief, the board unanimously voted to terminate the project. A number of Deere's agents in Europe and elsewhere in the world were upset by the decision; they had envisioned manufacturing in Europe as a potential breakthrough for Deere's products all over the world. V. C. Fischer, the chairman of Agar, Cross, the British trading house (and Deere's Argentine agent), wrote Wiman, expressing unhappiness. Wiman answered: "The fact that we finally reached the decision not to go ahead with the East Kilbride project does not mean that we have made a decision against a foreign manufacturing project as such. We will, of course, continue to study this important subject." But Wiman also mentioned the fertilizer project, to Fischer, further commenting that "increasing nationalism" on the part of various countries in the world was making a decision to manufacture abroad more difficult.

Wiman's shift in focus was sudden, and there is not a ready explanation for his seeming disenchantment with the rest of the world. The Scotland location may have been the wrong place to start. Many board members felt this way; perhaps Wiman was weary of bucking them.

MANAGEMENT SUCCESSION

Sometime in 1954, Charles Wiman learned that he had a fatal illness, one that would give him just a few months to live. One question then to be resolved was his succession at Deere & Company. He had been undisputed head of the company from 1928. The transition would be especially difficult because there had been substantial turnover in the rest of top management in this period of the early 1950s. In the four-year period of 1951–1954, the board of directors had lost a cumulative 148 years of actual board service (Dwight Wiman, 31 years; Willard Hosford and Theo Brown, 30 years each; Charles Stone, 27; Harold White, 14; Harold Boyle, 7; Benjamin Keator, 6; Virgil Bozeman, 3. With the exception of Dwight Wiman, these were all operating top management men).

The early Korean War years had given the company excellent financial results—on sales of $337 million in 1950, there was a $42.7 million profit; in 1951 sales rose to $433 million, with profits of $35 million. The next two years' performance gave everyone pause, however. Sales fell in 1952, though profits remained almost level with the previous year, but in 1953 sales again dropped back and profits fell to $24.7 million.

Several predispositions dominated Wiman's thinking at this juncture. Deere & Company had been a family endeavor for its entire 117 years; Deere was known widely as one of the country's preeminent "family companies." But the board had never perceived lineal descent as a sinecure, as a feudal right to leadership. Ineffectual members of the family had held directorships, but their weak ideas had been ignored; a few of these men even had been terminated as directors for poor performance. For senior management posts, the individual family member might have been accorded some marginal preference at the time of selection, but he had to pass muster to continue to hold the job (again, there had been significant instances of demotion, and even severance for low performance). In sum, there was a family ethic of achievement and performance. *Fortune* magazine, in a major article centering on the Wiman transition, elaborated: "Despite Deere's healthy condition, the wrong man could easily have wrecked it. For more than a century, Deere had been characterized by a self-confident operating style that allows managers an extraordinary degree of independence and puts a premium on initiative. By careful choice, more than luck, family leadership had preserved that style over the generations."

Wiman had a keen awareness of this record. If there was an outstanding descendant of the Deere family itself standing ready, this would loom large in selection; if more than one, the best would be picked, whatever the lineal relationship might be. There was inter-family rivalry, some push-and-tug about company matters among the three dominant lines—the descendants of the Deeres, the Webbers, and the Velies. But, contrary to what some thought at the time, there is strong evidence that Wiman would not have let Charles Deere lineal loyalties dominate in any decision. Similarly, if a non-family member of management were clearly the best candidate, he would be picked over any family member. This person would have to be found somewhere within management—it would have been almost inconceivable to Wiman, or to anyone in the Deere group, to have turned to the outside. Harold Boyle, the International Harvester executive brought back to Deere by Wiman at the end of World War II, had been for many years the only top management officer appointed directly. Deere was proud of its proven ability to nurture leaders and confident that it would develop an unending supply of fresh candidates.

As Wiman now surveyed the top management team he knew so well, his initial thinking centered on three people. Two were part of the Deere family—Pat Murphy (from the Webber side of the family) and William Hewitt, Wiman's own son-in-law. The third person was Lloyd Kennedy, a long-time confidant of the family, trustee for both the Charles Deere and William Butterworth estates. None of the extant evidence suggests that Wiman ever considered a fourth person who had been particularly influential in top management in the immediately preceding years, Budge Cook, who was married to Pat Murphy's sister.

Kennedy, at age sixty, had had more than forty-one years of company service, all of it in the financial side of the business. Murphy's service totaled thirty-one years at this point; all had been in the production side of the firm with the exception of a few months as assistant territory manager in the Minneapolis branch. He was fifty-four. Hewitt had just turned forty and had been with Deere & Company for just six years, all of it in the branch side of the firm.

Hewitt was the only college graduate of the three (Kennedy had attended a commercial college for one year, Murphy matriculated at the University of Iowa, staying for a few months). There had been ambivalence in the company over the years about whether a college degree for anyone other than engineers was a plus. Lip service had been paid to "hiring more college graduates," but actual practice seemed to be contrary to this injunction. Wiman, a Yale graduate himself, had tended to value the college degree more than others did in the company, an attitude closer to congruence with the postwar thinking in other major American companies.

Hewitt had worked two years between high school and college as a messenger in a San Francisco bank, then entered the University of California, Berkeley, majoring in economics. "Hewitt blossomed socially in college," the *Fortune* editor commented. "He fell in with one of those groups of young men who strike outsiders as all being destined for success. Among them were Robert S. McNamara; Walter Haas, Jr., now chairman of Levi Strauss; Willard Goodwin, formerly chief urologist at U.C.L.A.; and Stanley Johnson, who became an eminent trial lawyer in San Francisco. It was a close circle." Hewitt, McNamara, and Haas went on to Harvard Business School together. Hewitt had only enough money for one year, and he returned to San Francisco to work first in accounting at Standard Oil of California, then as an advertising copywriter in a menswear chain. Joining the navy shortly after Pearl Harbor, Hewitt rose to the rank of lieutenant commander on the battleship *California*; his cabin mate there was Gabriel Hauge, later chairman of Manufacturers Hanover Bank and from that time a lifelong friend and business confidant.

After V-J day, Hewitt returned to his native state to enter the farm machinery business as a territory manager for a Ford-Ferguson distributor. While on this job, he married Wiman's daughter, Patricia. Within a few months, he was asked to join Deere, and after some hesitation, he did so. He began as a territory manager for the San Francisco branch, then was brought in first as assistant branch manager, then branch manager, under Benjamin Keator. The latter was an excellent mentor for Hewitt. He was a Deere family member (from the Webber side) and a judicious and independent person.

Hewitt brought an unusual sense of style to the operation. The *Fortune* editor interviewed branch personnel about his early days with the company: "He conducted himself with grace . . . a good listener . . . did not throw his weight around . . . disarmingly cordial, and acutely sensitive to others . . . a good leader." Hewitt also demonstrated a coolness under pressure, evidenced particularly in the tricky assignment as one of the company investigators in the antitrust suit. After becoming a director in 1950, he was not only coached by his board colleagues on broader, corporation-wide issues, but was also inculcated with the company's special history and traditions. "No man could hope to serve as Deere's chief executive," commented the *Fortune* editor, "without having a deep knowledge of its rather singular character. Deere is a culture as well as a corporation, and the company's officers understand its strengths as reflections of its past."

Still, there was no overt evidence that Hewitt was being uniquely marked. "Contrary to what some of Hewitt's colleagues thought," wrote the *Fortune* editor, "Wiman was not taking any special trouble to help his son-in-law along." But the onrush of events in 1954 stemming from Wiman's ill health now forced him to bring the question of Hewitt's eventual role into his consciousness.

At this point, Wiman turned to Keator for advice. Keator, always close to Wiman, had just resigned from the board, thus putting himself more at arm's length from the sensitive upcoming decision. Keator immediately wrote Wiman his prescription for the presidency: "We should have a young man—this is most important—who is a leader, who knows the business and has vision for the future of our line, who is cognizant of the problems to be met and how to handle them." Keator felt that Kennedy and Cook were "good in their own fields" but were "getting along in age and also have no sales, field or factory experience." (Cook, at fifty-one, was just a bit younger than Kennedy, at sixty.) Keator felt Pat Murphy too old, too. At fifty-four, Murphy was clearly one of the most respected men in the entire organization, but most observers of this period seem to have agreed with Keator that "while he does know factory operations he has no vision or field experience and lacks progressiveness and enthusiasm." Keator put Murphy as his last choice.

After briefly discussing several other members of the board, Keator turned to Hewitt: "I have felt, of course, that Bill has been pushed along pretty fast but I must say he seems to be able to absorb it in his stride without its affecting his personality or his association with others. He has had enough Branch House experience to know that problem pretty well and he is alert to the requirements in the field. He has youth, he is smart, catches on to things quick and has vision, enthusiasm and courage to get things done . . . he has had a broad field to operate in and had the authority to act. This is important if he is to be able to face up to the task. . . . To be sure, he has had little or no experience in factory operations, which is of course very important in our Company, but couldn't someone be appointed to his staff who could tutor him in these decisions?" Keator concluded, "Bill is a very dignified and personable young man and you would go a long way before you would find his equal for that job."

Wiman wrote back immediately: "I find your comments very keen about the V.P.'s. I had thought a lot of LEK [Kennedy]. I wish I could feel better about Pat Murphy. Your conclusions about Bill H. are what I am now thinking of—I think he'd fill the bill best of all! However, gosh, I wish he were not my son-in-law—for it will look to so many that I am trying to push 'one of the family.' It would be so easy to push for his presidency otherwise—on all counts." Wiman concluded enigmatically: "I will go back to Moline to work on this problem and how it comes out I do not now know."

This worried Keator. Was Wiman going to waver because of the possible criticism of nepotism? He wrote back immediately: "I can't see why if Bill is the best material and best qualified in every way to take over, both for the present and for the future, as I believe he is, why it should make any difference if he is related or not. After all, everyone concerned should be thinking and wanting to do the best for Deere & Co., both for the present as well as for the future."

Wiman talked further with Burton Peek, but probably with no one else—the subject was too personal. Wiman now was spending most of his time secluded at his Arizona ranch, with only occasional trips to his hospital in New York City; he had been there to chair the annual meeting on April 27, 1954. Finally, in late June he came to Moline from Arizona, having called a special meeting of the board for June 22. With Burton Peek in the chair, Wiman placed in nomination the name of Hewitt for election to the newly created post of executive vice president. The motion carried unanimously, and the die had been cast for management succession.

Hewitt chaired most of the board meetings through the fall and winter of 1954–1955, receiving major support from most of the officers. "The prospect of taking over the company at an unexpectedly early time galvanized him," the *Fortune* editor concluded. One of the officers recounted to him: "You could just see him grow after he got to Moline." On May 12, 1955, Charles Deere Wiman died; twelve days later, William Alexander Hewitt was elected the sixth president of Deere & Company.[14]

Though many of Charles Wiman's personal characteristics differed from those of his grandfather, Charles H. Deere, the two men were strikingly similar in their drive to be both successful and innovative. Charles Wiman's legacy to the company could be favorably compared to that of his grandfather for he had inherited some of the same vision that Deere exemplified, as well as the latter's entrepreneurial bent. Wiman often could hardly wait to make changes and move ahead, relishing the challenges; yet sometimes he would turn sharply backward toward caution. The eulogies for Wiman were properly laudatory, but somehow failed to capture this Janus-like quality of his personality—charismatic, enthusiastic, positive, yet sometimes conservative, even pessimistic. His qualities of leadership were outstanding, for he had had a unique ability to fire others to action. He had weaknesses, to be sure—his overemphasis on engineering and development and an unsure grasp of finance. He was not the keen marketing man that both Charles and John Deere had been. His attitudes toward employee relations were progressive, and he enjoyed high respect among management personnel and line employees. He was restless and always searching for something new; as Theo Brown wrote in his diary, "Charlie Wiman said at the table, 'If I wake up in the morning and know where I am I want to take a train and go somewhere else.'" In the last half-dozen years of his tenure, this restlessness had almost become idiosyncratic; the sharp shift in interest from the international sphere (the Scotland effort) to the dominating preoccupation with the chemical company was one example among many.

Earlier, in 1946, Max Sklovsky had been writing a book (never published) on the company, in which he emphasized Wiman's "volatile spirit, which some consider a sign of unreliability," and noted that "his driving is not always steady." There was truth to this, a negative on the balance sheet. But the positive qualities overbalanced this: "He possesses a combination of realism and vision," concluded Sklovsky. "Charley Wiman turned into a remarkable executive and became the spark plug of the Company."

Endnotes

1. For the history of the War Production Board's efforts in farm machinery, see War Production Board, Policy Analysis and Records Branch, *Farm Machinery and Equipment Policies of the War Production Board and Predecessor Agencies*, Report 13 (November 10, 1944). For the threatened Department of Justice action, see ibid., 49; for the Ford-Ferguson proposal for building 100,000 lightweight tractors, see ibid., 86–89. For the Massey-Harris harvest brigade, see Merrill Denison, *Harvest Triumphant: The Story of Massey-Harris* (Toronto: McClelland and Steward, 1948), 314–17.
2. For Deere's efforts in World War II, see Neil McCullough Clark, "John Deere War History: An Account of Deere & Company's War Production During World War II," (Moline, 1944), DA. Establishment of the Iowa Transmission Company is discussed in Deere & Company *Minutes*, April 29, 1941. For the Cleveland Tractor Company contract, see ibid., March 3, 1942; for the merger proposal, see ibid., April 28, 1942; for its rejection, see ibid., July 28, 1942.
3. The controversy over the War Production Board statistics for civilian farm machinery production in early 1944 is discussed in War Production Board, *Farm Machinery and Equipment Policies*, 145–53. See *Chicago Tribune*, May 15, 1944, for the quotation regarding "glamorized reports"; see also *Congressional Record*, 78th Congress, 2nd Session, vol. 90 (May 19, 1940), 4,849; *San Francisco Commercial News*, May 13, 1944.

CHAPTER 12

4. The Killefer case was decided by the National Labor Relations Board on March 30, 1940 (22 NLRB 484), the Dain case on September 30, 1939 (15 NLRB 779). The Farm Equipment Workers (FE) began as the Farm Equipment Division of the Steel Workers Organizing Committee (SWOC); in July 1938 the SWOC authorized the formation of the Farm Equipment Organizing Committee; this group in turn became the United Farm Equipment Workers of America, having been given a separate charter by the CIO as an international union with jurisdiction in the farm equipment industry. For the battle between the United Automobile Workers and the Farm Equipment Workers, see Irving Howe and B. J. Widick, *The UAW and Walter Reuther* (New York: Random House, 1949); Philip Taft, *Organized Labor in American History* (New York: Harper & Row, 1964) devotes chapter 47 to the left-wing issue in the CIO. The definitive book on International Harvester's labor relations is Robert Ozanne, *A Century of Labor-Management Relations at McCormick and International Harvester* (Madison: University of Wisconsin Press, 1967); see also R. Ozanne, *Wages in Practice and Theory: McCormick and International Harvester, 1860-1960* (Madison: University of Wisconsin Press, 1968). The most comprehensive study of labor relations in the agricultural machinery industry is Arlyn John Melcher, "Collective Bargaining in the Agricultural Implement Industry: The Impact of Company and Union Structure in Three Firms" (PhD diss., University of Chicago, 1964). The quoted newspaper headline on the postponement of the Army-Navy E award is from *Davenport Democrat*, June 20, 1943. See L. A. Murphy to Capt. D. C. Hanrahan, June 21, 1943, DA, 19526. George Wilson's remarks are in Deere & Company *Minutes*, October 31, 1944.
5. For quotations on Charles C. Webber's death, see *Minneapolis Morning Tribune*, February 22, 1944; *Minneapolis Daily Times*, February 22, 1944. For the elections of Murphy, Kennedy, Curtis, and Gill, see Deere & Company *Minutes*, June 9 and 13, 1944; Blue Bulletin, June 13, 1944. For Cook's speech to the Factory Managers Meeting, July 27–28, 1944, see DA, 20157.
6. See Charles Wiman to Edward R. Stettinius Jr., November 30, 1944, DA, Overlook folders. Wiman's suggestions on airplane and deep-freeze production are in Deere & Company *Minutes*, October 26, 1943.
7. Decentralization of purchasing is discussed in Deere & Company *Minutes*, January 30, 1945; branch bulletin 1022, May 31, 1945. See also Edmond Cook's remarks to purchasing agents, ca. April 1945, DA, 20158. For the Cook–Boyle correspondence, see DA, 27000. For the quotation on "unrest," see "Organization Lines and Policies, Memorandum by Edmond M. Cook of Discussions . . . in July 1945 and of Discussions . . . August through December, 1945," December 21, 1945, DA, 27000. Minutes of the Branch and Factory Managers Meeting of December 12, 1947 are in DA, 27000. The first Wiman bulletins on decentralization policy are 562 1/2-A (Factory) and 1085 1/2 (Branch), December 12, 1947; "John Deere Decentralization Policy," bulletins 570-A (Factory) and 1094 (Branch), May 20, 1948. See also "Purchasing Procedures," factory bulletin 564-A, January 10, 1948.
8. See "Report Prepared by Mr. Kennedy . . . Deere & Company *Minutes*, October 26, 1937. For the Dubuque plant proposal, see ibid., January 30, 1945; for Wiman's remark on restrictions on the board in earlier years, see ibid., March 15, 1945. The financing proposal is outlined in ibid., March 22, 1945; the Lindeman purchase is in ibid., May 1, 1945. The Tulsa, Oklahoma, proposal is in ibid., January 28, 1947, and February 25, 1947; the Ankeny, Iowa, purchase is in ibid., June 19, 1947. For Wiman's remarks on "business truisms," see ibid., October 26, 1948. For Cook's quotation on the "sense of failure," see Edmond Cook to C. Wiman, April 1, 1947; Wiman's reply is handwritten, on the same letter, DA, Overlook folders. See Max Sklovsky to C. Wiman, April 9, 1945; the Cook memorandum on Wiman's draft letter is dated May 28, 1945.
9. For Deere's attitude toward centralized bargaining, see the confidential memorandum to all factory managers from G. K. Wilson, February 29, 1943, DA, 19-3555. The joint request by locals of the UAW and Farm Equipment Workers to bargain as a group was made in their letter to G. K. Wilson, September 20, 1945; Wilson's rejection came in the meeting with this group on September 28, 1945, "1945 UAW Negotiations," DA, 19-3563. The strike settlements with the UAW locals are elaborated in the November 4, 1945, issues of the Quad City local papers; for the Waterloo settlement, see *Waterloo Courier*, November 20, 1945; for the Plow Works settlement, see the papers of December 10, 1945. The International Harvester negotiations in this period are detailed in Ozanne, *A Century of Labor-Management Relations*. For additional analysis of International Harvester's industrial

relations, see US Congress, Joint Committee on Labor-Management Relations, 80th Congress, 2nd session, Senate Report No. 986 (Washington: Government Printing Office, 1948), 118–20; Benjamin M. and Sylvia K. Selekman and Stephen H. Fuller, *Problems in Labor Relations*, 1st ed. (New York: McGraw-Hill, 1950), 606–23; Robert B. McKersie, "Structural Factors and Negotiations in the International Harvester Company," in Arnold R. Webber, ed., *The Structure of Collective Bargaining; Problems and Perspectives: Proceedings of a Seminar Sponsored by Graduate School of Business, University of Chicago, and the McKinsey Foundation* (New York: Free Press of Glencoe, 1961), 279–303. The Allis-Chalmers strike is described in Melcher, "Collective Bargaining in the Agricultural Implement Industry," 131–39. Leadership battles in the national union of the UAW, and the attempted merger with the Farm Equipment Workers, are described in Howe and Widick, *The UAW and Walter Reuther*, chap 7. For the ouster of the left-wing unions from the CIO, see Philip Taft, *Organized Labor in American History*, chap. 47. See also Melcher, "Collective Bargaining in the Agricultural Implement Industry," 173–74.

10. Wendell Berge, "The Antitrust Program and the Farmers," speech before the Wisconsin Farmers Union Convention, October 25, 1944. For Cook's concern about marketing implications, see his memorandum of September 17, 1948, DA, 2573; Deere & Company *Minutes*, October 26, 1948. For Cook's reference to "elephants," see his memorandum of December 15, 1948. See "Deere & Company Antitrust Suit, Examples of Favorable File Material, EMC Memorandum, 1-12-50," Deere & Company *Minutes*, October 26, 1948; branch-house bulletin 873, November 8, 1938; J. M. Burt to H. B. Pence, October 25, 1948, DA, "Dealer Severance" file, 22217. For Leon Clausen testimony on Deere in the J. I. Case Company case, see *United States District Court, District of Minnesota, United States v. J. I. Case Company*, trial testimonies, 651–57, DA, 2834; Cook's comment on postponement is in his memorandum of April 20, 1950, DA, 2573. Quotations from the Nordbye decision are in *United States v. J. I. Case Company*, 101 F. Sup. 856, 866–67. See Warren Wright Shearer, "Competition Through Merger," 368, 402.

11. For the early history of Rudolph D. Diesel's efforts, see C. Lyle Cummins, Jr., *Internal Fire* (Austin, TX: Octane Press, 2021), chap. 14. Caterpillar Tractor and International Harvester diesels are discussed in Gray, *The Agricultural Tractor*, 24 ff.; Wendel, *Encyclopedia of American Farm Tractors*, 69, 159. See also Richard N. Coleman, "Milestones in the Application of Power to Agricultural Machines," in *An Historical Perspective of Farm Machinery* (Warrendale, PA: Society of Automotive Engineers, 1980), 76–83. Descriptions of Deere model changes are from Will McCracken, *John Deere Tractors, 1918–1976* (Deere & Company, 1976). The standard-setting process for the PTO is discussed in T. H. Morrell, "The Development of Agricultural Equipment Power Take-Off Mechanism," in *An Historical Perspective of Farm Machinery*. For the development of the self-propelled combine, see Quick and Buchele, *The Grain Harvesters*, 140–41; for the Deere combine corn head, see ibid., 163–64; for the hillside combines, see ibid., chap. 19. See also DA, 1864, 22577, 26692.

12. International sales percentages are from W. R. Klingberg to Charles Wiman, September 13, 1949, and July 28, 1950; see also Deere & Company *Minutes*, April 25, 1950. See H. J. Heinz II to Maurice Block, June 30, 1950. There are no extant copies of John Good's report on Europe of June 1950; there are verbatim minutes of the board discussions of July 5 and 8, 1950, DA, 20034; see also C. Wiman's report in Deere & Company *Minutes*, July 25, 1950. For the possibility of manufacturing in Brazil, see ibid., July 13, 1950; for a verbatim record of the debate of the final proposal and a summary, see ibid., February 1, 1951, DA, 20034. The export department's views are expressed in W. F. Haberer's memorandum, ibid., April 20, 1951. A board "conference" was held April 23, 1951; verbatim minutes are extant in ibid., April 23, 1951. The final proposal is in ibid., April 30, 1951.

13. For the first board discussion of the fertilizer project, see Deere & Company *Minutes*, October 10, 1951; see also ibid., October 30, November 16, and December 7, 1951. See *Forbes*, October 1, 1952. C. Wiman's remarks on an independent company are in Deere & Company *Minutes*, January 29, 1952; on working capital, ibid., April 29, 1952. The final vote on the project is recorded in ibid., May 16, 1952.

14. Benjamin Keator to Charles Wiman, May 26, 1954; C. Wiman to B. Keator, May 28, 1954; B. Keator to C. Wiman, ca. June 1, 1954, DA, Overlook file. Max Sklovsky to Charles Wiman, October 22, 1946, and attached ms., "Excerpts from 'Creators of Deere & Company.'" Theo Brown, "Theo Brown's Diaries," October 28, 1952.

PART V

WORLDWIDE CORPORATION: THE WILLIAM HEWITT YEARS (1955–1982)

CHAPTER 13

RISE TO INDUSTRY LEADERSHIP

All day long the big planes buzzed in and out of Dallas' Love Field. They carried passengers from New York and New Dorp (PA), from Paris, France and Paris, Ill., from Seattle and Sewanee (TN). When darkness fell on Monday, August 29, 1960, more than 6,000 passengers—the biggest industrial airlift in history—had been safely landed.

Forbes, *1960*

Innovations came rapidly in the first Hewitt decade. Some of them had been initiated by Wiman, in particular the final painful changeover from the two-cylinder to three-cylinder, four-cylinder, and six-cylinder tractors. Other important new directions were initiated by Hewitt himself, notably his ambitious ventures into Europe and Latin America. In addition to this expanded geographical focus, he brought a new set of management philosophies. By the end of Hewitt's first decade, Deere had become a major multinational corporation and the largest agricultural machinery producer in the world.

Hewitt had inherited a sound company from Wiman. Sales for fiscal year 1955 were up markedly from the previous year, as were profits. The company's

◄ Brochure introducing the new line of four- and six-cylinder tractors, 1960. *Deere Archives*

CHAPTER 13

working capital was at an all-time high, book value of the common stock surpassed all previous years. Prospects appeared excellent; the sound, conservative Deere organization seemed nicely positioned to take advantage of future opportunities.

But was it? A number of small signs in Charles Wiman's last three or four years as president intimated incipient weaknesses within the organization. Benjamin Keator gave a pungent valedictory as a board member in 1953, explicitly asking that his misgivings be put on the record in the board minutes: "I leave the service of the Company with regret and a mixed feeling for the future.... Certainly we have efficient management, but we have overlooked some very important things during these 'fat' years that we are going to need awfully badly in the coming lean years." He particularly urged better sales training programs throughout the organization, "especially of dealers, who C.D. Velie rightly used to say, are the weakest link in our chain." Keator also worried about the design of certain Deere equipment, "where in some cases we are woefully weak and getting pretty badly beaten by our competitors in the field." In a letter to Hewitt at this time, he reiterated the same theme: "All our emphasis for the past few years has been on production and now it is high time we put our emphasis on sales."

Indeed there were weaknesses in the marketing arm of the company, as well as in engineering and product development. Other signals also indicated that the overall organization was not functioning as well as it could. The company's leadership was quite senior in years, and the brand of conservatism that formerly had seemed to serve the company so well apparently was tilting toward excessive caution.

This posed a dilemma for Hewitt. He was a young man, with relatively limited experience in the company—a member of the family, to be sure, but representing the next generation. Though the situation seemed to demand change, organizations tend to have built-in rigidities, making a change agent's role always difficult. Entrenched attitudes in the predominately elderly Deere board made Hewitt's task a formidable one in establishing his credentials as the "chief executive officer" of Deere.

Despite these obstacles, Hewitt did establish his control over the organization quickly and effectively. Within months he had shifted Deere & Company in a major new direction, toward an evolving concept of a "worldwide" corporation. In the process, he made a series of organizational decisions that wrought a fundamental change in the philosophy of the organization.

A first step was a study of the company's organizational structure. Not since the far-reaching, two-year study that Charles Wiman had carried through right after World War II had this been done. Wiman's effort, as well as those preceding it, had been done "within house," by Deere employees themselves. This insular posture had characterized company thinking over the years—a feeling that the company was too closely knit, almost too idiosyncratic for any outsider to truly understand. Perhaps there were features so special and so valuable that they could only be fully appreciated by longtime Deere managers. Hewitt did not believe this to be so and startled the board, less than a month after he had taken office, with a request that the board approve an outside management consultant to analyze both organizational structure and the related problem of executive compensation. The board approved, with scarcely disguised misgivings expressed by a few of the senior board members. The firm of Booz, Allen & Hamilton was engaged, establishing a relationship that extended over two decades.

Hewitt also began to assemble his own personal judgments of the situation. In this he had to keep his own counsel, for while there had been no overt opposition to his election as president, he still was an unproven chief executive. Clearly, a "wait-and-see" attitude characterized the feelings of a number of the board members—some of whom were part of the problem.

Over the fall of 1955, Hewitt put together a frankly written personal document. He first wrote out a set of broad objectives, maintaining that the company should "lead the industry" in six key indices—sales, profit ratios, quality, new designs, safety of operations, and excellence in employee, dealer, stockholder, and public relations. This was a herculean set of goals. Had one queried Deere's then-top management, they most likely would have said, "We're a good No. 2 to International Harvester—let IH innovate." Hewitt challenged this thesis within the first two weeks of his election. In a speech to all the branch-house managers he said: "Yesterday I received a friendly letter from one of our most active competitors, and emblazoned across the bottom of the stationery in splashy red ink was imprinted: 'We are not aiming to be runner up.' Believe me, we in Deere & Company are not aiming to be runner up either—we are aiming to be first in all our business activities." The gauntlet had been laid down: "Pass International Harvester!"

Marketing mistakes and problems dominated Hewitt's thinking. The sales department "is not doing a satisfactory job—weak ... in many ways the production department is running the sales department." The marketing group made poor presentations at board and committee meetings—indeed, seemed to be "scared" in the way they acted. The advertising department "is too separate in its activities—the sales department guides it and confers with it too seldom." Advertising "is 'written selling'—should be planned more by the sales department, not only by the advertising department." Further, the marketing group should participate more in the basic design decisions. Too often designs were determined unilaterally and privately by the engineering groups in individual factories. Manufacturing design engineers needed to work with product design engineers "after a certain point" to make certain that new machines could be easily manufactured. Hewitt asked himself, "How do we decide to build a machine?" Should marketing be included more directly in such decisions? Were pricing decisions too often dominated by the factories?

Throughout, Hewitt exhorted himself on the need for more effective communication and coordination—better market analysis, more information to the directors, upgraded public relations to the outside, and expansion of foreign-language advertising, service, and parts literature. He wanted a heightening of executive-level training—"men to business schools, specific programs of shifting promising young men around." As to internationalization of the company, foreign manufacturing should be entered into "wherever justified."

There were omissions in the document, too; these gave some inkling of Hewitt's potential blind spots. On his entire 8-1/2-by-4-inch page devoted to "finance," there was only a single entry: "Retire preferred stock?" The company did stand in excellent financial shape at this time; nevertheless, net income had dropped in the year 1954, as had working capital and retained earnings. These danger signs carried important financial implications, but they appeared not as important to Hewitt at that moment. Finance was not Hewitt's strong point at the time.

One further item in retrospect turned out to implant a signal Hewitt mark on the entire company: "Build a new office building." Hewitt included two specific views on this. First, the office building should be on "a campus." Second, the engineering department should "handle power houses,

Exhibit 13-1. Dwight D. Eisenhower, former President of the United States, visits the Deere Administrative Center, 1965. Escorting him are William A. Hewitt (left) and Mrs. Patricia Hewitt.

foundries, supervision of factory layouts and purchase of machine tools *but not* architecture."

Aside from finance, the document was quite anticipatory; Hewitt had evidenced considerable sophistication about the company, given the rather short period that he had been assigned to the general company. If even a significant portion of the hundred-odd items on Hewitt's sheets came true, Deere & Company would be a markedly different company.[1]

EXPLORING OPPORTUNITIES ABROAD

The hallmark of the early Hewitt administration was the company's aggressive extension abroad, in contrast to earlier halfhearted moves. When the Wiman proposal for manufacturing in Scotland aborted in 1952, the board comfortably settled back to a "business-as-usual" emphasis on domestic production and marketing. Left in place, though, was a board committee on "foreign manufacture," chaired by Charles Stone. Stone had been one of the conservatives in the acrimonious discussions about the Scotland venture, and leaving the chairmanship in his hands seemed to imply merely lip service to the notion of foreign manufacturing. But in 1953 Stone, then ready to retire after forty-nine years of service with the company, surprised everyone with his final act. Just before stepping down, he took an extensive trip through Europe, accompanied by one of the senior Waterloo engineers. Reporting to the board on his return, Stone delivered an unsettling message. Deere's export sales were dropping; the foreign exchange problem in most countries was an obstacle to imports and an argument for developing overseas farm machinery manufacturing. "Since we abandoned the Scottish project," Stone continued, "we find ourselves with no definite prescribed course of action and with inadequate information on the subject." Stone's blunt report stirred the board, and he was asked to report back the following month with his recommendation for a next step.

Stone's last act as a company officer was this recommendation, and he did it astutely. Several of the younger members of the board—particularly Hewitt, Elwood F. Curtis, and George T. French—were interested in the international sphere; there were other new men on the board in 1953, too—Willard D. Hosford, Jr., Frank M. Dickey, Richard Edwards, E. C. English, and Maurice A. Fraher—each of whom might be more receptive to changes than their older counterparts. Fraher was the strong-minded head of tractor manufacturing; a Waterloo executive and provincial in outlook, he had never been out of North America. Nevertheless, Stone now made the seemingly incongruous proposal that Fraher become the head of the committee on foreign manufacture, a "small group of two or three good men," to keep in touch with developments around the world and "apprise the Board of opportunities internationally." Thus Stone had bypassed Pat Murphy, the most vocal opponent of the Scotland project and still the most conservative of the board members who were leery of manufacturing abroad. Fraher had never been global-minded, but he rose to Stone's bait and immediately dispatched a study team to Brazil and Argentina (without him).

In late 1953, another member of the board, Lloyd E. Kennedy, turned up another potential international involvement. Kennedy had gone to Europe in December to investigate a discreet inquiry from a West German banker who asked whether Deere would be interested in purchasing controlling interest in one of the old-line German tractor and implement companies, Heinrich Lanz, located at Mannheim. Lanz had been building farm machinery since 1859, and prior to World War II had been one of the two dominant German tractor manufacturers (at one point claiming more than 40 percent of all German tractor sales, with Deutz taking most of the remaining share). The postwar period, though, had not been kind to Lanz. A very large percentage of its prewar sales had been to the farmers in East Germany, a market that was locked behind the Iron Curtain after the war. Further, the Mannheim factory site had been almost completely leveled by Allied bombing during World War II, and the rebuilding had been helter-skelter, putting the firm back into a weaker manufacturing configuration. Its famous Lanz Bulldog still remained a simple, effective tractor (the name Bulldog had become synonymous in Germany with tractor, similar to Caterpillar's generic association with track-laying tractors). But the company's postwar efforts to upgrade the power of the Bulldog had weakened the machine and led to increasing failures in the field. Compounding the problems of Lanz were other product failures from new, unproven models, prematurely marketed. In sum, though Lanz was a great name in German manufacturing, the company's position in the 1950s was tenuous; it was manifestly going nowhere. Some 30 percent of the shares of Lanz were held by this time by a German bank, which proposed selling its share and securing additional shares so that they could deliver 51 percent of the Lanz stock to Deere. The bank's shares were offered at approximately $2.5 million; Ellwood Curtis analyzed the Lanz financial and operating structures and estimated that a total investment of $13 million would be required over the first five years.

Herein lay the dilemma for Deere. To become established in manufacturing abroad, it essentially had two alternatives. One was to make a "green fields" approach—start a new manufacturing unit from scratch in one or more countries. The other was to purchase an existing agricultural machinery manufacturer. Deere was unprepared to attempt the former; memories were long about the unwillingness of senior executives in the manufacturing, engineering, and marketing groups to depute first-rank subordinates to Scotland. The notion of starting on their own looked increasingly forbidding to the Deere executives, for not only had the price tags for such ventures risen considerably since the Scotland proposal, but Deere did not seem to have the expertise in international efforts that would be required. Perhaps it would be better to buy an existing organization, with manufacturing facilities, personnel, and established marketing channels already in being.

But the Lanz proposal aptly illustrated the difficulties of the alternative approach. Buying an industry leader in a major country for a reasonable price would be ideal, but this was not a realistic prospect. Only the weaker companies were on the market block; moreover, competition had become much more vigorous. Where there had been a few dominant companies in the prewar period, now there were new domestic manufacturers all through the developed countries of Western Europe. In Germany, for example, which had been dominated by Lanz and Deutz, there were more than a dozen tractor and implement manufacturers fighting each other for a larger market share. The situation had degenerated in the early 1950s to aggressive, predatory price cutting and sharp competitive practices. Inevitably, some companies were forging ahead; others, including Lanz, were dropping behind. The Lanz performance in the 1950s was not at all impressive. Yet it was a great name in the agricultural world; perhaps this past reputation could be traded upon to build a new organization that could be infused with Deere methods, products, and esprit de corps.

An extended debate on Kennedy's proposal ensued in a special board meeting on December 23, 1953. Charles Wiman was in the chair; the focus was heavily tilted toward the export implications of North American business. Finally, the vote came: the board turned the proposal down. Compounding the problem, Fraher's study team in South America now reported back and after some initial enthusiasm by some board members about Brazil, the board finally decided that conditions in neither Brazil nor Argentina were favorable at this time for a Deere effort. Once more a Deere initiative abroad had sustained a severe setback.[2]

THE MASSEY-HARRIS-FERGUSON PROPOSAL

In Hewitt's first month of office a startling proposal had come forward to him, on a confidential basis, from a surprising source. James S. Duncan, the president of Massey-Harris-Ferguson, approached Deere about the possibility of merging the two companies. Massey was then in the midst of a complicated (and as later evidence showed, acrimonious) internal squabble growing out of a recent merger. In mid-1953, the worldwide operation of the Massey-Harris Company, Limited, had been merged with the Harry Ferguson Company, with the new organization carrying the name Massey-Harris-Ferguson, Limited. Earlier, Harry Ferguson had broken his relationship with Ford, a bitter severance that had led afterward to a messy lawsuit brought by Ferguson against the Ford organization. The suit finally had been settled in 1952 (in favor of Ferguson), and the future of the Ferguson organization had seemed reasonably bright. But, in truth, its product line had slipped and there were many management tensions, not the least being that Ferguson himself paid less attention to the organization (especially the United States divisions) than others in the group felt necessary. Massey, in turn, brought to the merger of 1953 a set of its own internal problems, not as severe as those of Ferguson, but nevertheless troubling. Though Massey had strong operations in its divisions around the world, and had become one of the first truly multinational corporations in agricultural machinery, its United States operations had lagged, and there were unresolved differences within management about what to do.

The vested interests of managers in each of the Massey-Harris-Ferguson organizations, and the unwillingness of top management to restructure the new firm, soon led to increasing difficulties. The two companies' properties were reasonably compatible, so there was not much duplication of manufacturing facilities. But there was overlap in product lines and further complications in quality control. Exacerbating all of this was the duplication of the distribution systems. Duncan himself had advocated a "two-line policy," whereby there would be some exclusive Massey-Harris dealers, some exclusive Ferguson dealers, and a third set of Massey-Harris-Ferguson dealers. The mixed, often clashing lines of distribution were further complicated by an equally disparate set of products from the two organizations. Over the first two years of the merger, while the overall parent organization was performing reasonably, the North American operations began to sag severely. As if this was not enough, there were tensions among Massey's top managers, with Duncan heading one faction and W. E. Phillips, chairman of the executive committee of the board, leading the other (and backed by an Argus Corporation shareholder bloc).

When Duncan first approached Hewitt, in the spring of 1955, much of this internal dissension was not public knowledge, though Massey's lack of success in the United States was known. Hewitt outlined the pluses of such a merger in a private memorandum to himself: Deere could acquire an already established, large-volume overseas manufacturing and distribution system and thus "avoid 'growing pains' from trial and error type form of expansion." The merger could help solve some of Deere's problems on the West Coast of the United States, where it seemed likely that Deere would be severing most of its relationships with Caterpillar in the joint Caterpillar-Deere dealerships. (Deere now made a crawler tractor, directly competitive with Caterpillar.) Massey had strong West Coast dealerships. But as Hewitt ticked off the pros and cons of the merger, his list of minuses became much longer. It would not be easy to merge the two groups of officers, directors, engineers, and so forth. There might be adverse public reaction in Canada if one of its largest industries were taken over by a United States corporation. Important problems loomed in consolidating the tractor and implement product lines, particularly because Massey itself never had been able to solve this problem within its own organization. There would be personnel adjustments, difficulties in meshing the high-quality standards of Deere with the somewhat lower standards at Massey, and there would be some dilution of voting power in the merged corporation.

Most important, there would be enormous difficulty in merging the two distribution systems. Massey already was weakened by conflicts among its own dealers; Deere's dealers would certainly look askance at some of the implications of a Deere-Massey merger. Finally, there was a cloud hanging over the negotiations—antitrust implications. In the United States, Massey was quite strong in combines (stemming in good part from the striking effort in World War II with the Massey "Harvest Brigade," a story recounted earlier). Deere's combines, too, enjoyed a large market share. The total of the two would be a dominating powerhouse in harvesting machinery in the United States. Clearly, the Department of Justice could choose to investigate the merger and might require the merged organization to divest part of its harvesting operations. Such a divestment typically is not on a company's own terms, and Deere might end up having a less strong domestic harvesting business than it did before it went into the merger.

Two meetings were held with the Massey group, after substantial financial analysis by both companies, to project how such an arrangement could be brought about. The negatives were too strong for the Deere group, however. The complexity of the distribution problem alone seemed daunting. The company did not need another antitrust investigation; the experience in the late 1940s had been graphic evidence of the debilitating effects that such an effort brings. Massey also wanted to back off; its divided and bickering management was close to a final break. (Within a few weeks Duncan's resignation was asked by the Argus group, Phillips took over as the new chief executive officer, and A. A. Thornbrough was confirmed as executive vice president and, later, president.)

A Deere-Massey consummation would have given Deere an instant worldwide operation, with an organization that had strong roots in more countries than any other agricultural machinery manufacturer. Deere might have shortcut what was to be a tortured, difficult road to full-scale multinational status. Yet the disabilities of a merger with Massey seemed to far overbalance the advantages of gaining quick international reach. There was evident relief when Hewitt and his colleagues returned from their last meeting in Toronto, without a foreign operation.[3]

THE ORGANIZATIONAL SHAKE-UP

By January 1956 the Booz, Allen & Hamilton report was on Hewitt's desk. It was in two sections—a comprehensive "public" document for all of the board members and top management, and a private document for Hewitt on implementation. In the latter, the Booz, Allen analysts were dispassionately frank as to which executives should retain their incumbent positions, which should move up, and which should be shifted away from their present responsibilities. They were keenly attuned to the sensitive issue of management transition, for the election of Hewitt could have sent two markedly different signals to the organization. It could have been interpreted as an isolated act leaving in place the strong senior management, particularly the powerful triumvirate of Murphy, Kennedy, and Cook. On the other hand, Hewitt's presidency could signal a significant shift in management, bringing into central policy-making roles a new generation of leaders. The Booz, Allen document took the latter course.

The consultants proposed an unusual strategy, the establishment of a new "policy committee" at the top-management level. This group was to be manned by top executives, representing the major divisions of the company—production, sales, and finance. Its members were to spend full time on committee activity and were not to have any direct line of authority. The committee was to report to the president—not to the board—and only the president could issue directions to the line organizations. The committee's primary assignments were to be threefold: to develop objectives, to aid in defining policies, and to conduct special studies. The Booz, Allen consultants also envisioned an additional set of secondary functions: aiding the president in public relations, counseling with the president on administrative affairs, and providing guidance to principal operating executives in the company.

The Booz, Allen analysts avoided any mention of the already existing advisory committee instituted in 1952 by Charles Wiman. The advisory committee had not been effective, perhaps in part because its composition included not only the senior top executives but other lower-level line and staff officers of more limited perspective. The active members of top management had seldom felt it helpful to solicit advice from their older peers, to the evident chagrin of some of the latter. The proposed policy committee was to become a senior advisory group of direct value to the president. The consultants suggested that it initially be composed of four senior executives—Murphy, Kennedy, Cook, and Bruce Lourie. The first three, who had been at the center of the power structure of the organization, would become advisors to the president, rather than senior operating officers. In turn, the suggestion of Lourie recognized Hewitt's desire to make some changes in the operating side of the marketing arm. A further set of personnel moves would fill the line positions that would be vacated by the establishment of the policy committee, bringing younger faces into high-level management positions.

The Booz, Allen analysts put particular emphasis on the concept of decentralization in their public document for the board, in the process paying homage to Deere's longstanding canons of rugged individualism (they even cited the first formation of Deere, Mansur & Company in 1869). But the Booz, Allen authorities had misgivings about how well the concept really was understood; in their view many executives in the company had an overly simplistic view, confusing decentralization with almost complete autonomy. In a private memorandum, Hewitt had intimated similar unease, suggesting that the company "should keep all advantages of decentralization *but* strive for more coordination." (In his first public speech to the organization after assuming the presidency, an address to the branch managers in June 1955, however, he had given no indication of concern about the concept: "My best advice to each one of you is that you keep on doing what you are doing under our decentralized system. . . . The only change I am anticipating right now is that you should expect to be given more responsibility as our Company continues to grow. . . . As more and more authority is delegated to you, it is expected that you will delegate more authority to the men under you.")

The Booz, Allen authors warned of the "limitations of decentralization," noting that there were many groups—stockholders, dealers, farmers, the UAW, the general public—that were not at all interested in a highly fragmented, decentralized operation. "Rather, they demand unity; they require that the Company act with a high degree of uniformity in its relations with each one of them." While the company could readily conduct its operations "with maximum freedom given to the head of each manufacturing and distributing unit," there still was a critical need for "the responsibility of the General Company to provide leadership to the entire enterprise. The decentralized branch houses and factories must be brought together to achieve uniformity."

This was heretical thinking to many of the older generation at the company. While Charles Wiman had sensed the need for additional coordination and control when he instituted the two separate studies of decentralization in 1945 and 1947, in both instances he had pulled back from any decisive elaboration of the relation between the general company and the branches and factories. So the period from that time to the election of Hewitt had remained marked by a high degree of autonomy for branch and factory managers, coupled with a more than occasional unwillingness to coordinate and cooperate with other arms of the company. Now the Booz, Allen report gave Hewitt the opportunity to promulgate a more sophisticated version of decentralization, one that would put the general company more directly into a mode of leadership. This inevitably would mean some loss of autonomy at the branch and factory.

In this same month (January 1956), the Booz, Allen survey of executive compensation also reached the board. The consultants found that Deere's compensation of its top management officers was significantly below that of almost all of its competitors. Booz, Allen recommended significant increases in the base rates for management personnel at the company, so that a somewhat smaller percentage of each executive's total compensation came from year-end bonuses. The report cleared the air for the development of a comprehensive salary structure that put the company's executive compensation in line with that of the rest of the industry. Though the plan has gone through a number of iterations over the years, this report has continued to serve as a basic conceptual structure for Deere's executive compensation planning ever since.

In total, the Booz, Allen reports of January 1956 were seminal ones, far reaching in their changes. Booz, Allen clearly recognized that the company was in a period of critical management transition and needed decisive actions, particularly by Hewitt himself, to establish the new management philosophy of a new generation of Deere managers.

Hewitt's treatment of the Booz, Allen report and his own decisions reveal his evolving management style in this first year of his tenure as chief executive. He moved quickly on the report, and by the annual meeting of April 24, 1956, major organizational changes were announced. Though Hewitt was intrigued with the policy committee notion—even roughed out a hierarchy of assignments for it (including a study of "electronic computers" and a set of overseas visits)—he finally dropped the policy committee suggestion altogether and adopted a more cautious alternative—the appointment of Murphy and Kennedy as the company's first senior vice presidents.

In these new positions, they were to assume policy-oriented roles. The elevation of French to vice president of implement manufacturing and Curtis to vice president, finance division, and treasurer, clearly signaled the transference of line responsibility from Murphy and Kennedy to their successors. This new role was comfortable for Kennedy, not quite so for Murphy. The former had continued his responsibilities as financial advisor to the family, and the new job now freed him of onerous day-by-day financial decisions. Murphy already had evidenced some concern about his relationship with Hewitt and privately had discussed with Hewitt some changes in his own job assignment. At that time he had written Hewitt, "Whether I decide to remain with Deere & Company after April depends on a number of things, primarily whether there is any need for me here anymore, and what standing I will have in the future, etc." Thus Murphy's role as senior vice president was not wholly unambiguous. By nature, he always had immersed himself in operating details, and he now sometimes found it difficult to rise above the day-by-day to assume the policy perspective that the new job implied.

Cook, the other member of the triumvirate, opted for a different direction. He was the most independent and self-assured of the three in personality and by his professional role as a lawyer, which included outside board memberships. Cook had long advocated more rapid upward movement of personnel within the organization and had identified his successor several years before—Lewis D. Wilson. Now Cook chose to retire as general counsel, and at the annual meeting of April 1956 Wilson was elected to that post. In turn, Cook assumed a unique role; the three incumbent members of the advisory committee—Maurice Block, Virgil Bozeman, and J. L.

Deffenbaugh—retired from the committee (and the company) and Cook was appointed to it.

The character of the advisory committee now changed. No longer was it a sinecure for senior executives, whether they were directors or not. Rather, it became a one-person committee, Cook being its only member for the years 1956–1962 (except for a short period when Maurice Fraher took an assignment on it). In this special role, Cook provided the kind of advisory counsel to Hewitt that the Booz, Allen analysts had envisioned, and he was particularly helpful in one of the most important Hewitt moves in this period, the building of the Deere Administrative Center.

Other important changes were announced at the meeting of April 24, 1956. Burton Peek resigned as chairman of the board of directors at age eighty-four, after an incredible sixty-eight years of service with the company. Peek had always been very supportive of Hewitt but had taken himself out of the mainstream of management decision-making several years before. Apparently this seemed an appropriate moment for his disengagement to be made formal. Another of the "old guard," L. A. "Duke" Rowland, head of Deere's chemical company, also resigned. He did it reluctantly, only with the prodding of Hewitt. One other officer-level appointment was made at this meeting, too—Frank N. Dickey was appointed vice president, industrial relations and personnel.

Thus the management reorganization announced on April 24, 1956, became one of the most important in the company's history. A number of other Booz, Allen & Hamilton recommendations were put into effect, tightening the organizational structure, the communications channels, and the operations policies, particularly in marketing and finance, of the company. There was little doubt in anyone's mind after this critical meeting, just eleven months after Hewitt had assumed the office of chief executive, about "who was in charge." Hewitt had put his mark on the management structure decisively and with sensitivity and aplomb. In September 1957, when he persuaded the dynamic, strong-minded C. R. Carlson to leave the Minneapolis branch and come to the general company as vice president of marketing, Hewitt completed his set of initial moves. Bruce Lourie was appointed vice president of special projects and moved out of direct responsibility for line marketing. At this point Hewitt had his own full management team in place.

Right from the start, one person among these, Elwood Curtis, had the closest compatibility with Hewitt. The two men were remarkably complementary to each other. Hewitt's strengths in general management, marketing, and public and community relations were balanced by the financial acumen, analytical skill, and realistic perception of projects that Curtis brought to every situation. Hewitt's post as president was an unusually isolated one; most of the people in senior management who ostensibly were to be Hewitt's advisors were candidates for shifting, and yet most of the younger top management people were untried. Only time would determine whether they could successfully assume the key senior management posts of the future.

As subsequent events proved, Hewitt was perceptive in sensing that Curtis was the person he needed to balance his own thinking. Curtis shared Hewitt's enthusiasm about the challenges of building a new, expanded organization, and Curtis was profoundly supportive of Hewitt's thinking. Yet Curtis also brought a leaven of caution and conservatism from his meticulous training as an accountant. The correspondence between these two men in the first years of the Hewitt presidency chronicles a warm and growing mutual respect; one or the other was often in the field, frequently abroad, and these written bonds served both very well. When Lloyd Kennedy retired from the company as senior vice president in the fall of 1959, Hewitt established the position of executive vice president and requested the board to confirm Curtis as its incumbent (in the process moving him ahead of Murphy, the only officer remaining in the senior vice president position).[4]

MOVING INTO MEXICO AND GERMANY

Within days of the collapse of negotiations with Massey-Harris-Ferguson, Hewitt moved to rekindle Deere's international interest. This time he decided that the senior decision makers themselves should do the field work abroad, rather than follow Fraher's pattern of dispatching subordinates. He first urged Fraher himself to go to Europe. Fraher remonstrated that he had never been there, simply had no notion of what to do on such a trip, especially on such short notice. Hewitt replied, "Maurice, go visit with the European representatives of our auditors, go meet with our bankers' European correspondents—just go sit at a sidewalk cafe and observe—but go!" Fraher did. At the same time, Hewitt sent Curtis and three colleagues to Mexico.

Fraher and Curtis reported back in July. An amazing transformation had been wrought in Fraher—he came back an enthusiastic convert, very optimistic about the potentials he saw in Europe and worried about Deere's backwardness in not being there. Indeed, Deere's arch-competitors had been there for years; International Harvester had started a small, experimental manufacturing plant in Sweden back in 1905 and was into substantial production in German and French plants by 1910; Massey had purchased its French operation in the late 1920s. As Hewitt later told a *Forbes* editor, "I'd say our competitors had a 50-year jump on us . . . literally." Fraher now had the litany himself: "If we are to maintain our competitive standing in the industry we *must* have this foreign market. Not to enter it will mean the eventual loss of much of our export business . . . giving up a principal avenue for major growth of the Company." He not only wanted to enter the European market as soon as possible, but to do it on a surprisingly large scale, with the capacity to make sixty to seventy-five tractors a day.

Curtis, too, was positive about the Mexico trip—the Mexican government seemed enthusiastic about a John Deere plant there, either for assembly or for manufacture, "provided its products did not cost substantially more than imported units." Mexico's agricultural needs were substantial; its nearness to the United States might give the company a compatible venue in which to "practice" for further overseas ventures.

New implement companies had already started at several locations in Mexico; the government had promised an umbrella over them by refusing import permits to other manufacturers. Curtis urged early action: "If another company establishes a plant there we probably will lose a substantial part of our participation in that market because of the Mexican government's policy of protecting local industry by import licenses and tariffs." The company's Mexican distributor, Wells Fargo, had not been very assiduous in pushing Deere products, and it was beginning to curtail its operations by closing some of its branches.

Curtis and the Deere territory manager in Central America, Robert A. Hanson, both felt that if Deere wanted to increase its visibility (the company only had about 8 percent of the market at this time), it must set up its own branch in Mexico. If the board were to agree to this, Mexico would have the first sales branch outside of the United States and Canada.

There were some potentially ominous complications, however. The Mexican government had begun to pressure the automotive industry to increase the percentage of "local content"—that is, the proportion of components and parts in the Mexican automobile actually manufactured in Mexico. How would this decree affect tractor manufacturing?

This question did not deter Deere; the board seemed genuinely eager to act, and it immediately implemented the suggestion for a Mexican branch operation. A Venezuelan corporation, John Deere, C. A., was established as a subsidiary to own the Mexican operation and others that might be established abroad.

Harry Pence was appointed executive vice president of the subsidiary and, after a whirlwind investigation, recommended to the board that a plant be built in Monterrey, Mexico. Skilled labor and engineering talent were available there, as well as ready supply lines for the procurement of raw materials. Monterrey was one of the gateways to the agricultural part of Mexico, a logical manufacturing site. By late in the year (1956), the board had appropriated just under $2 million for the purchase of land near the city and for the construction of an assembly plant for some John Deere tractors and a few of the simpler implements used in Mexico—wagons, harrows, and disc plows. The manufacture of parts was to be introduced as soon as it proved economical to do so.

Meanwhile, the Lanz proposal in Germany surfaced again. In July 1956 Hewitt, Kennedy, and Pence returned to Mannheim to negotiate once more with the bank officials holding the block of Lanz stock. Hewitt's sense of urgency soon generated a formal proposal for Deere to purchase a block of 51 percent of Lanz stock for approximately $5.3 million, subject to a Deere audit of both the Lanz financial statements and the actual machinery and equipment in the plants (the tractor works in Mannheim and an operation in Zweibrücken, where harvesting equipment was manufactured). At the ensuing board meeting (on July 31, 1956) Hewitt expressed concern about certain imponderables. Though Deere had been tendered Lanz balance sheets and income statements, "the information they give is considerably less than is customary to publish in this country. . . . It seems to be the European custom to make certain charges against sales before arriving at a gross profit figure. These charges are credited to certain reserves or 'silent assets' which do not appear in published statements. For these reasons, it is difficult to appraise the true financial condition of the European firm."

Lanz receivables and inventories troubled the board, too—both had grown considerably in the two years since Deere had first approached the company. Still, the assets seemed to be somewhat understated—earlier in the year, the company had sold DM 10 million of new stock to purchase machine tools, at the same price that Deere was being offered the stock by the bank. Hewitt was particularly concerned about the outmoded, high-priced Lanz Bulldog: "It is doubtful whether such a tractor can successfully compete with the Ford-Ferguson, International Harvester tractors of modern design." Hewitt proposed that if the company did acquire Lanz, it bring the Dubuque line of diesel tractors to Europe to manufacture in place of the existing Bulldogs. Thus a substantial portion of the Lanz inventory would be subject to shrinkage, so that operating losses for the first few years would be probable. Hewitt estimated that the total investment in Lanz would likely reach a figure of $10–$12 million, not including immediate needs for working capital. The latter would probably balloon the project to a total of some $25 million. Still, with all these reservations, Hewitt strongly supported the purchase.

After lengthy debate, the board acquiesced to the acquisition of the 51 percent block of stock, and the relationship between Deere and Lanz was a fait accompli. Curtis was immediately dispatched to Mannheim to supervise the audit of the financial and physical assets; his preliminary work indicated that the Lanz machine tools were "better than those in our own implement and tractor plants." Curtis placed a value of $15 million on these machine tools, "on a conservative estimate."

Balanced against this were serious disabilities, some of them only incompletely known even at the time of the final commitment. First, preliminary figures on Lanz sales for 1956 showed a figure of $39 million, with a loss of some $2 million—"quite a surprise to some of the German Lanz directors," Lloyd Kennedy reported, "but it includes substantial write-downs of inventory." Kennedy, Pence, and Curtis were further surprised to discover a substantial number of tractor failures in the field and complaints concerning some of the Lanz models. "It had been believed that while the Lanz tractor design was outmoded, it was sturdy and quite acceptable in the areas in which sales were being made. The fact that this is not quite true makes it imperative to change the design," Kennedy admitted. The Lanz inventory of finished goods was quite high, and the study team also worried about relatively high production costs and a shortage of working capital. On the positive side of the ledger, the company appeared to have excellent personnel and still held a fine reputation in Germany and in some of the traditional Lanz export markets.[5]

There were striking similarities in the corporate histories of Lanz and Deere. Heinrich Lanz had founded the German company in 1859, first selling English and American machinery as a distributor. Within a decade he was making his own machinery, concentrating at the start on harvesting equipment. When steam threshing came into prominence in the 1870s, Lanz began making stationary steam engines—Locomobiles—and soon had a substantial hold on the world market. Heinrich Lanz himself was a paternal father figure, a stickler for detail, operating his factory in "absolute strictness" (according to one of his employees). His walks around the plant floor—every day at precisely the same moment—became legendary; for those who were late, he stopped and held a watch under their noses, stating that company employees must start on time. More than once it was reported he sat at the desk of an employee who hadn't arrived yet and took care of his work until the man showed up a few minutes late. He missed nothing on his factory tours. When he noted any nuts or bolts lying on the floor, he lifted them up quietly and took them over to the foreman, often making a note on his shirt cuff of the incident.

The loyalty of the Lanz employees in this earlier period was also legendary; the company paid well and there was a pride of product and an esprit de corps that made Lanz one of the most respected employers in southern Germany. In 1902 Lanz visited the United States and in the course of his travels met and talked with Charles Deere. (There is no record of where the meeting occurred, however, or whether Lanz visited Moline on the trip.) It is likely that Deere must have seen some similarities between his father, John Deere, and this precise German.

When Lanz died in 1905, the firm remained within the family and was operated by his widow and four children. By 1912 the firm was building agricultural tractors, the first model, the Landbau-Motor, a strange machine with a permanently attached rotary cultivator on the rear. In 1916 the company also expanded its harvesting machinery product line by acquiring an old Zweibrücken firm, founded in 1863 by Christian Wery.

It was the Lanz Bulldog, though, that came to epitomize the Lanz name around the world in the twentieth century. In the same year that Lanz

CHAPTER 13

Exhibit 13-2. Top left, Lanz Bulldog; bottom left, Heinrich Lanz factory at Mannheim, Germany, after Allied bombings, 1945; top right, Interior of the Mannheim factory, late 1970s; bottom right, Factory from the air in the early 1980s. *Deere Archives*

had acquired the Wery firm, an engineer named Fritz Huber joined Lanz. Huber developed a single-cylinder, two-stroke engine of extraordinary simplicity and exceptional economy. It was a slow running, low-compression engine using low-grade crude oil; the machine had no valves, no camshaft, no carburetor, no water pump. The first commercial model came out in 1921; those earlier versions had an external vaporizing chamber and for ignition used a device on the front called a "glow bulb" (also known as a "hot bulb")—thus a "semi-diesel." The Lanz glow bulb was constructed with ventilation shafts that looked like eyes on the front of the tractor—perhaps the reason that Huber and his colleagues settled on the name "Bulldog." The first production models were 12 horsepower, but the company found that larger versions were becoming so popular for land clearing and other heavy jobs that the firm soon concentrated on the so-called Big Bulldog, in 22 and 28 horsepower. When pneumatic tires from imported American tractors came to Europe in 1929, Lanz adopted these for the Bulldog. By the late 1930s a smaller version had been developed, a Farmer's Bulldog at 15 horsepower that cost only DM 2,750, about $1,100. By World War II, Lanz had produced more than 100,000 tractors out of the Bulldog configuration.

World War II was a disaster for Lanz. More than 80 percent of the Mannheim plant was leveled in a series of Allied bombings in 1944 and 1945. The firm did receive permission from the American Occupation Forces to rebuild, but then the Lanz management made a signal mistake. Rather than razing the whole area and building from scratch, they attempted to reconstruct the damaged buildings piecemeal. The bombings had not left any logical production patterns intact; indeed, as the company rebuilt it compounded its production problems of materials flow, shipping, and so on, and left itself with a legacy of scattered, ill-adapted buildings that plagued it through the postwar period. The Bulldogs were still the main product (along with the harvesting machinery from Zweibrücken, the plant there less damaged than the Mannheim operation). Emphasis was still on the old Lanz tradition, and efforts were made to rebuild the worldwide sales organization that had characterized the prewar period.

There had been important links with Australia and several countries in South America. Lanz tractors had also been assembled at a plant in Getafe, Spain, outside Madrid. Lanz Iberica, S. A., had been founded by Ricardo Medem in 1953; Lanz itself owned 8 percent of the manufacturing company. The products were the 28- and 36-horsepower Lanz Bulldogs.

At the heart of the onrush of difficulties that Lanz experienced in the early 1950s was its decision in 1951 to build a new small tractor model called the "Allround Tool Carrier" (later named the "Alldog"). Postwar agriculture in Western Europe encouraged a number of smaller farmers to experiment with mechanized agriculture. The Lanz Bulldog was too large a machine for them; what was needed was a small, general-purpose machine. The Allround/Alldog had a small 13-horsepower gasoline motor mounted above the rear wheels, with a saddle seat for the driver. In front, between the engine and the front wheels, was a long metal frame that allowed various tools to be mounted between its sides. The machine also could be used as a truck-like transport vehicle. In conception, it was innovative and might well have made a major dent in the small-farm market. Unfortunately, Lanz hurried the decision to get into production, deciding to purchase the motor, rather than developing its own. This was a mistake, for the outside contractor made an ineffective engine, resulting in many months of product failure (the frame itself had weaknesses at the early stages, too).

What then had Deere purchased in 1956? First, there was the badly outmoded, failure-prone product line. The Bulldog itself remained a reliable machine, but it was a single-cylinder tractor of limited horsepower that could not be expanded in size without major developmental expenses. By November 1956 the 200,000th Lanz Bulldog had come off the line, certainly testimony to its endurance. Its future, though, appeared limited. The Alldog was a major product disaster, offending a new group of farmers that the company could ill afford to alienate.

The product-line problem was worsened by outmoded methods and heavy personnel costs due to overstaffing and rigid production rules. Financial problems, shortages in working capital, and other miscellaneous difficulties added to the disabilities of the old firm. Though Lanz had very able people throughout the organization, Deere nevertheless was saddled at the start with a management structure that needed considerable honing. An American plant manager, James Wormley, was given the overall responsibility, but only a few able colleagues were sent to him in those first months. Once again, the entrenched managements in many of the Deere factories balked at releasing key personnel that would have eased the transition.

In general, Deere was naive in these first months of the Lanz endeavor. The company was just beginning to learn about the special characteristics of agriculture in the countries in which it was hoping to sell. Many farmers called on their tractors for uses not common to North America—in particular, there was a great deal of over-the-road vehicular use of the tractor. Often the loads were light and the higher-powered diesel engines introduced by Deere frequently were overpowered for such use. Indeed, if a diesel engine is not put under some considerable stress it can develop its own problems from under-utilization—crank case oil dilution by diesel fuels, for example.

Another example of misreading of European conditions was Deere's experience with the dealerships inherited from Lanz. The American marketers made a misguided effort to reconstruct them in the North American tradition of exclusive dealerships carrying a full line of products manufactured by one company. In Western Europe, though, the market was much more fragmented. After World War II large numbers of new, smaller agricultural machinery manufacturers came into being. Most of them were not full-line manufacturers and tilted, if anything, toward relationships with many manufacturers. Massey-Ferguson and International Harvester had been in Western Europe for many years and had been able to develop a cadre of exclusive dealers. But most of the rest of the Western European agricultural equipment dealers were fearful of an exclusive dealership, feeling that there was safety in numbers. They not only wanted to handle several companies' products, they also wanted to carry two competing tractors, two different combines, and so forth, all lined up together in their yards in a kaleidoscope of company colors. Company marketers persistently pursued the goal of exclusive dealerships, but frequently it was only the weaker dealers who were willing to do so. Thus the dilemma: Should the company stay with a strong dealer with only partial Deere loyalty, or go with a weaker one wholly dedicated to Deere? The company was not as sophisticated at the start in making this decision as it was in later years.

Another misjudgment was Deere's attitude toward European cooperatives. The cooperatives in Western Europe historically had been much more important than in the United States, and a great many Deere people had prejudices, growing out of the nineteenth-century experiences with the Grangers, that made it more difficult for them to understand the role of the cooperatives in the European scene.

Overall, the Deere group was very proud of its own name—in North America, it had become synonymous with quality and dependability. They were not used to dealing in an environment in which their company (in this case, Lanz) was losing its reputation and being upbraided for product failures and other slippages. It was a sobering experience for the marketing group out of North America to cope with weak dealers, poor product visibility, product failures, and other irritants.

Was the purchase of Lanz a mistake? In retrospect, it is clear that the company had rushed into the project without full information (albeit such information was not always readily available). Also, its hopes about how quickly it could "turn around the Lanz operation" and become a viable, major producer in Western Europe proved overly optimistic. It was to take almost two decades of very hard work and many disappointments to bring about this turnaround. Thus a short-run assessment would have been premature. But the answer to one important question was clear almost immediately. The Lanz purchase did accomplish in an unequivocal way the basic goal that Hewitt had envisioned from the start—it put the company into the international arena in a way that permitted no turning back. By no stretch of the imagination was the Lanz project just the dangling of a foot in the water—the company was totally immersed in international operations. Perhaps it was a naive new entrant, but it was there, and there to stay.[6]

THE DEERE ADMINISTRATIVE CENTER

Often it happens that a single symbol or achievement comes to stand as a personal mark for a lifetime or an era. Theodore Roosevelt's image as a Rough Rider and Henry Ford's Model T automobile both come to mind as examples. For Hewitt's period of leadership at Deere & Company, the initiation and completion of the new office building—the Deere Administrative Center—epitomizes what he hoped to stand for. Far more than just a new, indeed noteworthy, office building, it sought to define the feeling that Deere & Company employees had about themselves and their organization.

At this point, Deere & Company was 120 years old. Throughout this period, it had employed outside architects only in a limited way in constructing new plants and office buildings. Most of the work was done in-house by the company's own engineering departments. The result was that aesthetics took a back seat to utilitarianism—the buildings and their locations were conventional to a fault. Hewitt now decided to take a different approach, to find a nationally respected architect to mastermind the new building and then to assemble the proper data about that person so that he could persuade the board to authorize the expenditures. Robert McNamara, Hewitt's close friend and classmate at the Harvard Business School, was at this time a top executive at the Ford Motor Company and had just overseen the construction of an administration building in Detroit. Hewitt immediately contacted him, and McNamara sent him a packet of some two dozen architects' prospectuses. Henry Dreyfuss, the design consultant who had been so intimately involved with Deere during the Wiman period, suggested that there were two buildings that he (Dreyfuss) felt would be superb models to emulate—the new auditorium at the Massachusetts Institute of Technology and the General Motors Technical Center near Detroit. Hewitt visited each site and while at the latter first met the architect for both buildings—Eero Saarinen. Many years later Hewitt recounted this experience (in the *AIA Journal*). "If you want to work with an architect's architect," Hewitt quoted Dreyfuss, "put Saarinen on your list." Hewitt had found his man. "Then and there I decided Eero Saarinen was the man for the job."

By August 1956 Saarinen was in Moline with Hewitt, quietly looking over possible sites for the new building. One was a piece of land owned by the company along the Mississippi River, a mile or so from the enormous complex of offices and factories that had been the original location of John Deere. This particular property was not very attractive, though. A junkyard faced it from the rear, and it was bordered on one side by railroad tracks. The unsuitability of this river site (the 34th Street property) was quickly apparent to Saarinen and Hewitt. The other possible locations were on properties not then owned by the company. Saarinen observed that "he had never seen a

CHAPTER 13

Exhibit 13-3. Top left, William Hewitt and Eero Saarinen discuss the Deere Administrative Center, 1959; bottom left, the Center just before its opening, 1964; top right, the Center under construction, ca. 1963; bottom right, the Deere Administrative Center in 1983. *Deere Archives*

community that offered so many problems in regard to having potential sites marred by the nearness of shacks, trailer camps, cemeteries, cheap commercial buildings and other unattractive blight." After the two had traveled by auto and airplane through the area, they identified a particularly attractive possibility on the edge of the town of East Moline, along a set of bluffs overlooking the Rock River. There were some 120 acres, with enough level land to take care of the necessary properties, yet with trees and attractive views throughout—truly a potential "campus," to use Hewitt's earlier word. Hewitt and Saarinen borrowed a truck from a local utility company with a telescoping tower that allowed them to see out over the smaller trees, in order to envision a multistory building at the location. Both were convinced that this was just the right place.

Now Hewitt had the sensitive task of selling the whole project to his board peers. Hewitt prepared for it carefully. His own verbatim notes remain, and they are revealing in several particulars. A meticulous case was made to reject the company-owned land and to use, instead, the wooded property. It was much too early for any configuration of the building to be discussed, so Hewitt stayed away from the subject. Rather, he concentrated his remarks on the selection of the architect. Saarinen was Hewitt's choice, and he proceeded to build his case, first with a description of a number of the Saarinen buildings around the world, many of them renowned. Then Hewitt posed a question to the board that seems in retrospect to epitomize more than any other single statement or act the direction in which he wanted to take the company. His query: "Is he too 'fancy' for us?" His answer, "No."

Part of the reasoning behind Hewitt's rhetorical question was the expenditure itself. An outside architect of the talent and reputation of Saarinen was not going to be inexpensive. But that was not the real issue, for Hewitt was asking the board to enlarge its concept of the company—to see itself as a major, nationally important entity, on its way to becoming the preeminent firm in the industry. In effect, a statement was to be made by the company, one that Hewitt hoped would reverberate through the organization at all levels and out among the dealers, the customers, and the general public.

It is probably not excessive to say that at this meeting Hewitt turned this insular, Midwestern farm implement company in a new direction, pointing it toward a new role as one of the world's best-known multinational corporations. For that is how the Deere Administrative Center did turn out—a superb Saarinen building that came to be known architecturally as one of the finest corporate administrative centers in the world. Most important, though, was its effect on the organization, which would not really be known until well after the completion of the building, all too many years into the future. At this point, it was only an article of faith for Hewitt, who had truly put his personal credibility on the line in a way that was almost irrevocable, once initiated.

The appropriation was authorized by the board in January 1957, with only grudging acceptance from certain board members. Of the senior triumvirate of Kennedy, Murphy, and Cook, only Cook was an enthusiastic booster. Burton Peek already had resigned his post as chairman of the board of directors a few months earlier; Peek had been very supportive of Hewitt ever since the latter had arrived at the general company and he gave a grudging blessing to the new building. Yet Peek could not be truly reconciled to leaving the old offices along the river, where he had gone for so many years. Peek often told a story of a hunting trip he had taken many years earlier in northwestern Nebraska, where Peek's personal hunting companion was General John J. Pershing. During the day of the hunt, Peek and Pershing had come upon a large herd of buffalo in a field on one side of the road, and across the road in another field a single buffalo, "a ruff, mangy, ugly looking old bull." Peek inquired of the local farmer why the single animal was in the field. The farmer replied, "You know, originally when the herd was smaller, the field was smaller and all of it lay on this one-third side. Later, as we increased the herd, we had to increase the field, so we acquired this two-thirds on the other side. We drove the herd over there—all except this one. Try as we would, we couldn't get him to go, we had to leave him where he was, and there he has stayed to this day." Peek now publicly applied this story to the office change: "Someday, if any of you encounter a ruff, mangy ugly looking old buffalo around the present site of Deere & Company's office, please deal with him gently, because he may be Burton F. Peek, reincarnated."

The architectural integrity of the new structure was to be Saarinen's—that much was evident—though Hewitt worked personally with Saarinen (with the help of Nathan Lesser, the head of the engineering department). At the start, Hewitt laid out a comprehensive "letter of intent" for the architect, making clear that the building design "will be in harmony with our functions and additions, and also be indicative of the objectives and progress that we envision for the future." Hewitt characterized the men who had built the company as "rugged, honest, and close to the soil," men who had always put a central emphasis on "quality of product and integrity in relationships with other persons." Inasmuch as "the farmer wants and needs the most efficient and durable tractors and implements," therefore "we also want and need a headquarters building that will utilize the newest and best architectural and engineering concepts." Hewitt concluded, "The several buildings should be thoroughly modern in concept but should not give the effect of being especially sophisticated or glossy. Instead, they should be more 'down to earth' and rugged."

Hewitt spoke here to the concern that several of the more conservative board members had enunciated, that the farmer customers in particular would look upon the building as pretentious and somehow too "urban." Hewitt's instincts were that just the opposite would occur, and he was heartened in this by receiving an articulate anonymous letter from an employee shortly after the building plans were made public. The latter said, in part, "Some persons condemn as foolish the plans as announced; I have heard the opinion advanced that farmers will hold our 'extravagance' against us—raise bitter complaints about prices, etc. Personally, I believe that these farmers will be in a small minority; there would be more danger—more progress to be lost, both for us and our customers—in conforming exactly to the notions these customers might have as to what our Company's character and direction should be. We are in a position where we cannot only manufacture quality equipment—but provide leadership in farming and farmers' thinking." The employee concluded, "There's no longer any reason to hide the fact that we are a large company, alive and virile. . . . Our obvious enthusiasm and faith in the future will tend to attract better personnel to our Company." No one really knew which of these points of view would tend to dominate; only when the building was up and occupied would these subtle nuances of employee and public reaction become clearer.

The design of the new office building and the assignment of offices also concerned the board. These matters were not just part of an inconsequential decision about office layouts and locations of file cabinets, but were seen as clues to the recommendations of the parallel study of the management structure of the company being conducted by the outside consultant. For this reason, the board gave Cook the charge to rationalize the building concept with the management concept, and for most of the remainder of the Deere Administrative Center project, Cook worked closely with Hewitt in fulfilling this assignment. Buildings always take time, though—sometimes more time than the purchasers feel necessary—and the Deere Administrative Center was not to be occupied until more than eight years after Hewitt had first visited Saarinen outside of Detroit in March 1956.[7]

A NEW GENERATION OF PRODUCTS

Innovation burgeoned in the agricultural machinery industry in the 1950s. In tractors, the big news was greater horsepower. "In 1950, few wheel-type tractors exceeded 50 maximum belt horsepower, and most farm tractors were considerably below that size," wrote Lester Larsen in his book on the tractor history of this period. But the year 1950 heralded what was soon to be an all-out "horsepower race" (though few engineers designing tractors at that time would have dreamed that horsepower would reach the levels that today's tractors have achieved).

Exhibit 13-4. *Trend in Field Speeds (miles per hour)*

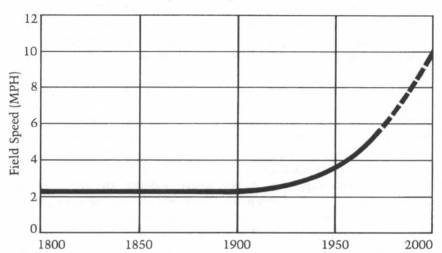

Source: Lester Larsen, Farm Tractors, 1950–1975, *(St. Joseph, MI: American Society of Agricultural Engineers,* 1971), 4.

Tractor field speeds turned up sharply during this period, from about 2.5 to 5.5 miles an hour (exhibit 13-4). Tractor transmissions also were going through a marked change, necessarily so with the heightened motive power of these larger machines. In 1954 International Harvester introduced a torque amplifier (TA), which allowed a regular five-speed transmission to become a ten-speed version (when in TA, the travel speed was reduced about 32 percent and the pulling ability increased about 48 percent). In 1957 J. I. Case introduced a torque converter that replaced the conventional clutch. When it was locked in, the tractor had an exceptionally good ability to start with a heavy load and, after gaining requisite operating speed, the torque converter could be locked out. The converter could be reengaged at any time to provide as much as a 100 percent increase in pull over that of direct drive, thus allowing the tractor to pass more easily through rough spots in the field.

Though distillates had continued to be a popular and economical fuel for many years, by the mid-1950s tractors burning distillate fuels were being phased out (the last distillate tractor tested by the Nebraska Tractor Testing Laboratory was the John Deere 720 All-Fuel tractor, in 1956). The most striking feature of the evolution of fuels, though, was the rapid rise in diesel fuel, a particularly strong trend after 1958. Efforts to provide more operator comfort were also widespread in the 1950s. Somehow, safety considerations had progressed at a slower pace. (Larsen, discussing the experiments with the mercury ignition switch, which would automatically shut off when the tractor reached a certain degree of slope, commented, "Most farmers didn't want to be bothered with such devices, which they termed 'foolishness,' and usually removed them after a short trial period.") It was not until 1966 that the first tractors became available with devices to protect the operator with a roll-over protective structure (ROPS). The pioneer was Deere, with its "Roll-Gard."

Deere's decision to move from the two-cylinder tractor to a four-cylinder model was taken in 1953, after years of arguing within the company, much push-and-tug, and probably a bit too much delay. But it took seven years more for the design engineers, the plant and production engineers, and the many others who had been laboring to bring into being a "New Generation of Power." Meanwhile, the company had to adapt and respond with its existing models to the onrush of new developments that other companies were introducing. Deere's popular 40-50-60-70-80 models that had served the company well in the early 1950s were all upgraded in 1956 to the so-called 20 series. These were successful models, with many refinements in hydraulics, fuel systems, and an engine design that allowed several models to establish new fuel economy records in the ten-hour runs at the Nebraska tractor test course. Operator comfort and convenience had been upgraded, and styling was more attractive (these models being the first to wear two-tone paint on the body, the combination yellow and green that has characterized the recent period). The 30 series that was introduced in 1958 had a number of cosmetic, face-lifting changes, one of which, new fenders for the protection of the operator, added considerable safety margins. Certain of these models were particularly popular, with more than 27,000 of the 720, 50-horsepower, tractor (PTO) selling in 1956–1958 and more than 29,000 of its counterpart 830, 57-horsepower, model in the following three years. Deere also startled the whole industry in 1959 with a four-wheel-drive tractor with more than 200 horsepower, introduced at a John Deere field day in Marshalltown, Iowa. To say that the enormous Model 8010 awed its onlookers is an understatement. It stood some 8 feet 2-1/2 inches tall, was 8 feet wide and more than 19 feet long, and weighed 19,700 pounds (actually, 24,860 pounds with its tires full of ballast). Its twenty-four-volt electrical system provided fast starting in any weather and its articulated power steering, and power brakes were regular equipment. It could pull a thirty-one-foot harrow or an eight-bottom plow at speeds up to seven miles an hour, readily lifting the integral plow off the ground for turns at the row ends. Only a few of the machines were built—the market in that particular period was not yet ready for tractors of this size. Within a decade, though, the situation would change, and tractors of this and greater horsepower range would become common.

In harvesting equipment, the company had held its position reasonably well during the 1950s. The combine corn head, first brought out in 1952, had been a great success almost immediately. The self-leveling hillside combine had also been well received. In general, though, Deere's harvesting equipment product development in the 1950s was directed toward minor improvements of existing, successful models. Not until 1963, with the Harvester Works' own "New Generation," did the harvesting machine change very much. Similarly, improvements and minor changes had been made all through the remainder of the company's products—the plowing, cultivating, and haying equipment, cotton pickers, and so on.

All thoughts of product development in the company would increasingly turn, though, to the change from the two-cylinder to the four-cylinder tractor, a story that came to a climax in Dallas, Texas, in August 1960. In what *Forbes* magazine called the company's "boldest venture of its 124-year career," Deere introduced its "New Generation" of tractors. The Texas event was a tour de force.

DEERE DAY IN DALLAS

Major model changes, always costly, can sometimes be devastating because of the time spent on conversion. The Ford Motor Company's overlong shift from the Model T to the Model A was a classic example of this. (All Ford's assembly plants, as well as its overseas factories, were completely closed for more than six months, and the company did not resume full-scale production until more than a year had elapsed; in the process Ford had irrevocably lost a startling amount of its market share.) Deere probably had spent too much time in the development stages of the four-cylinder tractor, but the conversion to its production at the Waterloo plant was spectacularly well done. The Waterloo shutdown lasted only about five months and the layoffs were staggered in such a way that as one group of employees was laid off another previously furloughed group would come back. By this time the harmonious results of the 1955–1956 contract negotiations with the United Automobile Workers had produced an era of close working relations, and employee-relations tensions during the changeover seemed at a minimum.

Downtime for changeover is typically reflected in production time lost on existing models, and the results of 1960 showed this. Sales dropped from 1959's $577.1 million to $511.7 million in 1960, and the net income took a sharp decline from $50.9 million to $20.3 million. Heavy outlays for the machinery and tooling for the new tractors had pushed total company capital expenditures to approximately $40 million in 1960—the net investment of the company, according to the annual report of 1960, jumped from $98 million in 1959 to almost $120 million in 1960. *Forbes* caught the essence of

CHAPTER 13

Exhibit 13-5. "Deere Day in Dallas," August 30, 1960. Top, Entrance to the Dallas Coliseum; bottom, William Hewitt and Stanley Marcus introduce the "New Generation of Tractors" at the Neiman-Marcus store. *Deere Archives*

the situation in its article on the new tractors: "As hindsight proved, Deere could hardly have picked a better year than 1960 to make the changeover.... The shutdown coincided with a marked fall off in domestic farm equipment sales as farmers felt the pinch of lower income in 1959 and early 1960.... Deere did not miss out on a potential bumper year. In fact, its indifferent profit showing was matched last year by competitors who had no such unusual expenses as Deere incurred." In other words, fortunate timing mitigated what might have been a far greater effect. The efficacy of the change still had to be proven by the success in the new models, but it was to this issue that Hewitt and his colleagues now turned with gusto.

"All day long the big planes buzzed in and out of Dallas' Love Field. They carried passengers from New York and New Dorp (PA), from Paris, France, and Paris, Ill., from Seattle and Sewanee (TN). When darkness fell on Monday, August 29, 1960, more than 6,000 passengers—the biggest industrial airlift in history—had been safely landed." Thus began the *Forbes* article on "Deere Day in Dallas." The next day's events opened with a preview of the entire line, enthusiastically received by the dealers as the tractors stood in a row in a huge open display area in the parking lot of the Cotton Bowl, near the Coliseum, on the outskirts of the central city. Perhaps the most spectacular single event of that day came at the stroke of noon, in downtown Dallas itself. Inside the prestigious department store, Neiman-Marcus, right next to its exclusive jewelry counter, stood a huge, twenty-foot box. Stanley Marcus, the ebullient impressario and president of the firm, walked up to the box and tugged away its wrappings. *Forbes* described the results: "Its contents: a rakish-looking, grass-green, farm tractor. From its sides myriad diamonds (hastily affixed with Scotch tape by willing [Nieman-Marcus] salespeople) twinkled the name of its maker: John Deere." A diamond coronet was even affixed to the tractor's exhaust pipe. Only slightly less spectacular were the fireworks that evening, the Al Hirt band that had been flown in from New Orleans, the Texas barbecue, and other hoopla. Laced through all of this were hard-sell speeches and informal contacts, with Hewitt himself personally acting as master of ceremonies for the evening events and C. R. Carlson, vice president of marketing, leading the Deere marketing contingent.

The choice of Dallas itself was noteworthy. Previously the unveiling of new tractor models always had been done in Waterloo (or Dubuque, if built there). They were in the "heart of the Midwest," logical places for the farmer customer; Dallas was not. But it was the *dealer* who was involved in the initial viewings, and dealers could be brought from all over the country to anywhere. Dallas was just a dramatic and offbeat enough choice to whet their appetites. (In more recent years, Deere has made extensive use of dealer "fly ins," shuttling them from all over the country to Moline in Deere-owned and commercial-charter aircraft).

The proof had to be the models themselves, and their enthusiastic acceptance was quick in coming. Over the next crop year their field results were excellent, and though a drought slowed down sales during the last half of 1961, the North American agricultural machinery results were up considerably in that year and the total sales for the company rose to $561 million (with income up to $36 million).

There were four models in the new line—the 1010, rated at just under 36 PTO horsepower; the 2010, at 46.6 horsepower; the 3010, at 59 horsepower; and the 4010, at 84 horsepower. The three larger models had Syncro-Range transmission, which provided eight forward and three reverse speeds, with synchronizers in the transmission permitting shifting on the go within a range between forward speeds or shuttle-shifting between forward and reverse. The closed-center hydraulic system of the three larger models provided up to three independent "live" hydraulic circuits to serve a rear rockshaft and one or two remote cylinders. The models were equipped with power steering, and the largest two models had hydraulic power brakes. All four tractors had high ratios of horsepower to weight, allowing operation with equipment loads at higher speeds, reduced lugging strain on the engine, and increased efficiency. All four were available with gasoline, diesel, or LP gasoline engines.

Operator comfort had never been a high priority with tractor manufacturers, but in the new line Deere pioneered, with the help of a perceptive outside consultant, Dr. Janet Travell (a posture specialist who was later White House physician for President John F. Kennedy), an orthopedically sound "comfort" tractor seat was developed. The enclosed Roll-Gard cab was introduced in the early 1970s.

The two largest models were relatively the most successful, some 45,000 of the 3010 being sold in the four years 1960-63 and more than 40,000 of the 4010 in just the three years 1960-62. Their successors, the 3020 (65 PTO horsepower) and the 4020 (91 PTO horsepower), developed in 1963, were even more spectacularly received. In the period 1963–1971, more than 86,000 of the former were sold and in the same period some 177,000 of the 4020 were made and sold. The latter was far and away the most widely sold single model of tractor ever built by the company.

Deere had dubbed the new tractors a "New Generation of Power" and *Forbes* thought this "somewhat grandiose." Yet, in terms of the models' eventual success and Hewitt's efforts to establish irrevocably a new company image, the pretentious name for the model group and the hoopla surrounding the Day in Dallas seemed to justify the superlatives. *Forbes* called the new models "symbols of a livelier, more dynamic company" and attributed this in part to Hewitt himself. "He has transformed the eminently successful but rather shy and conservative company," they concluded.

But it is never the chief executive alone. *Forbes* added: "Part of Deere's success, too, unquestionably is due to its long and careful study of the farmer, for the farmer's lot and Deere's are intertwined in a way which broader-based companies are not. Deere men pride themselves on knowing the farmer's wants. Not least of all, they accord him the respect which city dwellers do not always give him." (*Forbes* and other national media often enjoyed stereotyping Deere employees as "country bumpkins," almost as if all of them, top management included, came to work in overalls, pulling the straw out of their hip pockets; they came close to implying that Moline rolled up its sidewalks at night. Such remarks seemed gratuitous to the Deere group, particularly so after the emergence of the company as an international power. "Farmers are no longer hicks, and neither are we," asserted Hewitt to a *Forbes* editor in 1963.)

The company, incidentally, did continue to receive complaints from farmers who were diehards about the two-cylinder tractor; finally, C. R. Carlson developed a thoughtful, four-page form letter that explained the reasons for the change (probably even this not satisfying some!).

The outstanding success of the New Generation tractors fueled Deere's rise to industry leadership. The company's share of the wheel tractors sold in the United States stood at 23 percent in 1959; by 1964 it was up to 34 percent—more than one-third of all the tractors sold in the country. Even more important, in 1963 Deere passed International Harvester for the first time in history in total sales of farm and light industrial equipment. Deere's $762 million sales exceeded Harvester's $665.4 million. (The net sales of Massey-Ferguson were $636.1 million and of Allis-Chalmers were $543.9 million.) In 1963, Deere was also ahead of its farm machinery competitors in net income per dollar of sales and net income per dollar of total assets (only Caterpillar, mostly in heavy construction, scored above Deere in these two key measures.) Under Hewitt, Deere never relinquished its position as the world's largest producer of farm machinery. Hewitt's euphoric plan to be "No. 1" in the industry had come true; more important, over the next two decades it stayed true.[8]

MISCALCULATIONS IN GERMANY

Euphoria in Moline helped ease memories of depressing results in Europe. James Wormley, appointed president of Heinrich Lanz at the first board meeting in December 1956 and dispatched immediately to Mannheim to head the organization, was the only major top management Deere representative on the scene. Deere had taken a calculated risk to leave in place the entire management structure of the predecessor German organization. The decision could be viewed as naive, as if it had been said, "Let's buy this German company, change the color of the paint, call it John Deere-Lanz, and begin selling tractors in Europe." One could argue, alternatively, that there was wisdom in moving slowly, not uprooting an organization too rapidly. With probably a bit of both of these rationales in mind, Wormley left for Europe. He returned in July 1957 to apprise the Deere board of just what the company had really purchased.

Progress had been made in the months that ensued, though his central message was not reassuring. On January 1, 1957, Lanz had a total of 8,060 employees; by the end of July, total employment had been reduced to approximately 7,400 and Wormley expected year-end employment to be around 6,700. Still, the company was patently overstaffed. The long history of Lanz employee benefits and emoluments had certainly built employee loyalty—the number of second, third, and even fourth generation Lanz employees rivaled that of Deere. Yet there seemed to have been a slackening of employee productivity ever since World War II; Wormley estimated that the output per employee at Lanz was only about DM 18,000 per year; other similar German firms had outputs of DM 25,000 to DM 30,000 per employee for the same period.

The product line, Wormley observed, had been too broad at the time Deere bought the company; Lanz had been producing tractors with 12, 13, 16, 18, 20, 24, 28, 32, 36, 40, 50, and 60 horsepower. Only some of these sold in adequate enough quantities to permit an economic scale of production. Wormley immediately discontinued the 13-, 32-, and 36-horsepower tractors. All of the remainder, with one exception, were various sizes of the Bulldog—all one-cylinder, two-cycle diesel engines. The exception, an 18-horsepower model, was the Alldog (the forward tool carrier tractor), with a two-cylinder, four-cycle diesel engine (and a purchased motor of dubious distinction, as related earlier). Lanz was also making 4-foot and 6.4-foot pull-type combines and 6.3-foot and 8.5-foot self-propelled types. Wormley immediately discontinued the pull-types because there simply was no market for them. The implement line was composed of fourteen types of machines, including mowers, tedders, hay rakes, side delivery rakes, and potato and sugar-beet diggers. For the moment, Wormley decided to leave those in place.

There were other difficulties, too. For example, there was no central price-making authority; dealers did not know what Lanz would charge them. Indeed, record keeping as a whole was in a shambles. The company even had lent money to employees for personal expenditures without setting up a proper collection system.

Some of the problems hanging over from the earlier Lanz period could be readily ameliorated. There was, for example, a seven-year accumulation of chronic field complaints; to the chagrin of Deere, "previously such complaints were completely ignored" (as Lloyd Kennedy reported after a visit in October 1957). Wormley immediately appointed a new marketing director and insisted that field personnel follow up on all complaints. This paid off—whereas there had been complaints concerning 52 percent of all tractors sold, Wormley was able to reduce them to about 7 percent by April 1958.

The problem of dealerships was more intractable. Lanz's fading reputation in the field had begun to damage the distributorship structure. Many who had formerly sold mostly the Lanz name were still willing to sell the harvesting equipment and the tillage tools, but were less enthusiastic about the Bulldog. In truth, just as Deere tenaciously had held onto the two-cylinder tractor beyond the optimum changeover point, so too had Lanz stayed with a simple, traditional model of tractor at a time when other companies were marketing more attractive, more effective multicylinder versions. It seems incongruous that Deere had purchased a company with a single-cylinder engine at just the point when it was going through the lengthy, costly, and disturbing shiftover to the four-cylinder tractor. There were some differences, too—Deere was still able to sell its two-cylinder tractors very effectively in the marketplace, while Lanz was really heading downhill on its sales curve well before the purchase by Deere.

Deere's first three years as majority owner of Lanz were discouraging. The firm sold some $35 million in equipment in 1956, at a loss of $767,000; the sales fell to $33 million in 1957, with a loss of $1.5 million, and similar sales in 1958 produced a loss of $1.2 million. By the end of 1958 the cumulative deficit was $3.5 million. A more rigorous and effective sales program in 1959, while producing only $32.6 million in total sales, did result in a break-even year. By this time, the company had upgraded the outmoded Bulldog product line cosmetically to gain more appeal, at the same time selling the new Deere concept to the dealers and ultimately to the buyers. Henry Dreyfuss had been brought into the picture early, making a visit to the Mannheim plant in 1957, and his simple styling changes added greatly to the appearance of the Bulldog. The Lanz blue paint had been changed to Deere's green and yellow. Even here, though, the Mannheim group designed the paint patterns in a rather complicated, showy fashion; Hewitt wrote Dreyfuss, "The whole darn thing just looks too busy—it looks like a Persian carpet." Dreyfuss immediately suggested new painting combinations and the result was quite striking.

Still, it was painfully evident that Lanz needed new, modern tractors; Deere executives had seriously underestimated the sag in the Bulldog's reputation. Faced with a major engineering and manufacturing challenge, the company was forced into a crash program. The effort began in 1957; by late 1959 the models were already in prototype form. In retrospect, this foreshortened development path probably pressured both the North American and German engineers in the company into a set of too-rapid decisions. Compromises were made in attempting to meld American and European field needs. In essence, the Waterloo and Dubuque engineers working on the problem made the decision that the smaller Dubuque tractor was the prototype for Germany, and they proceeded to adapt this model for manufacture in Mannheim (with the engine itself built in Dubuque and shipped to Mannheim for assembly).

There was a serious flaw in this decision, however. The Dubuque tractors were sold in the United States and Canada as utility tractors, not designed in

Exhibit 13-6. Henry Dreyfuss, renowned industrial designer, who was instrumental in the styling and design of Deere equipment for more than three decades, 1966. *Deere Archives*

terms of power and ruggedness to serve a multipurpose set of needs. It was true that the German farmer wanted a tractor of this horsepower size, but the uses to which the German farmer put his tractor were strikingly different from those of his North American counterpart. He wanted a tractor that could do heavy plowing and other arduous field work, but also wanted to use it for front loading work, for transport and hauling, even for over-the-road travel. In sum, the German farmer wanted a small tractor, about the size of the utility tractors from Dubuque, but one that was built more ruggedly and with more wide-ranging capacities.

The Waterloo and Dubuque engineers missed the point. Indeed, this is just one example—certainly a serious one—of many misreadings of the European agricultural scene by the American contingent. There was insularity in the Deere views all across the board. Hewitt had sent Maurice Fraher to Europe "to sit at a sidewalk cafe" for just this reason. The learning curve to truly understand and be empathic with another culture is a more upward-sloping line than the Deere people assumed. There was substantial chauvinism, and in the case of the engineers more than a bit of arrogance in dealing with the European. The Lanz engineering group did have rigid ways of structuring itself and dealing with problems, and these seemed to the Americans to be too hidebound. On the other hand, the European engineers were well trained, intelligent, and used to making their own decisions. The fact that the Americans called the shots often irritated and riled the German engineers. Thus there was major difficulty at the start in melding the two engineering groups.

The two-culture problem evidenced itself throughout the organization. Lanz employees had some deeply embedded stereotypes about American business firms and were uneasy about being taken over by one. The German worker believed that "Americans hire and fire without asking"—that American business firms follow boom-and-bust employment patterns. Wormley's necessary reductions in the force had mostly been done by attrition—many employees were leaving Lanz for other employment in the Mannheim area, where there were many excellent jobs available, indeed a labor shortage, at this time. Still, Deere had to get over an initial skepticism by the German workers that Deere's personnel policies were less humane than those of their predecessor. In reality, Deere's policies were, if anything, more employee-oriented than those of the German company—the reputation that had been established by John and Charles Deere in the nineteenth century still held strictly at Deere & Company. This was a fact that could only be fully appreciated over time, though, and at the start Deere-Lanz employee relations were ragged. The infamous "beer strike" was a case in point.

Just about all German companies at this time allowed the sale of beer in their factories, to be quaffed at breaks and at lunchtime. At Lanz, this had been elevated to a science—there were four different locations around the factory where beer trucks would back up each day and deliver fresh kegs; designated employees were then deputed to bring back large mugs of beer. This was done in combination with some rather loose practices relating to the breaks themselves. For example, one of the workers would take fifteen or twenty minutes off to go to the keg for his beer, his assistant standing by. Then, sometimes, the assistant would go to get his beer, and the worker was not fully served. In sum, to the Deere executives at Mannheim the system seemed woefully loose and potentially dangerous, given the mixing of alcohol and high-speed machinery.

When Harry Pence came to Mannheim to succeed Wormley in 1960, he decided to tackle this system and in March announced a fixed breaktime, to be taken by all personnel at the same time and to be precisely fifteen minutes in length. Concurrently, he announced that only small beer bottles would be made available at the plant, not the generous steins of the past. One needs little imagination to guess what happened. Employees instituted a strike and demanded that the workers council take the matter before the Mannheim labor court for arbitration. The dispute hit the public press: "Tough Harry . . . would be well advised to pull his sleeves down," one editorialist vowed. "German workers dislike the sober and speedy mechanisms of American hustling," said another. Representatives of the Social Democratic Party at Mannheim quickly passed a resolution demanding that "American management should adapt itself to the conditions and laws of the host country." Even the mayor signed the resolution. Fortunately, the strike lasted only a few days—it was just at the time when the new tractors that had been developed out of Dubuque were ready to be manufactured, and a delay at this point would have been disastrous. Pence was able to tighten the slippage in the breaks, though his plan for fixed times had to be modified. Further, the beer remained—and has up to the present. The imbroglio was a good example of the need to respect cultural differences. Pence started with an assumption that productivity still lagged at the plant, and that industrial discipline was badly needed if Lanz was ever to compete with its aggressive competitors. But taking a man's beer away was too much!

Across the board, Deere's North American management now became more involved in Lanz. By 1959 Curtis told the board: "Earlier it was thought that the best policy was to interfere as little as possible with Lanz management. We have now determined that following this policy was not fruitful... we are now sending more American personnel to assist and advise Lanz. We shall continue following this policy of increased help in an effort to inaugurate so far as possible many of the policies and practices which have proved successful in our North American operations." In the same year, Deere management persuaded some of the German top managers to step down from their positions. This created uneasiness among the employees as they saw colleagues who had been in charge leave without much of an explanation, but this concern dissipated over time. Subtly over the period 1959–1961 the Lanz organization shifted more into the mode of a Deere subsidiary, rather than that of an autonomous German company that just happened to be owned by some Americans. In December 1959, at the annual shareholders meeting of Lanz, the name of the company was changed from Heinrich Lanz to John Deere-Lanz A. G.

In January 1960 the two new wholly Deere tractors were ready to be shown in Germany (prematurely, as we noted above). As the North American marketers had done before with the changeover from the two-cylinder to the New Generation tractors, the German marketing group now sold out the Bulldog line. The last Bulldog came off the production line in this year; it was number 219,253. The Bulldog had had an unbroken period of production from 1921 to 1960, one of the longest active model runs of any tractor company anywhere in the world. Thus in one year, 1960, Deere said farewell to two historic tractors—the famous North American two-cylinder "Poppin' Johnnie" and the ubiquitous Lanz Bulldog.

The two new John Deere–Lanz tractors were powered by four-cylinder diesel engines—the 300, a 28-horsepower model, and the 500, at 36 horsepower. With Dubuque engines, the remainder of the tractors (including the transmission) was manufactured in Mannheim. Within two years of the introduction of the two basic models, the company also began assembling the model 1010 crawler tractor, with a major portion of its components supplied from Dubuque. A larger model 3010—a 65-horsepower, four-cylinder diesel tractor from the United States—was brought over from Waterloo and assembled in Mannheim, giving Lanz a large-horsepower model to sell in Western Europe.

The exteriors of the two tractors were quite appealing—there were substantial similarities with the North American New Generation line. Most of the basic engineering design was well conceived, too. Still, the tractors had numbers of problems, stemming from the inappropriately wholesale adaptation from the utility tractors made in Dubuque. As horsepower was downgraded from the 40–50 horsepower range, the engineers made compromises that precluded later flexibility (although one always hopes to have a basic engineering design that can be downgraded and upgraded without redesign of certain key dimensions, in this case, many basic dimensions had been locked in). The engines themselves were an older design, the so-called "sleeve and deck" arrangement. This was an outmoded concept, and later the engines had to be changed to a new configuration. There were also transmission and hydraulics problems, along with minor difficulties with the brakes and the fenders. Most of these shortcomings were correctable, though the basic design problem continued to plague the engineers as they both downsized to a two-cylinder, 18-horsepower tractor and upsized to a four-cylinder, 50-horsepower tractor in 1962. The 100-300-500-700 models were certainly a boost for Lanz, for they phased out a Bulldog that had reached the end of its productive career. Still, the new models had serious problems that kept Lanz from putting its best face forward to a market that was extremely competitive in performance and price. Massey-Ferguson and Ford tractors continued to sell at prices below Lanz list prices, and continuing cutthroat competition characterizing the German market in this period made it doubly difficult to sell the more costly Lanz tractors, relatively expensive in their weight-to-horsepower ratio. Deere's German company remained a sick operation.[9]

DEERE MOVES INTO FRANCE

Hard as it was to look above the line of fire in Germany, company personnel could not help but be aware of several major, multicountry efforts occurring in the European environment. The most important, by all odds, was the institution of the European Economic Community (EEC). Its six original members—France, West Germany, Italy, Belgium, the Netherlands, and Luxembourg—had consummated the Treaty of Rome in 1957, and by the early 1960s they had effected an ambitious effort to link their economic policies through the device of a Common Market. Another seminal effort in 1961, the Organization for Economic Cooperation and Development (OECD) brought together the same six Common Market countries with other European countries, as well as the United States and Canada, again with the objectives of expanding trade and economic interrelationships and providing aid to less developed countries. The message to Deere was clear—there was an opportunity, and a need, to get more substantially involved in other European environments, beyond just West Germany. During this period in the late 1950s and early '60s, Western European agriculture had begun to expand; agricultural machinery potentials were attractive indeed. In particular, the use of tractors had accelerated in a major way (exhibit 13-7).

The formation of the EEC had great portent for companies all through its member countries. In France, three small agricultural machinery companies had become concerned about surviving in the new environment. Remy & Fils manufactured hay tools, disc rakes, cultivating equipment, and forage harvesters in Senoches; Ets. R. Rousseau manufactured low-density balers and threshing machinery at Fleury-les-Aubrais, near Orléans; Thiebaud Bourguignonne also manufactured low-density balers (their smallest version, the "Lady Bug," had achieved great success), as well as grain separators, rakes, and mowers at Arc-les-Gray. All three felt vulnerable because of their small size, lack of anything approaching a "full line," and fears about increased competition from the other EEC countries.

In 1958 the three small French firms formed a fourth company, Compagnie Continentale de Motoculture (CCM), with its central purpose the daunting task of developing a tractor. In effect, the three companies hoped in one fell swoop to jointly become a full-line company. In 1959,

Exhibit 13-7. *Number of Tractors on Farms by EEC and by Country, 1951 and 1962*

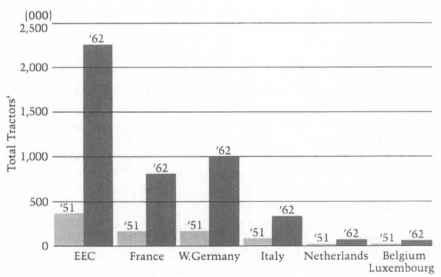

ᵃExcludes garden tractors.
Source: FAO Production Yearbook, *1963, United Nations, vol. 17, 1964 Rome.*

almost by chance, Deere made contact with one of the three companies' plants and learned of CCM.

Discussions soon led to Deere joining with CCM; the company would buy a 51 percent controlling interest in CCM and would also purchase small blocks of stock in each of the three companies, somewhere in a 15–20 percent range. A sum of $1 million was appropriated for the interest in CCM, and an additional $1.2 million was to be split three ways to purchase the interests in the three companies. All three companies agreed to adopt the John Deere green and yellow colors; at the start, each manufacturer was to maintain its own trade name. There already were some overlaps among the three companies' products; now Deere was required to further mesh the three sets of products with its own North American products, particularly those from Ottumwa (where Deere's North American mowing equipment, balers, and other related products were made).

The heart of Deere's interest in CCM was the tractor proposal. CCM engineers had already sketched preliminary ideas about how a tractor might be made, but these were far short of what was necessary to make a viable tractor. Deere's agenda was not to build a complete tractor, but, rather, a tractor engine. If this could be accomplished, the engines could be sent across the border to the Lanz factory at Mannheim, replacing imported engines from North America. This notion was especially attractive because the EEC tariff regulations would soon make importation from the United States relatively more costly. Therefore, the initial appropriation of the board for the purchase of CCM, Remy, Rousseau, and Thiebaud stock also included $4.2 million for "establishing or purchasing a tractor plant and providing necessary tools."

Perhaps it was the sobering experience at Lanz in buying a shaky company with multitudes of problems, perhaps it was the fact that there were no readily available engine companies to purchase. At any rate, Deere management assumed right from the start that in this case they would begin with a "green field." A land site was found near Orléans (in its suburb of Saran), and by 1962 a building was underway. When it was finished in 1965, the company began making three- and four-cylinder diesel engines for the German operation.

Two of the three companies in CCM did very well under Deere partnership. Remy in Senoches continued its success in forage harvesters and tillage equipment, while Thiebaud at Arc-les-Gray made all of the French-made balers and prospered with its excellent mower-conditioner, a model more suited to the natural grasses of Europe than the Ottumwa machines (the latter more widely used on legumes). Rousseau had lost its balers to Thiebaud when CCM had been founded; its product line of threshing equipment had not done well and the financial drain from an overly ambitious plant expansion in 1957 had put a severe strain on the company. In 1962 Deere increased its ownership to about 80 percent of the stock of Rousseau; the former managing director resigned and Deere began assembling some of its industrial equipment there, the components coming from Dubuque. (The French requirements for this type of machine were somewhat different than the North American standards, and the equipment had to have special French engineering built into it).[10]

MEXICO:
LOCAL CONTENT, LOCAL OWNERSHIP

The Mexican operation had begun as "one of the bright spots in overseas operations," said Curtis in his report to the board in 1959. The new factory building in Monterrey opened early that year; it was given an auspicious, fast start because the company had already been assembling the Model 630, 730, and 830 tractors, together with disc and rotary plows, tool carriers, and harrows in rented quarters. Sales stood at $1.5 million in 1956 and by 1960 they were up to $4.5 million. The Mexican government had promised to protect this new domestic industry by restricting imports from companies outside the country. Costs were higher than in the United States because of smaller production runs and lower employee productivity. Nevertheless, a break-even was expected for the year.

The euphoria of those first four years of operations soon was tempered, as the Mexican government began to reveal, in a rather convoluted way, its long-term plans for the Mexican tractor industry. Its goals appeared to be threefold: (1) to increase the Mexican content of all international companies in the country; (2) to encourage substantial—indeed, majority—participation of Mexican capital in them; and (3) to keep the price of a tractor manufactured in Mexico precisely equal to the cost of an imported tractor, were it to be allowed in the country. At this time (1961), four international companies were selling into Mexico in some quantity—Deere, International Harvester, Massey-Ferguson, and Ford (only Deere, though, had an assembly operation in Mexico itself). All three of these Mexican goals troubled the international companies. They saw Mexico's tractor market as a limited one, with low-quantity production runs needing to be spread over a wide range of horsepower sizes. Costs would therefore be high. Were there also to be rigid price control at an unrealistic level, profitability would be tenuous indeed. If control of the companies was exercised from within the country, pressures to satisfy short-term stockholder demands at the expense of longer-run growth might be exacerbated.

In the rumor-filled summer of 1961, the companies vied to ascertain the real agenda of the government. There were intimations that only two manufacturers were to be chosen—those with the "best" proposals—and that then the borders would be closed to any further entrants. The first candidates were two English companies, David Brown and Nuffield, but their plans were quickly rejected by the government. Deere, feeling under considerable pressure, finally submitted its own proposal in August 1962, an arrangement that would provide 40 percent local content by the fourth year and 60 percent content by the eighth year, all to be accomplished by a company that would have 40 percent Mexican ownership. Though Deere personnel waited with keen anticipation for an answer, none was forthcoming; they were left in limbo for months, their offer never acted upon by the government.

In November 1963 International Harvester submitted its proposal; whatever the particular chemistry involved, this time the offer was accepted. Once again, Deere submitted its own proposal and now it, too, was approved. Deere proposed to manufacture two models (tractors with 41 and 53 horsepower) and to increase its investment in manufacturing facilities from $1.5 million to $6.1 million (in the process, to add about 500 employees). The local content would reach 60 percent at the end of two years, rather than the eight years previously proposed. In regard to Mexican ownership, 40 percent of the company would be sold "to the Mexican investing public." Just who these investors would be was not clear. Deere pressed the government to reaffirm that the manufacturing program would include only two companies and that importation of all tractors up to 65 horsepower would be prohibited. Finally Deere exhorted the government to set tractor prices at such a level as to permit an adequate return on investment.

The government seemed to have approved these requests, so Deere proceeded to increase the plant size. Rising sales in 1964 and 1965 and modest profits in those two years seemed to justify the decision. Three unsettling developments soon occurred, however. First, imports of tractors by Massey-Ferguson and Ford continued to be allowed. As Deere estimated its unit costs to assemble in Mexico to be about 30 to 50 percent higher than in its United States plants, and as transportation costs were only about 10 percent and duties about 5 percent, the price squeeze was severe. (Deere continued to import its own larger tractors, and the differential worked in its favor there.) Complaints from Deere and International Harvester finally did succeed in shutting off most of the competing products, but the Mexican government more than once surprised them again with special barter arrangements with other international tractor manufacturers that allowed significant importations.

The second problem related to the "Mexicanization" of shareholders. As Deere contemplated ways to sell shares to the general public, it became apparent that the shares had little appeal. Not only were the financial results of the company still quite modest, but also the government required that only Mexican citizens or Mexican-controlled institutions be the purchasers. Sources of considerable capital actually in the country—other foreign nationals living permanently in Mexico, and so forth—were excluded by this constraint. Further, the free-wheeling Mexican investing community was skittish of most agricultural projects, given the power of farmers to exert political pressure on the government.

A third problem was even more serious. Just as Deere's manufacturing facilities were nearing completion, the administration of President Lopez Mateos was replaced by that of President Diaz Ordaz. Soon it was known that the latter had no intention of limiting tractor manufacturers to just the two that were already there; in mid-1965 the new administration announced that it was weighing approval of four additional manufacturers. George French, who now headed all overseas manufacturing for the company, pushed the board in mid-1965 to accelerate its efforts to sell shares to local investors: "The new administration does not feel committed to all the agreements of its predecessors. Nevertheless, if 40 percent of the shares were sold at the right time, we expect that the Mexican government would live up to its agreement for not less than two nor more than four years of its administration. . . . With strong support from Mexican shareowners, it is probable that the present limitation of two manufacturers can be maintained for a long time." Curtis came away from a meeting with one of Deere's local bank officers with much the same feeling: "His real recommendation was that we seriously consider 51 percent because this would go a long way toward buying peace with the Government—not only this Government but succeeding Governments."

Within weeks, though, the Mexican government undercut the French-Curtis thesis by approving, first, a Massey-Ferguson proposal and then a similar Ford offer. Despite this, the government still made it plain that it expected not 40 percent Mexican ownership, but 51 percent. Deere was under the gun—now it simply had to sell 51 percent of its outstanding stock. To its consternation, no investors stepped forward. Finally, 51 percent of the shares were trusted to one of Deere's lead banks there, the Banco Nacional de Mexico, S. A. ("Banamex"), which agreed to use its best efforts to sell at least half to the investing public. Despite the bank's promotional work, only 8 percent was placed and the bank was the trustee for the rest. At least technically, Deere was now "Mexicanized." Given the whimsical actions of the Mexican government, however, all of these developments were most unsettling to company officials. Fortunately, the operations did stay modestly profitable for the rest of the 1960s.[11]

ARGENTINA, SOUTH AFRICA, JAPAN

The old-line British house, Agar, Cross, which had been Deere's distributor in Argentina since the turn of the century, had become overstaffed and debt-ridden by the mid-1950s, with monumental arrears due from Argentine farmers. Yet the Argentine agricultural outlook was quite promising—the country seemed to be entering a period of stable economic growth, its Southern-hemisphere "green belt" one of the great agricultural areas of the world. It seemed a good time to part company with Agar, Cross, and to establish Deere's own bridgehead in the country.

This new Deere initiative included not only marketing but production; in September 1957 the board authorized $3.6 million for construction of a plant to assemble tractors in Rosario, with a capacity to produce 3,000 tractors per year for the Argentine farmer. William Klingberg, Deere's Mexican manager, was brought in to head the Argentine operation. Replicating Mexico, there now began a complex negotiation with the government about the percentage of local content of the manufactured products. Here, though, the definition itself was more complex. First, there were specified levels of local content, separately defined for each manufacturer. Next, there was a further specified list for imported components, some free of import duties, others carrying high tariffs. The Argentine authorities were adamant in insisting on precise explanations of each part's manufacture, an intrusion into the company's internal cost data that Deere officials found upsetting.

The first government decrees in 1959 seemed to allow about 55 percent importation, 45 percent local content. By 1961, the latter was up to 50 percent, a year later to 60 percent, and one year after that to 70 percent. By 1967, the local content required was 90 percent of the total product. A substantial plant addition was required in order to accomplish this; the company put it in place in 1964, and by 1965 Deere did indeed have its local content component up to 90 percent. (Later, in the 1970s, this went to 95 percent.) But there was an even more rigorous requirement: any part that could be made by an Argentine manufacturer was automatically prohibited from importation.

There were further aggravations. The company had had constant problems in clearing imported machine tools through the labyrinthine Argentine customs. Production costs in Argentina were much higher than those in the United States because the large fixed overhead was spread over too small a production run. Further, the company had to pay high prices for locally purchased materials. Finally, the local-content issue remained a central problem. As one of the Deere executives put it in 1966: "Forcing an industry such as ours to a 90 percent or higher local content is not always in the best interests of either the industrial economy or the customer. Undoubtedly, if the local content were reduced to somewhere between 50 and 75 percent, our costs would be greatly reduced and the selling price to the farmer customer would be lower." J. I. Case, a small importer and modest implement manufacturer, pulled out of the country at this time, apparently in part because of the local-content issue. Fiat, the other international company manufacturing in the country, stayed. Its investment was large, and its management was reluctant to jettison a substantial bridgehead, so it too lived with the draconian local-content percentage.

Deere's operation turned unprofitable by the early 1960s—there was considerable labor unrest at the plant, and the optimistic projections of economic stability had faded with renewed political unrest in the country.

CHAPTER 13

Deere was third in market share in the country for tractors (after Fiat and Deca S. A.), holding about 14 percent of the tractor business. But the bright promises held about Argentina as a country, and the equally optimistic hopes of Deere about its own operations there, had faded by the mid-1960s.[12]

Deere had been exporting farm machinery to South Africa since well before the turn of the century, but in the early 1960s its market share was still modest. In tractor sales, its 4 percent share was considerably behind those of Massey-Ferguson, International Harvester, and Ford (with only the latter assembling in the country at that time). There had been only a trickle of implement sales—South African duties on such equipment were substantial.

In 1962, a study team was sent out from Moline to look into the possibility of direct manufacture in the country (for both International Harvester and Massey-Ferguson had already decided to do so). Contact was made with a South African implement manufacturer, South African Cultivators (Proprietary) Limited, which had grown from a small farm implement repair shop into a company with an annual volume of about $1 million, making tillage tools and ground-engaging tools, such as disc blades and plow points (some for their own products, some sold as components to other South African manufacturers). The plant was located in Nigel, east of Johannesburg.

The South African company had expanded operations so rapidly that there now were pressing needs for working capital. A proposal was consummated for Deere to buy 75 percent of the stock of the organization, with the founder, Nils Gregerson, to stay as its chief executive. Just over $1 million was appropriated by the Deere board for the purchase. The new company became known as John Deere-Bobaas (Proprietary) Limited; later in the 1960s its name was changed to John Deere (Proprietary) Limited.

Deere took over the tillage lines, introducing its own products. In addition, some local assembly was done on smaller Deere tractors (the larger sizes still being imported as whole goods). For most of the 1960s the South African company limped along unprofitably, though market share for tractors rose to 6.5 percent in 1966 and 8.2 percent in 1967. The changeover to Deere equipment was not easy. As was often true of equipment placed in a different environment, substantial difficulties were encountered in trying to use American tillage equipment in the South African soils. The harsh, high-silica nature of the latter brought extra wear and some product failures.

Gregerson was a stubborn, production-oriented executive who sometimes overestimated his sales volume and built up too large an inventory. On the other hand, he and his staff had a perceptive ability to deal with South African labor. In the mid-1960s in South Africa the tension between blacks and whites had not yet become as high profile an issue as it later was. Curtis, in his presentation for the initial purchase, told the Deere board: "It appears unlikely that racial upheavals will occur in the foreseeable future—probably not for at least 10 years. This uncertainty, though important, has not deterred our competitors and major automotive firms from investing." A year later, though, he had increased misgivings and wrote the board: "We believe the racial problems in South Africa have tended to become worse rather than better since we decided to buy the interest in John Deere-Bobaas last July. The possibility of all the other independent nations in Africa banding together with the help of Russia to try to force the 3-1/2 million white people out of South Africa is a definite possibility within the next one or two decades." But another Moline staff member, in a month-long visit in late 1964, played down the issue. "There does not appear to be great concern on the part of the whites that the situation is explosive in nature. They realize they have a problem but are doing something about it."

The nagging question of just what kind of political and social environment really characterized South Africa became more important to Deere in the mid-1960s as the government began to make overtures about encouraging more domestic production. In the past the government had allowed importation of assembled tractors from abroad with only modest duties applied upon entry, and South African agriculture had become increasingly mechanized. Smaller tractors were most popular, owing to the ready availability of inexpensive labor. The initial purpose of the policy to encourage domestic output had apparently been to utilize more local labor. But it soon became clear that the dominant reason was South Africa's increasing preoccupation with self-sufficiency. Pressures from abroad concerning its apartheid policy were mounting by the year; a boycott by other countries that would shut off the flow of materials and products for the country was always a threat. The South African government was quite sophisticated about the implications of pushing "local content" too fast; a forced indigenous industrialization tended very often to result in much-increased prices to the local users, and the ruling officials were keenly attuned to the farmers. The Deere managing director in South Africa wrote Moline in 1967: "This country is determined to become self-sufficient just as quickly as possible for at least two reasons—one, the fear of trouble over South West Africa and the other over the Rhodesian situation. Some months ago, all industry was advised of the need to increase stocks of goods on hand—largely in the spare parts category. Our own Marketing Division was asked to increase their supplies of goods (spares) obtained outside this country and the government even offered to agree to allow the banks to extend lines of credit for their purchase if present lines were exhausted."

Thus, over the remainder of the 1960s, Deere and the other international companies had to grapple with the increasingly insistent proposals of the government to require more manufacturing within the country, with tractors as the central issue. As early as 1963, the government attempted to promulgate a tractor manufacturing protocol for the industry and solicited proposals from the four major international companies. Deere, Massey-Ferguson, and Ford presented a joint proposal in April 1963, after International Harvester had surprised everyone with an offer to manufacture a tractor, including engines, with 70 percent local content within five years. The joint proposal of the three other companies was for 50 percent content, to be balanced by modest tariff and subsidy provisions from the government. In the subsequent bargaining, the government backed off on its effort to pressure the companies, and the tractor manufacturing protocol stayed in suspension. The government raised the matter again in 1967 with a formal "Enquiry into the Tractor Industry," and at this point Deere made a second formal proposal (in March 1968) for local manufacturing, suggesting two alternative options—40 percent and 60 percent local content. Again the proposal was not acceptable to the government and the issue remained a moot one as Deere moved into the 1970s, still searching for a profitable and stable mode in the country.

In 1963 Deere also built a link to Asia through a licensing agreement with Hitachi, Ltd., one of Japan's largest manufacturing concerns. Hitachi would produce certain Deere farm and industrial tractors and equipment for the Japanese, South Korean, and Okinawan markets. Deere would furnish technical assistance and would also sell some machines and component parts to Hitachi. The bright promise of the collaboration did not come true. Hitachi had taken on agricultural tractors only at the insistent urging of the Japanese government, which was worried about the lack of mechanization on Japan's predominantly small farms. When pressure from the government eased, Hitachi lost interest—agriculture was not its prime target. Though Hitachi did sell some Deere industrial equipment, it seemed more and more that Hitachi was mostly interested in Deere's technology, not in selling its products. A desultory effort was made in 1969 to put together a new joint venture in industrial equipment, but, as one of the Deere representatives on the scene put it: "There are broad differences in customs and language . . . communication with Hitachi involves the utmost in patience and understanding." Such understanding was not to be, though, and in October 1970 the two companies agreed to go their separate ways.[13]

THE "WORLDWIDE" TRACTOR

In mid-1961 Deere had increased its ownership of Lanz Iberica in Spain to 42.5 percent, at the same time buying a similar percentage of Ricardo Medem's sales company; the latter at first kept 42.5 percent ownership of the manufacturing company and controlling interest in the sales company, but soon sold controlling interest in both to Deere.

Spain was not scheduled to join the EEC for a number of years; meanwhile, the ability to move parts across the border into France and on into Germany was not as propitious as if Spain were already within the EEC. Spain did have high tariff walls and did require substantial local content (90 percent within two years of establishing a company), thus making some of the movement of components and parts from North America and other countries in Western Europe more difficult. In sum, Spain at this particular time was almost the textbook description of a "pocket" market.

With this Spanish addition, the company was now manufacturing in six countries (the others: Germany, France, Argentina, Mexico, South Africa). It was clear that the market for tractors around the world was large, yet a company such as Deere could not produce everywhere or, indeed, satisfy all the special needs of the world's farmers. If the company could achieve some standardization, whereby components manufactured in one location could be utilized by others, it could improve its flexibility and reduce costs.

The initiative for developing a standardized product came during a critically important discussion in September 1960 on the concept of a "worldwide" tractor. A group of thirty-four top executives in the company assembled around the Moline board room table; the group was so large that it had to be divided. Those around the central table were to carry the burden of the conversation and a second group at the outer perimeter was to listen and absorb. Hewitt stated the issue succinctly: "Deere & Company leads competition in the United States; however, competition is ahead of us in the foreign markets. . . . Investments for foreign manufacture will be rewarded if properly planned, and we can contribute to the world welfare by making equipment available. . . . We should evaluate this market, the facilities required, and production capabilities to assess the potential for a worldwide tractor, bearing in mind that farming customs and conditions differ in various countries, and it would be uneconomical and impractical to attempt to design a tractor for each country. Therefore, the final design must represent a composite of world requirements."

Subsequent speakers laid out the market potentials for such a tractor, elaborated on competitor models, and sketched out what such a tractor—"A utility or general-purpose tractor, of low profile"—might look like. A number of suggestions were made about horsepower; the consensus seemed to be that it should be no smaller than 25, nor larger than 50 (PTO). The participants ranged over issues of engineering requirements, interchangeability of components and parts among different countries, provisions for a so-called "basic design," and standardization potentials (including the query of whether the British metric system should now be adopted and concerns about how such a piece of equipment could be tested on a worldwide basis). In total, it was a formidable challenge to the assembled Deere executives; many later recounted the electric feeling that this meeting generated.

Out of these early discussions about the worldwide tractor came detailed plans for two prototypes that would become the heart of the program—the X-21 at 37 horsepower (PTO) and the X-22 at 55 horsepower. As plans progressed in North America on the design of these two tractors, Frank T. McGuire was sent by Curtis on a survey of the European plants, accompanied by Clifford L. Peterson and T. S. Keisling. The rubric laid down by

Curtis was that "all Deere factories everywhere in the world were to design their products with the maximum possible standardization." This would allow interchangeability of parts, would ensure the maximum use of design talent, would raise worldwide quality standards, and in the end "should result in a worldwide 'family look.'"

McGuire's recommendations took into account manufacturing costs for the two tractors (compared to competition) and the most likely locations for manufacture of various components of the tractors. McGuire felt that Mannheim should make the engines for both tractors, the new Saran factory in France should make the transmissions (the opposite of the original plans for Germany and France). The tractors would be assembled both at Saran and Mannheim. In turn, all of the company's European combine manufacturing would be concentrated at Zweibrücken. (Mannheim had continued to make some combines up to this point.) The Deere operation in Spain, Lanz Iberica, would be licensed to make products specially adapted to the Spanish market, but Spain would not at this time be part of the larger Western European manufacturing complex. McGuire concluded, "The big Deere effort in Europe should be concentrated in France and Germany and devoted to tractors, combines, industrial equipment and balers."

McGuire, too, was preoccupied with the issue of local content, but from the perspective of EEC members. To be sure, a number of the Western European countries had had substantial tariff barriers. With the advent of the EEC, however, there was now to be a lessening of tariffs and eventually no barriers whatsoever. In other words, there likely would be no "local content" requirement between, say, Germany and France. There might well be subsidies paid by individual countries to their own producers, as was subsequently adopted by several of the "Six" for agricultural products. In terms of manufacturing, though, everyone would be in the same basic situation. The difficulties of meshing Deere's operations in Spain illustrated the implicit height of the borders being put around the EEC.

By the time the X-21 and X-22 models were introduced in 1965, a third model had also been developed. The three prototypes were the 310, a three-cylinder diesel at 32 horsepower (PTO); the 510, also a three-cylinder diesel at 40 horsepower; and the 710, a four-cylinder diesel at 50 horsepower. There had been modifications in McGuire's production planning—the Saran factory was in operation by this time—and it was decided that engines for both French and German tractors were to be made there. In turn, the transmissions and a number of other components were made at Mannheim. Subsequent history proved the efficacy of the switch from the McGuire proposal.

Deere thus took a giant step toward a worldwide manufacturing program, with significant components coming from North America but with major manufacturing shared uniquely by West Germany and France. Deere executives had been overly naive at the beginning of their European venture; indeed, there were some conceptual problems in trying to be a multinational, worldwide corporation even at this point. Nevertheless, the Americans had accomplished a signal victory in realistically bringing the West Germans and the French together in a joint manufacturing venture. Nationalism might have made such a cooperative venture much more difficult had either one of the two countries alone initiated the project. Deere had also thrown an umbrella over the three small French companies and had shown that a large multinational corporation could indeed assuage the fears of smaller firms in Western Europe that were concerned that they were too small to hold their own against evolving competition from the large companies.

Market share for Deere-Lanz now began to climb a bit (Appendix exhibit 25), though Deere's dent in the European market was still modest—only 3.4 percent of all Western European sales. There were to be many more heartaches after the introduction in 1965 of the new models. The European operations were still in a tenuous position, and the following several years would prove even more difficult. One could not have said in 1965 that Deere had "made it" in Europe. What could have been said, though, was that a major conceptual change had been effected in a way that seemed to bridge the gap between the American and European views of management.

Europe was not the "world," the modifier used by that important meeting in September 1960 to discuss the "worldwide" tractor. In 1965, Spain, South Africa, Mexico, and Argentina each remained essentially a "pocket market." The challenge of trying to bring these four country operations more into the mainstream of Deere remained to be met.[14]

HEWITT'S FIRST DECADE IN RETROSPECT

On April 27, 1965, Deere & Company held its quarterly meeting of the board of directors, followed by the annual stockholders' meeting. At the director's meeting, William Hewitt was presented a key to the Deere Administrative Center, to mark his tenth anniversary as chief executive officer of the company. Curtis, in presenting the key, pointed out that it had been especially designed for the occasion by Henry Dreyfuss and "opens all the doors—including the door of the wine cellar." (In truth, wine was served only at state occasions for foreign business visitors; at Deere there was a rigid rule of no alcoholic beverages in any of the United States offices, branches, and factories.)

During Hewitt's first decade of leadership the organization had assumed a new role, a multinational corporation that had "arrived" as the number-one farm machinery manufacturer in the world. Without forgetting its roots in Midwestern conservatism, it had found a new style and verve that gained it widespread national attention. Hewitt now had in place his own management team and had realigned the organizational structure in important ways. After the major shakeup stemming from the Booz, Allen & Hamilton report of 1956, he had tightened the structure more effectively. Factories were consolidated in Moline with the transfer of various assembly operations of the old Wagon Works to other factories in Moline and Des Moines (the Wagon Works as an operating entity was then assigned to oblivion, as had been the wagons it made). In October 1959 an anomaly in the operation of the board itself—interim meetings of the local Moline members of the board of directors in place of an executive committee—was ended. This "local board" had had no official status and had suffered because the quorum requirements for the full board had to be met in order to conduct official business. At the meeting of October 1959 a formal executive committee was reconstituted, with a quorum of five. The new system gave more flexibility and also centered power more strongly in the hands of the Moline members of the board. (Over the years, this sometimes has been a source of irritation on the part of the outlying members.)

In early August 1961, with the death of Pat Murphy, the company lost a loyal and influential executive. He had been ill for a number of months but characteristically had worked until a week before his death. Murphy had been a director of the company for twenty-four years, one of the key leaders of the manufacturing arm of the company for an even longer period. His dedication to the company was unstinting, his leadership qualities significant. His innate conservatism and focused perspective sometimes constrained his willingness to expand his horizons and movement into new areas. Yet an organization must have individuals on all sides of the spectrum, and Pat Murphy's honest and open representation of his point of view served Hewitt and the others on the board very well indeed.

With Murphy gone, George French and C. R. Carlson assumed the key roles as heads, respectively, of manufacturing and marketing (with Curtis as the key financial officer). On the board itself there were also new faces: D. C. Glover, related by marriage to the Wiman side of the family, came on the board in 1955 (but abruptly resigned his management position as industrial sales manager in 1958). Joseph Dain was elected to the board in 1960, himself a third-generation link to one of the founding companies. Frank T. McGuire came on the board in 1962, as did Ed W. Ukkelberg, from the marketing side of the business. A. B. Lundahl joined a year later and in April 1964 three additional board members were elected—John H. Graflund (manufacturing, with a focus on overseas), Clifford L. Peterson (finance), and Comart M. Peterson (marketing).

Hewitt also asked for a significant change in his personal status in the company in April 1964. Explaining to the board that "business people outside the Company feel that 'President' carries more weight than 'Executive Vice President,' even though the responsibility might be the same," he asked that Curtis be appointed president of the company. In turn, Hewitt's title was changed to "chairman." Hewitt made it explicitly clear that this was not synonymous with "chairman of the board"—that he remained the chief executive officer of the company.

There had been persistent problems in organizing for the overseas operation. At the start, only a few American personnel were posted abroad; a number of top-line people who might have profited from an overseas assignment were not sent, their continuing presence jealously guarded by the heads of the domestic operations. As the company learned it needed to become more involved in the foreign operations (particularly as Lanz in Germany and the new French operation continued to have difficulties), more senior management people were deputed abroad.

In 1961, the international group was reorganized. The Venezuelan holding company, John Deere Intercontinental, S. A., was phased out, and a new Canadian company, John Deere Intercontinental, Limited, was formed, not to handle the Canadian operation but to be the rallying point for the rest of the world. Two divisions were set up—a Latin America–Pacific–Far East division under Harry Pence, and a Europe–Africa–Middle East division under Comart Peterson. Still unclear, though, was the way that all of this fitted into the top-management structure at Moline.

Late in this same year, Hewitt asked Booz, Allen & Hamilton to consider the restructuring of the international operations. When the Booz, Allen analysts reported back, they suggested a radical departure. While they felt that the functional basis of organization still made sense on the domestic side of the business (with all manufacturing being centered in one group, all marketing in another), they concluded that a number of factors peculiar to Deere's international operations argued for a different structure there. These factors were the wide variety of ownership situations, the special complexities of international business itself, the development of the "worldwide" tractor concept, and the particular needs for coordination between France and Germany. Their conclusion: the company should be organized abroad in a geographical mode. Two senior vice presidents would be assigned responsibility for managing the two major segments of the business—the North American operations and the international operations, respectively. In turn, the international operation would be divided into two groupings, each to be headed by a vice president. One would embrace all of the European operations, the other everything outside of Europe (i.e., Argentina, Australia, Mexico, and South Africa).

Hewitt adopted most of the report; by early February 1963 a company-wide bulletin announced a "Region I" to include Latin America, Australia, South Africa, and the Far East and a "Region II" to take in all of Europe, North and Central Africa, and the Middle East. In addition, there would be a separate organization for export sales to dealers in countries where no Deere sales branches had previously been established. This concept of dividing the world made eminent sense, and the pattern has stood to the present as Deere's way of operating abroad. (Later there was one modification: as the South African business climate was affected most particularly by patterns in Europe, it was shifted in 1966 to Region II.)

There was one striking modification of the Booz, Allen report, though—the historic functional basis of organization was continued abroad. The strong, sometimes opinionated personalities of the two individuals who were to take the senior vice president positions, C. R. Carlson and George T. French, finally persuaded Hewitt to remain with the functional definition of the company, worldwide. Carlson, as senior vice president of marketing, was given complete charge of marketing throughout the world; the directors of marketing in the two regions reported to a vice president–overseas marketing (Comart Peterson), who was directly under Carlson. In turn, George French, as senior vice president–manufacturing, was given overall supervision of all manufacturing throughout the world, with his vice president–overseas manufacturing, John Graflund, also supervising manufacturing in both of the regions. Hewitt explained his reasoning to the board: "Where previously we had one man in charge of both marketing and manufacturing in each of our various overseas areas, we now have separated the responsibility for marketing from that of management in each of these areas. . . . Carrying this separation of responsibilities up and down the line enables our management people to devote their efforts to their own specialities in which they are well trained. . . . This new organization parallels the proven and successful structure of the John Deere North American organization."

This turned out to be a signal mistake. There had always been an insularity in both the manufacturing arms and the marketing groups in the domestic company. Though this had been a historical source of strength by vesting a high degree of decentralized authority in the hands of, for example, factory managers, it also had been a perennial problem, for this dichotomy of thinking between the two groups of people often made each unresponsive to the other. In the overseas operation the difficulties were compounded—the lines of communication were longer, the differences in each country very striking, and the sophistication and understanding of the Deere personnel about foreign environments much less than that of the domestic operation. For example, if the marketing manager in a foreign operation felt that a product needed to be re-engineered to fit a local condition, he could not make this request directly to the manufacturing man in his country. He had to go through Moline to his own senior vice president of marketing, who then would deal with his counterpart in manufacturing. Obviously there was some communication horizontally between the two functions throughout the organization. Yet the uneasy fact remained that the formal linkages were vertical, and the practical effect of this was to compound insularity. Within three years, the company was forced to change its international organizational structure, turning back to the original Booz, Allen concept.

If there were indeed some organizational difficulties, there was no uncertainty about the overall mission and goal. Hewitt had been able to dramatize and then carry through a wide range of new initiatives. One of his key devices for accomplishing change was a periodic goal-setting process (on an approximate five-year cycle). Each board member participated, developing his own extrapolations of company sales and profitability, as well as his own suggestions for new initiatives. When the process was renewed in February 1964, the evident success of the company over the previous years dominated individual comments. Hewitt had given a ten-point set of his own goals, which focused particularly on the overseas operation and the need for product development and reliability improvement. Permeating Hewitt's own personal presentation and the contributions of each of the directors was a sense of excitement about the challenges and opportunities of the moment. Yet there was a grasp of the future throughout all of the goal-setting process—a realization that short-term growth might not always produce the wisest long-term pattern. A number of years earlier, Frank Silloway had stated this succinctly in one of his most enduring speeches: "We are in business for the long term. Natural human impulses stem from short-term motives. Therefore, resist your natural impulses or, at least, count ten before you yield to them."

In this goal setting of 1964, deep concern about international losses intruded repeatedly into the discussions. The need to look to the longer run was stressed by Frank McGuire in a letter to Hewitt, which was read at the meeting: "For as long as I have been aware of your thinking and observed your leadership, you have emphasized that Deere must seek growth. In this, we may be forced to take short-term losses but will accept these provided the long-term growth will yield satisfactory profit levels. . . . The advantage of this is that no one then could possibly be under any illusion that you are interested in what I describe as 'empty' growth, namely, sales increase with excessive dilution of profits or, worse, a continued lack of profit."

McGuire certainly had nothing to worry about in the ten-year results, 1955–1964. Profits more than doubled, reaching nearly $60 million in 1964, and the increase in sales was even greater (exhibit 13-8).

The uses of these funds in the ten-year period give many clues to the increasing strength of the company. Retained earnings were high all through this period (with the exception of the new-model year of 1960); the average for the last four years (1961–1964) was more than 60 percent. After $40 million had been put into capital expenditures in the new-model year of 1960, substantial capital outlays also were made in each of the next four years (reaching $65 million in 1964). Beginning in 1960, the comptroller's report included precise figures on product research and development as a percentage of sales; the figure was well over 4 percent for each of the years from 1960 onward. Thus one of the signal marks of the Hewitt era—the high percentage of company resources invested in new product efforts and new manufacturing facilities—was put in place in this early period. At the same time, organizational changes in marketing found effect in a more aggressive and effective marketing team throughout the Deere organization itself and, especially, among its dealers. The latter group was greatly strengthened during 1955–1964.

Perhaps the centerpiece of the first ten years of Hewitt's tenure was the completion of the Deere Administrative Center, which opened in June 1964. Eero Saarinen had died shortly before the construction of the building began; his colleagues, Kevin Roche and John Dinkeloo, finished the architectural planning and oversaw its construction. The concept of the building remained very much that of Saarinen. The site was a beautiful one and it was complemented by a surprising feature that Saarinen incorporated into the building; the exterior was constructed of unpainted steel girders, called "Cor-Ten," intended to weather and rust to blend into the countryside. (Some of the Deere engineers were most alarmed about this: "We've been warning farmers against rust for 120 years, now we are building a big, rusty building.")

The structure, once in place, was indeed striking, and it received dozens of awards. It was one of Saarinen's greatest triumphs, probably his most important administrative building, certainly a worthy companion for many

Exhibit 13-8. Deere Sales and Profits, 1955–1964

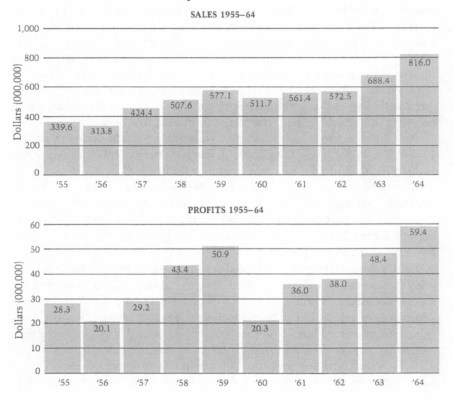

of Saarinen's beautiful public buildings around the world. When articles are written in architectural journals and general-interest publications citing the best of the world's architecture, the Deere Administrative Center is often among those buildings pictured. As an architectural and aesthetic success, the Deere Administrative Center is outstanding.

Beyond this, though, the building was more than just a physical structure to win awards from *Architectural Forum*. It was a statement by the company that it strove for the best in quality, style, and effectiveness. C. R. Carlson, one of the more conservative of the older board members, penned a delighted letter to Hewitt on his first day in the new building. "I have spent the biggest share of my years with Deere working in old buildings. At Welland, as you well know, they were cheap and badly constructed to start with. At Waterloo during the period I spent there we lived in an old wooden second story office that had been built on as a makeshift, and even with water on the roof we cooked all summer long.... At Minneapolis, after we put in all new lighting, moved the men's and women's washrooms from right out in front of the sales department offices, and opened the whole office area up, they were comfortable and livable, but of course they were old and nothing like this. I must say that the surroundings are wonderful and the building itself is an excellent one in which to work.... I am sure that it is going to be conducive to better work on the part of all of us who have the privilege of working here."[15]

Carlson's words were a particularly appropriate summary of the first ten years of Hewitt's leadership. The "old guard" had indeed been won over; with considerably more unanimity of purpose than had characterized 1955, the company seemed ready to take on the challenges of the still-sagging international outreach. This key segment of Hewitt's plans had not taken hold.

Endnotes

1. Benjamin Keator to Charles Wiman, reprinted in Deere & Company *Minutes*, July 28, 1953. B. Keator to William Hewitt, May 25, 1955, DA. The Booz, Allen & Hamilton study was authorized at the board meeting of June 20, 1955; Hewitt's remarks are from Deere & Company *Minutes*. Hewitt's planning document is undated; internal evidence places its completion by December 1955. For Hewitt's speech to the branch-house managers of June 8, 1955, see DA, 46332.

2. For Charles Stone's memorandum to the board on foreign manufacture, see Deere & Company *Minutes*, June 4, 1953; for the recommendation for a new chairman, see ibid., July 28, 1953; for Maurice Fraher's reports on the Brazil and Argentina surveys, see ibid., October 27, 1953, April 27, 1954, July 27, 1954, and October 26, 1954. For Lloyd Kennedy's report on Heinrich Lanz Mannheim A. G., see ibid., December 21, 1953.

3. The best source for the merger discussions between Harry Ferguson and the Massey-Harris group is Colin Fraser, *Harry Ferguson: Inventor and Pioneer* (London: John Murray, 1972). The internal problems of Massey-Harris-Ferguson are documented in Edward P. Neufeld, *A Global Corporation: A History of the International Development of Massey-Ferguson Limited* (Toronto: University of Toronto Press, 1969). The Deere merger negotiations are discussed in ibid., 200–201. James S. Duncan's view of his severance from the company is in J. S. Duncan, *Not a One-Way Street: The Autobiography of James S. Duncan* (Toronto: Clarke, Irwin E. Company, 1971). See also Peter Cook, *Massey at the Brink: The Story of Canada's Greatest*

CHAPTER 13

Multinational and Its Struggle to Survive (Toronto: Collins Publishers, 1981). The Hewitt document in preparation for these meetings is in DA, Hewitt ms.

4. The confidential report of Booz, Allen & Hamilton, dated January 1956, is titled "Program for Implementation of the Proposed Plan of Organization," DA, 45515; the overall report, "Top Management Organization Survey," is dated February 1, 1956, DA, 45514; their "Survey of Executive Compensation" is dated January 1956, DA, 45510. See L. A. Murphy to William Hewitt, September 30, 1955, DA, Hewitt ms.; L. A. Rowland to W. Hewitt, February 15, 1956, DA, Hewitt ms. For Cook's explanation of his decision to take early retirement, see Deere & Company *Minutes*, April 24, 1956.

5. Establishment of the Venezuelan subsidiary, John Deere, C. A., was authorized by the board on July 26, 1955; for Curtis's report on Mexico and Fraher's report on Europe, see Deere & Company *Minutes*. For Harry Pence's report on the Monterrey manufacturing proposal, see ibid., February 24, 1956. For Pence's suggestion of Lanz, see ibid., April 24, 1956, and June 29, 1956. For the proposal for purchase of Lanz stock made by Hewitt, see ibid., July 31, 1956; for the appropriation for the Monterrey plant, see ibid., October 30, 1956; for Curtis's remarks on machine tools, see ibid., August 30, 1956. Reconstitution of the Lanz board and concerns about integrating Lanz and Deere dealers were discussed in ibid., October 30, 1956. For the quotation on the lack of sturdiness of Lanz tractors, see ibid., April 15, 1957.

6. For an "official" version of earlier Lanz history, see Paul Neubaur, *Heinrich Lanz, Fünfzig Jahre des Wirkens in Landwirtschaft und Industrie, 1859–1909* (Berlin: Paul Parey, ca. 1910). See also Wolfram Fischer, "Ein Jahrhundert der Landtechnik die Geschichte des Hauses Heinrich Lanz" (1958). There are two company-written histories of Lanz, also covering the recent period: *Geschichte der John Deere Werks Mannheim* (Mannheim, 1979), and *Lanz Und die Land Wirtschaft* (Mannheim, 1960).

7. For Hewitt's first contact about Eero Saarinen, see Henry Dreyfuss to W. Hewitt, January 17, 1956, DA, Hewitt papers. Hewitt first discussed an outside architect in Deere & Company *Minutes*, June 15, 1956. Saarinen's visit to Moline was in August 1956; for the board's authorization of him as architect, see ibid., January 29, 1957. For Burton Peek's buffalo story, see *Moline Dispatch*, October 29, 1953. Saarinen's fees are discussed in E. Saarinen to W. Hewitt, September 17, 1956, and W. Hewitt to E. Saarinen, October 5, 1956, DA, Hewitt papers. For Hewitt's statement of goals for the building, see W. Hewitt to E. Saarinen, August 23, 1957. The anonymous employee letter to Hewitt was dated August 21, 1957; Edmond Cook to Joseph W. Lucy (of the Saarinen organization), April 19, 1957.

8. Lester Larsen, *Farm Tractors, 1952–1975* (St. Joseph, MI: American Society of Agricultural Engineers, 1981), 2. "Generation unto Generation," *Forbes*, March 1, 1961, 19–22. The C. R. Carlson form letter explaining the decision to drop the two-cylinder tractor is dated September 15, 1960. For the quotation on "hicks," see *Forbes*, April 1, 1963.

9. For James Wormley's report on Lanz, see Deere & Company *Minutes*, July 10, 1957; for Lloyd Kennedy's report on complaint backlogs, see ibid., October 29, 1957, and April 29, 1958. Report on Deutz-Porsch and Fahr-Guildner relationships is in Curtis's report to the board, ibid., June 12, 1958. Force reduction at Lanz is discussed in W. Hewitt to J. Wormley, March 25, 1957, DA, Hewitt ms. 719. The "beer strike" is discussed in E. Curtis to W. Hewitt, March 8 and 11, 1960, DA, Hewitt ms. 705. See *Der Spiegel*, March 16, 1960; *Badische Volkszeitung*, March 15, 1960; *Bild*, March 8, 1960. For the Henry Dreyfuss involvement in Lanz, see W. Hewitt to H. Dreyfuss, November 5, 1957, and July 8, 1958 (the quotation about "Persian Carpets" from the latter), DA, Hewitt ms. 714. For Curtis's remarks on North American management becoming more involved in Lanz, see Deere & Company *Minutes*, April 27, 1959. The difficulties in changing German officers of Lanz is discussed in E. Curtis to W. Hewitt, June 10, 1960, DA, Hewitt ms. 705. For Harry Pence's report on the slowness of change and low morale at Lanz, see Deere & Company *Minutes*, July 27, 1959. For a discussion of the "semi-diesel" principle and the function of the glow bulb, see Cummins, *Internal Fire*, 267, 277.

10. The Compagnie Continentale de Motoculture was established in November 1958 and was converted to a joint stock company on September 22, 1959 (see "Protocol," July 21, 1959). The Curtis-Lesser-Wilson negotiations are discussed in Deere & Company *Minutes*, July 27, 1959; for the major project proposed by Curtis, see ibid., August 17, 1959. For Deere's investment in the three companies, see ibid., November 22, 1960; for the additional purchase of Rousseau shares, see ibid., June 26, 1962. Difficulties at Rousseau are discussed in E. Curtis to W. Hewitt, October 3, 1961, and March 24, 1962, DA, Hewitt ms. 705.

ENDNOTES

11. Curtis's remark on Mexico as "a bright spot" is from Deere & Company *Minutes*, January 27, 1959. The local-content issue was first discussed in ibid., April 25, 1961. See also ibid., October 30, 1961. The David Brown and Nuffield proposals are discussed in ibid., January 29, 1962. Deere's proposal that was accepted is dated September 30, 1963; see ibid., March 24, and April 27, 1964. Local participation is discussed in ibid., April 27 and July 27, 1965; George French's quotation is from the latter. See E. Curtis to W. Hewitt, August 13, 1965, DA, Hewitt ms. 705. For a useful summary of the years 1958–1966, see "John Deere de Mexico, S.A. de C.V.," Harvard Business School case 9-313-239 (1968). The overall problem facing American industry in regard to percentage of ownership is discussed in "Curbing Yanqui Cash: Mexico's New Regime Increases Restrictions on Foreign Companies," *Wall Street Journal*, June 17, 1965.
12. The best overall description of the Argentine government's decrees on local content is "Una Breve Reseña Sobre la Industria Nacional de Tractores," prepared for Dr. Victorino O. Rodriguez, Secretario del Consejo de la Industria de Tractores (ca. December 1967), DA, 79/17/10. The quotation on local content is from "Problems Pertaining to our Manufacturing Operations in Argentina," June 30, 1966, DA, 79/17/10.
13. The initial feasibility study for South Africa by R. C. Deffenbaugh, O. E. Hintz, and D. M. Willis was dated June 1962, DA, AN/80/43/2. For the proposal to purchase an interest in South African Cultivators (Proprietary) Limited, see Deere & Company *Minutes*, July 1, 1962; see ibid., September 13, 1962, for a modification of the appropriation. The land purchase in Nigel was authorized in ibid., November 8, 1962. For Curtis's remarks on race relations in the project proposal to the board, dated July 24, 1962, Hewitt ms. 705, see ibid., July 1, 1962, and E. Curtis to H. B. Pence and selected board members, May 15, 1963, DA, AN/80/43/2. See J. H. Goddard, "Report of South African Investigation of Cost and Availability of Materials . . . (September 29, 1964), DA. The joint proposal of Deere, Massey-Ferguson, and Ford was "Memorandum to the Board of Trade and Industry, South African Tractor Manufacture" (April 1963), DA. See also Pence memorandum, February 8, 1963; George T. French memorandum, July 28, 1964; "Proposal for the Manufacture of Tractors in the Republic of South Africa" (March 1968), DA. For the quotation on South African government efforts to increase self-sufficiency, see H. I. Martin to G. French, January 16, 1967, DA. The government "Enquiry into the Tractor Industry" was its Report 907 (1967). For Deere's relationships with Hitachi in Japan, see Deere & Company *Minutes*, October 27, 1970.
14. For the situation in West European agricultural machinery in this period, see Cynthia A. Breitenlohner, *Structural Changes in West European Agriculture, 1950–70*, US Department of Agriculture, Foreign Agricultural Economic Report 114 (November 1975); *The Grain-Livestock Economy of the European Economic Community: A Historical Review, 1951–63*, US Department of Agriculture, Foreign Agricultural Economic Report 31 (1966). See also Robert Trow-Smith, *Life from the Land: The Growth of Farming in Western Europe* (London: Longmans, Green & Co., Ltd., 1967); Henri Mendras, *The Vanishing Peasant: Innovation and Change in French Agriculture*, English trans. (Cambridge, MA: MIT Press, 1970). The meeting to consider the feasibility of a "worldwide" tractor was held on September 9, 1960, DA, AN/80/10, 4157. The McGuire-Peterson-Kiesling report, "Manufacturing Survey-Europe," is dated December 15, 1961; Curtis's charge to this group of October 2, 1961, is reprinted there.
15. See W. Hewitt to Henry Dreyfuss, May 3, 1965, DA, Hewitt ms., concerning the latter's role in the presentation of the key to Hewitt. Glover's resignation is discussed in D. C. Glover to W. Hewitt, October 8, 1958, DA, Hewitt ms. An excellent discussion of the consolidation of the Wagon Works is in K. W. Anderson to Curtis L. Oheim, August 9, 1956, DA, Hewitt ms. Hewitt's remarks on the reorganization of the international operations in 1963 is in bulletin B-103, February 5, 1963. The Frank Silloway quotation is from DA, 20165. See Frank McGuire to W. Hewitt, January 28, 1964, DA, Hewitt ms. 707; C. R. Carlson to W. Hewitt, April 20, 1964, DA, Hewitt ms. 750.

CHAPTER 14

OBSTACLES ABROAD

Advanced farm equipment and skills have played a positive role in building a modern state farm. It has emancipated people's minds, broadened their vision and made them realize that mechanization is the fundamental way to increase agricultural output. . . . It has battered down management methods that are inappropriate to agricultural mechanization and brought about many changes in farm management.

Guangming Ribao
People's Republic of China, 1979

The year 1964 was "the best ever." Deere's net sales totaled $816 million, its net income more than $60 million (7.3 percent of sales). Yet within this satisfying company-wide performance, difficulties in the foreign operations were a constant source of frustration to management and a drain on company resources. Though the Mexican and Argentine operations each had a small profit, all the rest of the major foreign operations registered unsettlingly substantial losses. Most serious was the German loss of $5.2 million. The French operation dropped more than $1.6 million, Spain was in the red for more than $1 million, and South Africa had a loss of almost $500,000. The pattern continued unmitigated in 1965, 1966, and 1967, with Germany alone losing almost $40 million total in those three years; France

◀ Balers awaiting shipment at Arc-les-Gray, France, 1978. *JD Journal* lost more than $9 million in this same period; the Argentine operation lost $7.6 million alone in the year 1967.

Not until eight years after that euphoric year of 1964 was the international operation able to turn a profit. It took a full decade for the pivotal German operation to shuck its persistent pattern of losses. A number of problems conspired to produce what turned out to be almost two decades of difficulties in Deere's bold decision to "go multinational." Many new ways of producing and marketing farm machinery in differing environments had to be mastered, and the learning curve for these turned out to be steeper than had been first imagined. In addition, two sets of more fundamental questions about international operations also had to be resolved before the international operation became a balanced component of the business.

First was the problem of structuring the organization itself. Could the domestic operations just be extended intact beyond the shores of North America, as Deere originally thought, or was a totally new structure needed? Further, could Deere go it alone, or was it necessary to join with another strong company, presumably a foreign one, and likely to be European? A second set of questions related to the rampant nationalism in a number of the countries Deere had entered. Could the company live with some of the increasingly stringent demands put upon it by individual countries, particularly relating to import restrictions and local-content restraints? Further, Deere found itself working in, or dealing with, countries whose political and economic philosophies differed sharply from those of the United States. The Eastern bloc and other Socialist countries espoused particular viewpoints that made business arrangements additionally complex; some countries, particularly South Africa, posed complications for Deere outside of the business realm, relating to political ideologies. Finally, questions were being raised about the very appropriateness of Western technology in the developing world. These additional enigmas, beyond the already difficult business situation, made Deere's international outreach a complicated one in the period 1965–1980.

REORGANIZING FOREIGN OPERATIONS

To this point, Deere had carried all of its major foreign subsidiaries on its books as equity investments, not consolidated in the company financial statements each year. This had the signal disadvantage that losses could not become offsets in the calculation of United States corporate income taxes for the company. In Germany, there was a further complication that went beyond any narrow financial concern: the German public held a minority of Lanz shares, about 3 percent; under German law this required outside business and labor members on the company's board. All the company's reports and performance instantly became a matter of public information, in a country well known for a relentless business press and a watchful public very keenly interested in business developments. For a fresh, untried competitor, this fish bowl atmosphere was very disconcerting. "Publication... has been a detriment of an undeterminable magnitude to our business because of this public knowledge," Frank McGuire told the board in April 1967. McGuire, who was managing director of Region II, which included the European operation, recommended that Lanz revert to a branch operation, similar to a domestic branch. The Deere board concurred and authorized purchase of the minority shares. The offer was a fair one and the outside shareholders sold with alacrity, probably glad to be out of what they perceived to be a star-crossed company.

Unfortunately, the German press reported the sale in unflattering terms. One headed its article (as translated into English), "Heritage Staked and Lost: The Story of an Unsuccessful American Business Venture"; another had a nostalgic article on the long history of Lanz, ending its article, "This name 'Lanz' must not be allowed to be forgotten." Still a third chimed in, "Lanz Stockholders Did Not Wear Mourning." Deere felt the spillover from an inchoate anti-American attitude, and the shift to being a "branch operation" of a foreign company did not help.

Though the branch modus for Germany improved Deere's financial figures, persistent trouble spots remained in the company's handling of foreign operations. It finally was becoming clear that Deere's overall organizational structure was impeding company growth, rather than facilitating it. The greatest difficulties loomed in the international side of the business, where the functional approach and the lack of an on-the-spot management presence had led to substantial confusion. But there were also problems in the North American organizational structure: still too many constraints on marketing's participation in product planning, the failure to resolve questions of overall responsibility for engineering, and uncertainty over the precise role of the rapidly growing industrial equipment division.

Early in 1966, Booz, Allen & Hamilton was summoned again for another organizational study. It turned out to be a major endeavor, extending over seven months and involving five separate progress reports. The Booz, Allen analysts were particularly critical of missteps in the international operations stemming, they believed, from wrong choices in organizational planning back in 1963. They uncovered strong evidence that European design requirements were being given secondary consideration in the North American planning and engineering process. It seemed to take too long to get technical assistance from the United States, and specifications were reportedly changed with little or no advance notification. Further, the unit heads and the regional directors abroad had not exercised the authority inherent in the decentralized concept; this had led to poor performance and excessive turnover in key jobs abroad. There appeared to be a "lack of urgency . . . a lack of leadership" (the consultant's words) in the foreign operations, compounded by confusing signals from senior management in Moline, most of whom evidenced an unsettling degree of myopia about the individual foreign operations. There were especially severe problems in the coordination between France and Germany: "Plant and branch managers and their subordinates are expected to resolve their mutual problems with their counterparts. This has not worked well nor is it likely to when many of the people really do not understand the Deere philosophy, have been in their positions only a short time, often cannot communicate in a common language, and have inadequate written descriptions of responsibilities and procedures."

Resolution of this apparently widespread mismanagement could be effected, the Booz, Allen analysts maintained, by organizing the international businesses "on a general management concept"—with one senior executive in charge of all manufacturing and marketing abroad. For Europe, the dominant problem area, the consultants urged the appointment of a strong vice president and general manager, to be resident in Europe, with "authority and stature to straighten out the present unsatisfactory situation." The analysts made their point crystal clear: "The Company cannot afford to continue to rely on Deere & Company management resident in Moline to provide this leadership. It has not worked well and the continued financial deterioration in Europe attests to the need for taking drastic action."

For the domestic side, the Booz, Allen analysts were satisfied that the North American farm equipment business should continue to be organized on a functional basis, but they did recommend that the North American industrial equipment business be split off into a separate division "as soon as practical." To strengthen product planning, its executives were to be given a reporting relationship directly to the general-company marketing executive. The domestic factories would continue to have the basic responsibility for product development and design, where the Booz, Allen analysts felt they had done an outstanding job.

This time the board adopted the Booz, Allen recommendations almost in toto. The organization plan, as announced in a major memorandum in late November 1966, provided that George French assume the position of senior vice president, overseas operations, in charge of all manufacturing and marketing abroad. Reporting to French were John H. Graflund as vice president, overseas administration; Francis T. McGuire, managing director, Europe and Africa (Region II); and William R. Klingberg as managing director, Latin America and Australia (Region I). C. R. Carlson retired the following March, and Edgard W. Ukkelberg then became senior vice president, North American operations. French's successor as head of manufacturing and engineering, also at the senior vice president level, was A. B. Lundahl. The rest of the senior officers reporting to Curtis and Hewitt were: Frank M. Dickey as vice president, industrial relations and personnel; Joseph Dain Jr., vice president, corporate planning; Lewis D. Wilson, vice president and general counsel; and Clifford L. Peterson, vice president, finance.

The most critical ingredient of this plan was sending Frank McGuire to Europe as managing director, the first time that a senior executive of the company had actually been physically stationed overseas. One major exception was taken to the Booz, Allen report—the industrial equipment division was left in place. Robert J. Gerstenberger, who was the director of industrial equipment marketing, reported to Ukkelberg, and the industrial equipment manufacturing was combined with that of tractors under the leadership of H. B. McKahin. By 1969 the industrial equipment division had its own separate marketing organization and in 1970 the original Booz, Allen recommendation that there be a separate division was carried through (George French heading the division for a year before his retirement, Delno W. Brown replacing him at that time).

Domestically the system preserved most of the strong features of decentralization. Some people, though, viewed the reorganization of 1966 as an incipient effort to draw together more authority at the top. Hewitt was concerned enough about any misreading of the plan to comment to the board on the need "of ensuring that the Company's policy of decentralization is in fact being implemented at all levels of the corporate structure.... This policy needs constant reiteration so that personnel will not lapse into centralization modes of operation." Hewitt gave practical effect to this by taking this occasion also to remind the board of his strong feeling that personnel should be promoted on the basis of ability, without regard to seniority or age. He stressed particularly the need for "giving promising young employees as much responsibility as they can take as soon as possible.... Young people want and need challenges to develop latent abilities."

Hewitt's bulletin to the organization about the new assignments chose not to elaborate on the profound change that had been made in the conceptualization of the overseas structure on a "general management" basis. He warned only that "in many instances they will require substantial changes in familiar habits and relationships." The careful, staged approach of the consulting effort, specified by Hewitt at the start, ensured that everyone concerned had been personally involved in the process, and so there were no surprises at the end. This caution paid off, for the organizational realignment was well received. Deere already was a full-scale multinational corporation; in 1966 for the first time it had the organizational structure to fit that status.

SEEKING A PARTNER: TALKS WITH DEUTZ

The unremitting flow of red ink in the foreign operations persisted through the remainder of the 1960s, despite the reorganization. There had been a short-lived upturn in Germany in 1967 with the arrival of the new Mannheim tractors, the first models designed specifically for the European market. But the agricultural market remained depressed, the old negative Lanz image was exploited by rival firms, and competition continued to be cutthroat, with numerous firms selling tractors below cost. Germany alone lost $19.6 million in 1968, $20.6 million in 1969, and a very unsettling $25.7 million in 1970. The worldwide sales of Deere & Company had risen steadily in the six years 1965–1970 and profitability was substantial, though it dropped toward the end of the decade. But serious losses from the overseas operations continued to be heavy (Appendix exhibits 26 and 27).

There were abundant criticisms of the company during this period. Moline management seemed particularly stung (perhaps more than they should have been) by a highly negative article in the German newspaper, *Die Zeit*, in May 1969. The title (translated), "The Gambled Heritage: The History of an American Failure," telegraphed its uncomplimentary message. "Region II is still a weak point. Here the shield animal, the leaping deer, goes lame on the hind legs," the author began. "The new uncles from America came with their pockets full of money" but "their European ambassadors were not good publicity makers. . . . They still do not master the German language . . . they confirmed Jean-Jacques Servan-Schreiber's conclusion that the Americans made more mistakes in Europe than their competitors. At the same time they refuted his opinion that they could well adapt to local conditions."

Deere certainly was not the only American firm to be raked over the coals by the Europeans at this time. The earlier worship of American management superiority, in the late 1940s and early '50s, had long since dissipated; not only had European recovery brought renewed strength in its business sectors (with concomitant pride and chauvinism), but the unpopular American involvement in the Vietnam War negatively affected United States business firms abroad. Still, one cannot gainsay the fact that Deere itself continued to be viewed as an inept newcomer even after fifteen years in Europe. Most of the "bad press" was a direct function of the company's unremitting losses. It was hard for Deere to take, and by the fall of 1970 there was a siege mentality in Moline about the overseas troubles.

At this point, McGuire wrote Hewitt a blunt letter outlining his thoughts on possible alternatives. The most critical issue, he felt, was the Mannheim tractor factory: "Unless we can solve this problem we will continue to have poor performance. It cannot be solved by lower-cost designs alone. We need this, and in addition the very strong benefits to be obtained by economies of scale." McGuire sketched out six possible alternatives: (1) make all the Dubuque tractors in Mannheim; (2) form a joint venture with Klockner-Humboldt-Deutz, the Cologne, Germany, truck and tractor company; (3) consider essentially the same notion with Fiat, the Turin, Italy, conglomerate; (4) purchase tractors with or without equity involvement from one of the British manufacturers, David Brown or Leyland; (5) purchase tractors from Japan; (6) close down and liquidate the Mannheim operation.

Each of the six alternatives had danger signals embedded in it. If all the Dubuque tractors were made in Mannheim, McGuire foresaw a break-even operation, but with serious problems back at Dubuque and Waterloo. Either of the joint ventures had promise, but there would be major difficulties in meshing the product lines and manufacturing facilities, as well as rationalizing the two managements, the marketing structures, etc. Either one of the British proposals, said McGuire, "would be a business of retreat in the eyes of people, but in view of our loss situation it is not totally unpalatable." The Japanese had had little experience with agricultural tractors: "Our experience with Hitachi did not give us a base of experience on which to build." The closing of the Mannheim operation "means simply getting out of the tractor business except perhaps for the pocket markets.... We could remain in the pocket markets of Argentina, Mexico, Spain and Australia, and be a short-line implement combine dealer with the EEC." But this had some strong negatives: "Our dealer organization is not strong, but in all probability this action would wreck them and cause chaos for a number of years.... It's a solution which is almost impossible to reverse at a later point in time should our successors seek to do so."

A wide divergence of thinking emerged in Moline about these alternatives. Some of the directors did not like any of the six proposals, preferring to continue frontally attacking the market for increased share, just as in the past. But this seemed to others a counsel of despair; the losses, if continued, would eventually erode the strength of the domestic operation. There was some support for exploring a joint venture. At this point, surprisingly without a formal board vote, conversations were initiated with Deutz.

Secret negotiations had been conducted between Deere and Klockner-Humboldt-Deutz AG as early as 1966. At that time the suitor was Deutz, through its chief executive officer, Dr. Karl-Heinz Sonne; Deere's representative was Curtis. Deutz wanted to increase its volume, and Sonne proposed that his firm supply Deere its tractor engines, with Deere providing transmissions for Deutz. In addition, Sonne intimated a possible relationship on combines (Deutz owned the Fahr organization, one of Germany's better-known combine manufacturers). There were also suggestions that Deere might buy Deutz forgings and gray iron castings, Deere to reciprocate with some sourcing for the Deutz Argentine operation through Deere's foundry operation there, a company called Cindelmet. Nothing came of the negotiations, however; Deere had just put its French engine factory at Saran into full operation, and the losses from the German operation were not yet so fearsome as to pressure the company to make a change. Indeed, Deere appeared a bit cavalier in the process of rejecting the overtures.

Four new years of astronomical losses in Germany changed the tune. In 1970 Deere was the aggressor; Curtis and Sonne again were the key negotiators. The two met in Cologne, Germany (the home office of Deutz), in April; once more the negotiations were highly confidential, for fear that premature public notoriety via the always assiduous German press might unduly upset employees, shareholders, bankers, and other interested parties.

The scope of the proposal became much larger. Indeed, Frank McGuire, who handled the on-the-spot contacts, sketched a breathtaking vista in a letter to Hewitt and Curtis in August: "When we were together with Dr. Sonne and his associates, we touched on a full collaboration which in our context would have evolved as an end case into a joint venture up to a full merger." (Sonne, though, did not react as enthusiastically to the notion of a full merger; apparently, he felt that he could not sell it to his stockholders.)

At this point, Deutz was a company of approximately the same size as Deere, with total sales in 1969–1970 of just over $800 million; its profit, though, was only about $5.9 million. Deutz also made trucks and buses and was in the seemingly unrelated business of planning and constructing industrial factory buildings.

As the delicate negotiations continued through the fall, both parties envisioned combining all of the tractor production of Deutz and Deere in Europe, as well as combining distribution in many other countries (excluding the United States, Canada, Mexico, Argentina, Spain, or Iran). There were further possibilities, such as a joint venture in Algeria.

At the same time, the hidden agendas of the two wary parties began to surface, exposing three dominant problems. The first was an expected one: how could the two marketing organizations be meshed? Both Deere and Deutz were proud names; neither management group seemed particularly willing to phase out its own colors, let alone face the unpleasant vista of meshing the two dealer organizations in the field. This problem might have been surmounted had not two other contentious issues surfaced: what should be the equity positions of the two parties in the final company, and how would management itself be shared, both at the board level and the operating level? Sonne proposed a fifty-fifty sharing of ownership and voting rights, with Deere taking 70 percent of the profits (the assets Deere would contribute were more than 70 percent of the total); Hewitt felt that the profit sharing was too low for Deere. The second issue was even more sensitive: Sonne did not want any constraints on the ability of the joint venture to eventually sell in the United States and Canada. Deere was not at all interested in subjecting its highly successful North American operations to any intrusive competition from a jointly managed outside entity. On this point, the negotiations began to seriously founder. The "chemistry" between the two management groups had just not been right.

At just about this time, Hewitt happened to meet again an old friend, Giovanni Agnelli, the chairman of Fiat, the huge Italian conglomerate. Agnelli was addressing a symposium in New York City, sponsored by the *Financial Times* of London. His speech held some prophecy for Deere: "Creating Europe's Industrial Giants." When Agnelli suggested that Fiat and Deere explore a joint venture for both agricultural and industrial equipment (but to exclude all United States and Canadian operations), Deere decided to break off the negotiations with Deutz. McGuire notified Sonne in May; the latter's gracious reply preserved the amicable relationship, Sonne even suggesting further talk about the Algerian project. But this never came to fruition and the Deutz-Deere negotiations were irrevocably terminated. Later that summer, the public finally was apprised of the aborted relationship when Sonne, answering a question from the floor at his company's annual meeting, replied that the project "had failed because KHD was not willing to go along with Deere's demand to desist from exporting tractors to North America."[1]

SECOND EFFORT: THE FIAT VENTURE

The Fiat negotiations, begun in December 1971, proceeded with astonishing rapidity. By late January Curtis and McGuire headed a task force of some forty Deere people from both Moline and the German headquarters, "operating under very tight security." Curtis exhorted all the participants that "absolute security is essential" and warned them not to do any trading of their Deere stock "until such time as the project has either been abandoned or been the subject of a public announcement." Harold J. Berry, Deere's outside investment counselor (who by this time was with Merrill, Lynch, Pierce, Fenner & Beane), had been brought into the discussions, as were Deere's bankers in London, its auditors in Chicago and Milan, and lawyers from both Chicago and Washington. Additional counsel in France and Italy were also employed. By mid-spring, Henry Dreyfuss and his colleagues had been called in, to be queried about incompatibilities in design between some of the products of the two companies.

Fiat was the largest industrial enterprise in Italy, one of the largest companies in Europe. About 88 percent of its overall sales were automotive (automobiles, trucks); it was the second most important automobile producer in Europe, outranked in sales of units only by Volkswagen. In addition to its namesake car, it owned Alpha Romeo, half of Ferrari, and 30 percent of Citroën,

Exhibit 14-1. Suggested logos for the proposed joint venture of Deere and Fiat. *Deere Archives*

the French auto producer. The remaining 12 percent of the company's sales were accounted for by an incredible variety of operations—production of iron and steel, shipping, trading, financing and insurance operations, production of railway rolling stock, marine engines, aircraft, nuclear energy, engines and turbines, machine tools, domestic appliances, and lubricants.

Most important to Deere, Fiat also made agricultural tractors and industrial equipment. In wheel tractors, Fiat produced just over double what Deere was making in Europe in 1969 (Deere at that time had 4.2 percent of the market, Fiat 9 percent). Fiat also produced wheel tractors in Argentina, where it had 57.3 percent of the market (to Deere's 19.8 percent). Both companies also operated in South Africa; there they were quite close (Deere's 10.4 percent of the market balancing with Fiat's 13.7 percent). Deere also produced combines, its market share in Europe about 7.2 percent. Fiat was a major producer in Europe of crawler loaders and dozers (Deere was a very small force there); the joint venture would have about 35 percent of Europe's market. Fiat was also substantially involved in wheel loaders, back-hoe loaders, and hydraulic excavators.

One particularly important asset from Fiat was the mechanical front-wheel drive assist in its 60-90 horsepower sizes. Upwards of 20 percent of Fiat's sales in Europe were in four-wheel-drive vehicles. Curtis wrote Hewitt in early 1971: "We have never produced a four-wheel drive tractor at Mannheim because the standard mechanical four-wheel drive is not a practical possibility due to the design of the transmission of our tractors. . . . Some two years ago Mannheim was granted a significant appropriation to develop a hydrostatic front-wheel drive patterned after the Waterloo type. . . . This effort is not yet successful." The lack of this front-wheel-drive capability had hurt Deere's European thrust.

Measuring the two companies side by side was difficult—the complex negotiations attempting to meld the two organizations had to grapple with these comparisons in a minute way. Deere's and Fiat's European agricultural and industrial equipment—when combined with the two companies' South African and Argentine operations, together with Fiat's Brazilian operations—added up to be two units of approximately the same size, close enough that a joint-venture could be contemplated without too much adjustment in working capital or other compensations. The Fiat agricultural and industrial enterprises were only a small part of the total Fiat empire, and not one of its most successful, either. The dominant size of the total Fiat operation was especially important because the incredibly complex Fiat operations had a great many merged costs, a complication that plagued the joint-venture discussions from start to finish. Fortunately, from the beginning, Fiat made it clear that it had no intention of eventually pushing the joint-venture company into the North American tractor market, à la Deutz.

The charisma of the Agnelli family was legendary, and its titular head, Giovanni ("Gianni" or "Johnny"), was the most legendary of all. (His brother, Umberto, was managing director of the family enterprise and also an important figure in the story.) In a fascinating article in the magazine *Car and Driver*, just at the time Deere entered negotiations with Fiat, Giovanni was called "the Prince" with an undisguised allusion to Machiavelli. The article spoke of his "multifarious" activities, the author clearly intending an unflattering implication. The writer, perhaps carried away by his own prose, called Agnelli "a beautifully combed and perfumed gray mouse running breathlessly on the inside of a spinning drum as scenes of technicolor utopia are flashed on its mirrored inner surface. . . . Gianni Agnelli is executive jets, double-breasted shirts, holidays at the world's glamour spots, contacts with hyper-isolated beautiful people." After painting such a broad brush, the article then enumerated the amazing variety of activities that Agnelli presided over and called him "the commander-in-chief of Italian industry." Both *Newsweek* and *Time* had also featured major articles on him at this time; he was even on the cover of *Time*, which called him "the most widely admired and envied Italian industrialist—the *Numero Uno*." British journalist Anthony Sampson, in *The New Europeans*, commented, "Agnelli has a mythology not unlike President Kennedy's. . . . Clearly his presence fills some kind of psychological gap."

Superlatives sometimes overstate, but in this case Giovanni Agnelli appeared to be in real life as impressive in stature as were the accounts in the press. It was a heady opportunity for Deere, far and away the most exciting new venture in the post–World War II years. In substance and style, the notion of putting together Fiat and Deere was a compelling proposition; both were major producers, well known for their products, both had outstanding and distinctive managements, particularly at the top. In sum, the combination boded great promise.

Major study committees were formed by the two companies—a manufacturing group, a marketing group, and a financial group, each to report back to a central negotiating body. The latter, in turn, amassed all of the detail and incorporated it into an instrument called the "Heads of Agreement." It ran only thirty-five pages, including two appendixes, and yet in this short space incorporated the essence of the joint venture proposal in such a form that it could be acted upon operationally by the two boards. The new joint entity would include all of Deere's European holdings, as well as those in South Africa, Argentina, Iran, and Turkey; Fiat would contribute its agricultural and industrial equipment operations in Europe, as well as similar operations in Brazil, Argentina, Turkey, and Australia. All of Deere's European wheel tractors and all of Fiat's agricultural wheel tractors would be included, and Fiat would include its agricultural crawler tractors, back-hoe loaders, and crawler dozers. Deere would include certain of its crawler dozers, too. Deere and Fiat were to include their four-wheel-drive loaders, Fiat would contribute its hydraulic excavators, and Deere would include certain models of scrapers, graders, and skidders. All of

Deere's combines made in Europe would be included, as well as Deere and Fiat balers and forage harvesters. The capital of the joint-venture company would be held 50 percent by each of the parties, with the agreement providing ways of bringing each company's total contributions of working capital, equity, and borrowing to an equal amount. The agreement was to be carried out by July 31, 1971, if possible.

As this document was to be formal authority to proceed apace, the Deere board, when it met on May 17, 1971, knew the meeting had great portent. A number of misgivings surfaced in the many hours of discussion. Hewitt laid out the rationale for bringing the "Heads" agreement to the board: "A public announcement of some sort would be required before completion of these negotiations because too many people in both organizations would be involved in them to permit effective security. Since both parties believe it is unwise to make such an announcement without reasonable assurance that the venture will in fact be consummated, a commitment like the Heads seems a necessary step in the negotiating process. Further, a definition of the broad outlines of the venture . . . will be necessary to support certain discussions with governments, which it is desirable to complete before execution of a definitive agreement." Several members of the board then presented the economic rationale for the project; the "fit" appeared excellent, the case seemed strong.

There were risks: such a step might foreclose all other alternatives, there were knotty details in several important areas, disclosure would itself make the roles of the two companies more complicated. There was significant opposition from at least three of the directors. George French abstained, stating that "while he felt intuitively that the venture should be a useful solution he lacked sufficient information to make a reasonably prudent decision." D. C. Glover not only felt that information was lacking, but argued that the "Heads of Agreement" was a firm commitment that could not be turned around were there to be additional compromising information; he voted against the motion. A third board member, Frank Dickey, abstained, though not putting his misgivings on the record. An overwhelming majority of the board did want to go ahead, and the motion approving the "Heads of Agreement" was passed. Fiat, meanwhile, also had agreed to the proposition.

Now a public announcement of the joint-venture-company proposal was inevitable, and the press coverage (particularly in Europe) was substantial. The journalists commented extensively on the scope of its endeavor, its competitive implications for agricultural machinery in both Western and Eastern Europe, and in general seemed to applaud the move. There were some caveats, though; one influential French magazine asked how a fifty-fifty arrangement could work "by a Latin Expansionist and a Puritan Mid-Westerner." The German press seemed less complimentary to Deere, one paper editorializing: "When such a centralized company strives for a joint solution in an important part of the world market and in doing so allies itself with such a significant partner as Fiat, which is far stronger, then a great deal must have occurred to change the proud minds of the Americans." The Italian press, understandably, concentrated more on Fiat. The agreement did honor to the "enterprising Italian spirit" and would open up new work possibilities and social benefits. One paper alleged that Fiat was "in a state of extreme confusion," and some of the press implied that the venture was a labor relations ploy, that there would be strikes by "Leninist cells." A British correspondent asked whether "the 'new giant' would panic others into 'marriage.'" Much of the press commentary suffered from a lack of details—both Fiat and Deere had much work to do before any further public announcement.

Large teams from the two companies continued to work over the summer and fall of 1971. Agnelli paid a personal visit to Hewitt in Moline, and the latter reciprocated with a visit to Turin, both significant events in that they telegraphed to the organizations the continuing goodwill and desire for agreement on the part of the two chief executives. Cordiality was the hallmark of the meetings for the operating groups, too. Out of the enormously complex negotiations leading to the many drafts of the "Agreement for Closing," the essential features of the original proposal stayed in place. One Fiat plant (in France) was eliminated, there were complicated questions about Fiat's Argentine operations that had to be answered, there were concerns about timing and about transfer prices for components. Even more difficult to assess were misgivings about the political stability of the project in Italy and in some of the Latin American countries.

Undoubtedly the most difficult problem throughout the negotiations was the projection of costs and profits. Deere was long used to being able to generate precise figures for such proposals; its American accounting practices, established by Curtis and his predecessor, T. F. Wharton, had given the company an analytical financial capability that demanded equally high caliber analysis from other participants. Fiat's accounting practices were truly different from those in the United States. The Fiat accounting and finance system permitted far greater discretion and the Deere executives found it frustrating not to get exact figures comparable to their own.

There was a further problem, unique to Fiat: its many other manufacturing endeavors (particularly those in automobiles and trucks) generated a wide range of common costs and intracompany interchanges that were almost impossible to quantify precisely. From the start, the Fiat projections of the joint-venture-company costs and profitability were quite optimistic, but Deere could not be certain just how realistic they really were. By November, it did appear that some of the projections needed to be downgraded, and the profitability of the joint-venture company seemed less attractive than it did back in May—they were "quite unsatisfactory . . . considerably less optimistic than those contained in the feasibility study," said one top Deere executive. The Fiat negotiators were game, though, and now suggested certain changes—particularly the exclusion of their French manufacturing plant, at Bourbon Lancy, to eliminate some of the overcapacity that had crept into the joint-venture company. In addition, some of the components that were to be supplied by Fiat to the joint-venture company were shifted over to the joint-venture company itself. "We'll guarantee a profit," one Fiat executive promised. The Deere negotiators, however, estimated that only about one-third of the improved financial projections of the new plan represented benefits that would have fairly certain consequences; the remainder represented only a more optimistic assessment of projected economic conditions in the near future.

Fiat at this point began to press Deere for a quick decision. Apparently Fiat felt the pressure of time; they had already spent the previous two years studying a joint venture with Allis-Chalmers, now they had invested another year with Deere. Given Fiat's skittishness, the Deere board decided to meet in an extraordinary session to be conducted over two days, January 11 and 12, 1972. The negotiating team presented all aspects of the proposal, including the new Fiat suggestions. Frank McGuire summarized the conclusion: "The likelihood of profitability which had been demonstrated was insufficient to justify taking the risks and accepting the burdens which are inherent in the joint venture. While it is conceivable that a more thorough probing of Fiat costs, especially for components to be sold to the joint venture, might uncover further possibilities for improvement, such probing could not be completed within the time period for final decisions which Fiat insists on." McGuire cited the further problem of loss of control in a fifty-fifty partnership and commented on the rather unstable political situation in Italy. Finally, the matter was put to a vote; the final decision, to terminate.

Agnelli fortuitously happened to be visiting Henry Ford II in Detroit the following day (Agnelli was in the city to address a meeting of the Society of Automotive Engineers). Hewitt and Curtis, leaving directly from the board meeting, were at the Ford estate for breakfast the following morning. The evident friendship that the two chairmen felt for each other continued to the end. Agnelli asked only one question: "Was it because of our government?" When Hewitt replied in the negative, Agnelli responded, "I have been in your home, and you in mine. Let us continue our friendships that way." The two men shook hands and the Deere-Fiat joint venture had ended.

GOING IT ALONE

As company officials took a retrospective look at the Fiat endeavor, they counted many pluses: "We learned a great deal about ourselves in the process." There is much truth in this assertion; the project forced company personnel to think in a more system-wide perspective, and the company became more "multinational" in the process. Yet there were high costs to the project, too. The salaries and expenses of dozens of people working many months on the project were substantial, particularly for top management. Still, this is a normal expenditure for any alert, growing company. There is little evidence that the out-of-pocket costs of the Fiat project were excessive; profits in 1971 were $63 million—5.3 percent of sales—the second highest profit in the company's history (exceeded only by 1966, at $78 million).

The nonmonetary costs of the negotiations were more substantial. Perhaps no other issue in the post–World War II period had so polarized the board as did the Fiat project. These were serious divisions, going beyond the usual day-by-day differences of opinion. D. C. Glover, at this point no longer in active management, was the most vocal critic, not only of the project but of the management style behind it. Glover abstained in the final vote for termination in the meeting of January 12 but made two suggestions for the record. First, he wanted "management" to present to the board at the following meeting "some suggestions for a solution to our foreign business problems." Second, he advocated that "the Board consider the desirability of bringing in one or more outside Board members with experience in overseas matters." Hewitt agreed that bringing in outside directors should be considered, though he felt that these should not necessarily be limited to persons whose experiences were primarily related to overseas activities. Glover was also exercised about a critical article that had appeared in the *Wall Street Journal* a few days earlier, alleging that the company had incorrectly reported its earnings for the year. (The allegation was patently untrue, and Hewitt adroitly answered the accusation by sending a printed copy of the article, with his marginal comments, to every shareholder.)

George French, too, had been in the "loyal opposition" to the Fiat project; there again the dissent ran deeper than just the Fiat project per se. French had objected to the haste with which negotiations were conducted and to what he felt was inadequate preparation by the Deere team for final board approval. French wrote Hewitt a blunt private letter, advocating "other options" (but did not specify any particulars). Shortly after the Fiat termination, French submitted his request for retirement from his post as senior vice president of the industrial equipment division; he remained a member of the board until 1974.

At the meeting of January 25, Hewitt reminded the board that at the beginning of the Fiat negotiations there had been three alternatives for the company: (1) withdraw from overseas business; (2) find a partner; or (3) go it alone. After a morning of discussion, the dominant opinion clearly favored the third alternative. Hewitt urged each director "to raise a question if he had any misgivings," but none surfaced. The results for 1971 were available at this time; the losses in Germany alone amounted to $26.5 million, the overall losses in the foreign operations were more than $23 million. Was "going it alone" going to continue such a pattern?

CHAPTER 14

When news of the aborted Fiat negotiations became public, there was surprise and consternation among Deere's European dealers. Competitors were quick to exploit the malaise. Some Deere dealers were soon inveigled to defect to competitor products; there had been a particularly skeptical *Forbes* magazine article at the start of the negotiations that seemed to imply that the Fiat effort was a "do or die" hope by the company; one German competitor had the article translated to show to Deere dealers. Neel Hall, by this time director of all the European operations, was worried enough to urge Hewitt personally to attend the company's new product presentations scheduled for Zweibrücken in September. Referring to the *Forbes* article, Hall commented: "Taken out of context, these comments would make some people believe Deere's tenure in Europe would be short. I think that a statement from you to the dealers at this dealer meeting would be very helpful."

Hewitt not only agreed but, when the time came, brought all of the directors of the company with him for a major two-week visit to Germany, France, and Spain. The directors' wives were invited; even Charles Wiman's widow, Patricia, came for part of the sessions. A special meeting of the board was held in Heidelberg on September 11—the first time that a board meeting had ever been conducted outside the United States. Collateral meetings also took place with Pehr G. Gyllenhammar, the president of the Swedish automobile company, Ab Volvo. (Hewitt and Gyllenhammar were discussing a possible collaboration, though this did not later come to fruition.)

An international press conference was held early in the visit, with wide coverage by most of the key European business press. Hewitt took the opportunity not only to emphasize the company's North American strengths and the strong new product lines both in North America and Europe, but made it explicitly clear that Deere intended to stay in Europe. "We are carrying out this lengthy program . . . to make manifest to our employees, dealers, customers and competitors that we view our European operations as an integral and important part of our Company." Reactions from the press were reassuringly positive. *Le Figaro,* the French newspaper, headlined its article, "The American group John Deere takes up the offensive in Europe," and other papers commented on Deere "turning a new leaf" and deciding to "go it alone." An influential British trade paper emphasized "JD's aggressive market tactics." The directors' trip was a resounding success, reminiscent of the euphoric Day in Dallas in 1960; it seemed to defuse competitor inroads and certainly stopped the dealer raiding in its tracks.

Fiat resumed its negotiations with Allis-Chalmers and in 1973 did develop a viable joint venture, but limited in this case to industrial equipment. In 1973 Deere and Fiat once again considered a collaboration, a joint venture for the manufacture of diesel engines. Fiat also hoped to sell its trucks and forklift vehicles in North America. After a brief exchange of ideas, Curtis wrote Fiat that Deere had decided to continue alone in its diesel engine production, but he mentioned again that "the friendship and trust of the Fiat organization are important to us and are a source of satisfaction to us."

The judgment of the board about "going it alone" did indeed prove correct. Few would have predicted that a turnaround would come as quickly as it did, however. The company had put in place a sound organization overseas—the product lines were improved, the manufacturing operations had been realigned and rationalized (at considerable cost), and the dealer organizations were, if not excellent, at least adequate to the job. Further, there was a substantial upturn in European business in the early 1970s, already becoming evident toward the end of the Fiat negotiations. To the evident delight of everyone, the overseas results for 1972 were a net $2.9 million profit (after a $21.5 million loss in 1971). Australia was still in a start-up position and had lost $1.4 million, Spain had lost $2 million, and the German operation still showed a $9.3 million loss, but the other major operations were profitable enough to compensate. In 1973 the results were even more satisfying—the overseas operation as a whole netted $26.9 million profit. Though the Mannheim factory had still lost $3.7 million, the Zweibrücken factory and the domestic branch were profitable enough for the German operations to show a sizable profit. In 1974 the overseas operation registered a profit of $23.1 million, and this time Germany showed a profit of more than $13 million. The company's decade-and-a-half search for a profitable multinational status had finally been realized.[2]

ECONOMIC NATIONALISM IN MEXICO

In several countries in which Deere manufactured or assembled equipment, the ingredients of a "pocket market" were in place. Usually this term connotes a generally small, local market within a country, insulated from the outside by high tariffs or outright prohibition on imports. The cost of manufacturing and distribution by domestic firms is often higher than it is in the international market because of small production runs, inefficiencies in labor, inadequacies in machinery, and so on. Government officials in "pocket market" countries face complex choices about how much international competition the country can take; the strength of indigenous businesses and the economic fortunes of their employees are deeply affected by such competition. This insulation is perpetuated by requirements that foreign companies manufacturing or assembling in the country use considerable proportions of locally made components and parts and that portions of the foreign companies' local operation be owned by nationals. Economic nationalism in countries around the world varies widely from minimal to almost prohibitive. Deere now began to face increasing constraints.

By 1967 the Mexican government had begun to insist on 60 percent local content. With engines and most of the transmissions used by Deere still manufactured in the United States, almost all of the remaining components and parts had to be manufactured or purchased locally. Costs for these were typically higher than in the United States; the bugaboos of small production runs and less efficient manufacturing practices both held sway. But a pervasive populist sentiment, in place since the Mexican Revolution in 1910, made it politically unpopular to penalize farmers by "high" prices of the final product. So price controls were rigidly applied in the tractor sizes manufactured by the company in Mexico (the 60–120 horsepower range); they were particularly onerous during the administration of President Luís Echeverría Álvarez (1970–1976), when politics split the Ministry of Industry and Commerce into two separate groups. Industry lost the function of price control and import permits for machines and parts to Commerce, whose political constituency was the customer. The Ministry of Agriculture also wielded considerable power and seemed always to favor low import prices for farmers. Deere's operation in Mexico did eliminate losses after 1962, and even during the 1970–1976 period continued to make modest profits, the $1.7 million figure in 1976 being the largest (it was about 4 percent of its sales). But profitability only became possible because of good margins on imported tractors in the higher horsepower sizes and on imported combines. The Mexican-built tractors were losing money throughout this period.

In 1977, the Mexican government announced a new "Automotive Manufacturing Decree," which for the first time included tractors. Local content was to be reduced from a flat 60 percent to a lower percentage tied to the ability of the manufacturer to export some of its production from Mexico—in effect, a "compensatory exports" credit. Mexican tractor companies generally imported substantial amounts of larger horsepower tractors, and some also imported other expensive specialized machinery, such as combines, which might cost upwards of $50,000 each. Were the tractor companies to be forced to compensate for all of these expensive, specialized imports?

The formula, as eventually evolved by the government, was a complicated one. In essence, Mexican companies were to be required to export the equivalent of 70 percent of all their imports, the latter to be figured at cost price (70 percent of list). In turn, a company could drop its local content from 60 percent to 50 percent and could even further diminish these figures by increasing the actual growth in sales of its Mexican operation (that is, expanding companies were to gain further concessions).

These new requirements were double edged. On the one hand, the drop from 60 percent to 50 percent local content was a major break—some comparatively expensive operations could be eliminated. For example, in the early 1980s Deere was able to import fully assembled engines from the company's European operations, whereas in the past substantial local assembly of motors had had to be utilized. The negatives lay in the rigorous requirements for export. During the recession of the early 1980s it was particularly difficult to fulfill these requirements. Among the Mexican tractor manufacturers Deere came the closest to meeting the compensatory equation; choppers and shredders were shipped from Mexico to other Deere operations abroad, under the aegis of the Ottumwa factory (Ottumwa losing the production in the process). But Deere and the other Mexican manufacturers had increasing problems finding export outlets, for nationalism and protectionism were running rampant around the world and local manufacturers in other countries pressured their governments to protect them from outside competition. Finally, the Mexican authorities realized that they had to be flexible in applying the export requirements, and there was some informal letup in their pressure on the companies.

In 1978, under the government of President José López Portillo, the Ministry of Agriculture and the Rural Credit Bank announced an ambitious expansion program to increase areas under cultivation and to significantly mechanize agriculture. The government in earlier years consistently had been the largest purchaser of tractors; the substantial new government orders were to emphasize large tractors, those particularly built by Deere. Because Mexican manufacturers did not have enough capability to supply such an order as well as the rest of the market, the government initially intimated that substantial imports were going to be necessary. But this brought an outcry of criticism in the public press—why did the country have to import tractors when it had a domestic industry? The foreign companies found themselves in a "Catch 22" situation. If they were allowed to import, there would be intense local reaction; if they expanded domestically in Mexico, the stringent price controls and harsh requirements on local content and compensatory exports made the prospects quite unattractive. This, too, was not what the public wanted to hear.

There was another alternative—the government itself could take care of the increased production needs. By the mid-1970s the pressure for "Mexicanization" already had led both Ford and Massey-Ferguson to suggest the sale of a majority position to the government, and in the early 1980s Ford did consummate such a plan (with Nafinsa, a government bank, taking 60 percent ownership). International Harvester, meanwhile, had sold a majority of its tractor and engine operations to a private Mexican company and had a link with a private Mexican bank for its industrial equipment. Massey-Ferguson sold its entire holdings to a large private conglomerate, the Alfa group of Monterrey; it retained certain licensing and sales agreements in this process.

The private manufacturers were naturally ambivalent about the government involving itself in production. Political pressures on the government by the farmers often forced short-term decisions on price control, and so forth. On the other hand, a government company often acted as an umbrella over price, particularly if the government operation was somewhat less efficient than a comparable private one.

Deere itself had a further complication: it was under increasing pressure about its own Mexicanization. The earlier attempt to sell shares publicly had aborted, and 26 percent of its shares were still held in trusteeship; by this time, Banco Nacional de México—Banamex—had increased its holdings to about 17 percent, with an additional 8 percent held by other private investors. The government's Foreign Investment Commission now ruled that trusteeship was not true Mexicanization, and this was upheld by the courts. Apparently the government feared that if Deere could successfully use a trusteeship, other foreign manufacturers would immediately attempt the same route; the government felt that this would thwart their own goal of a more nationalistic base for Mexican industry.

Companies in the country that were not majority-owned by Mexicans were not to be allowed to expand in any way. There was a particular threat in this for Deere, for Allis-Chalmers had entered the country to manufacture combines with significant local content. Deere already had some 65 percent of the market with its imported combines; if there were to be further constraints on importation because of the lack of Mexicanization, Deere's market share in combines would shrink.

Several choices were potentially available to Deere. The company could arrange a direct sale of the 26 percent block through private placement, with the buyer either a private group or the government. But such a sale did not provide new financing for expansion and had negative tax implications in the United States. A more desirable approach would be to constitute a new joint venture with a partner and have the latter put enough new capital into the organization to bring the Deere percentage down to 49 percent. Fortunately for Deere, the latter approach became a reality, for Banamex agreed in early 1982 to provide such funding and to join with Deere in increasing the capital structure.[3]

"YO-YO" NATIONALISM IN ARGENTINA AND VENEZUELA

Nationalism was not only strong in Argentina and Venezuela, it was also more unpredictable than it was in Mexico. While Deere may have preferred stability, like most business firms, it encountered in these two countries wide swings in political leadership and sharp changes in economic policies.

Deere's Argentine operation had been on a considerably larger scale than in Mexico, but profitability had been more elusive. (With the exception of 1964, there were continuous losses until 1972.) The decade after the mid-1960s was one of great tension in the country. From 1968, during the regime of General Juan Carlos Onganía, a spate of terrorist atrocities by urban guerrillas subjected Argentina to a reign of terror. The aims of the shadowy groups were predominantly political, but they chose to kidnap (and often kill) industrial leaders, both businessmen and labor leaders. Fiat's managing director had been murdered. Many foreign companies pulled out all of their nationals, and even Argentine nationals in companies were threatened by kidnap, attack, or maiming. The clandestine terror seemed directed particularly at foreign companies, but no operation, Argentine or foreign, was without its fears. One could never be certain whether the union leaders across the bargaining table truly represented their employees; outrage from the left and repression from the right dominated the country. Accompanying this was galloping inflation. The political scene soon took a surprising shift with the return of General Juan Perón to take leadership of the country. Perón died after a year and his demise eventually brought back another military government. The pattern of guerrilla attacks finally tapered off in the late 1970s, but the decade of turmoil had done great damage to the country's business system. In the late 1970s an agricultural depression also hit. Interest rates had been freed from control and had risen enormously—indeed, far over the actual inflation rate, and interest payments combined with greater agricultural taxation to put the farmer into depression. Agricultural machinery sales declined as a result:

	Total	Deere	Deere Market Share
1976	20,674	3,423	16.5%
1977	21,699	3,506	16.1
1978	6,666	1,744	26.1
1979	7,962	1,508	18.9
1980	4,877	1,413	28.9
1981	2,800	821	29.3

During the 1970s, four major manufacturers—Deere, Deutz, Fiat, and Massey-Ferguson—shared roughly one-quarter of the market each. Each of the companies had been forced to develop full-scale manufacturing facilities in Argentina; the government's local-content requirement had risen as high as 95 percent. Deere even had found it necessary to purchase a separate Argentine company, Cindelmet, a manufacturer of rough castings, blocks, heads, machine tools, and windmills. A series of decrees in the late 1970s eased the stringent local-content requirement, but at the same time the barriers on importations were abruptly eased. In the process, the indigenous Argentine tractor companies suddenly found themselves competing with lower-cost imports. This development, combined with the collapse of the market due to the farmers' depression, convinced Deere in 1978 to cease manufacturing in the country. The Argentine legal entity was terminated and the net tax loss of $17.5 million was absorbed in the Deere & Company figures for 1981. Manufacturing virtually ceased at the Rosario plant; the Argentine operation was converted to a branch, with part of the Rosario plant continuing to be used for assembly of components manufactured at other Deere locations in Europe and the United States. Cindelmet, the Argentine subsidiary, was sold.

Thus Argentina lost a tractor factory capable of producing at least 6,000 units per year. Deere emphasized to Argentine authorities that it had no intention of abandoning the Argentine market—it was willing to compete in a market dominated by imports. In a letter to Argentina's minister of economics, José A. Martinez de Hoz in May 1980, Robert Hanson, Deere's president, did warn, though, that "proliferation of tractor makes in Argentina through indiscriminate importation of units from all over the world cannot work to the farmer's advantage. . . . Eventually the market will sort out the winners from the losers, but I would suggest that without restricting imports . . . guidelines should require the manufacturer to prove that he has sufficient stock of repair parts, trained personnel and on-site service points reasonably close to the farmer to adequately maintain the operating integrity of the unit he hopes to sell. . . . Argentina has a great deal to lose by allowing itself to become an uncontrolled dumping ground for farm machinery exporters." With Argentina's great agricultural promise, the world will surely be watching carefully to see what evolves in the 1980s, as Argentina pursues this new path. Meanwhile, the country had lost a significant portion of its domestic agricultural machinery capacity.[4]

Much the same abrupt change of policy occurred in Venezuela in the late 1970s, and Deere again was caught in the middle. Shortly after Carlos Andrés Pérez Rodriguez took office as president of Venezuela in 1974, he invited tractor manufacturers from around the world to submit bids for a project to manufacture (not simply assemble) tractors in Venezuela. These were to be in Venezuela's popular sizes—from 60 to 120 horsepower—and were to have 60 percent or so national content. Venezuela had long suffered from the syndrome of predominantly imported goods; the tractor project was seen as an opportunity to provide critically needed agricultural equipment within the country. The Pérez government did apparently recognize that such a domestic operation, with small production runs, could not compete frontally against low-priced imported tractors, so the government promised a 30–35 percent customs duty protection to overcome any differential.

Twelve different international tractor manufacturers submitted bids for the project. The winner (this word seemed applicable at the time) was Deere & Company. A new company was established, Fabrica Nacional de Tractores Motores, S. A. (Fanatracto); the Venezuelan government held 45 percent ownership (through its Guayana development corporation). Deere took 20 percent, its local distributor took an additional 15 percent, and a financial organization reserved another 20 percent of equity for subsequent sale to other local distributors and groups of retail tractor buyers. The government said it chose Deere because of Deere's excellent reputation, high market penetration in Venezuela, and low cost projections for the project.

Fanatracto was to manufacture diesel engines—some 5,500 per year—and these were to be used not only for the tractors but as marine engines, in construction equipment, and so on. The government stipulated that the plant for the project be located in Ciudad Bolívar, a developing area in the southeast of the country, some 150 miles upstream from the mouth of the Orinoco River. This was not the ideal location for an agricultural machinery factory—the natural markets were far to the north and west—but the government was keen on developing this largely jungle area. Deere projections did show a modest but economically viable market demand for tractors in the country; there were further export possibilities to other Andean Pact countries, particularly Colombia, Ecuador, and Peru. The industrial site in Ciudad Bolívar was also to house a planned Mack-Fiat diesel truck factory.

By mid-1978 Fanatracto had acquired land and begun the construction of a substantial 25,000-square-meter plant. At the same time a pilot plant nearby had been opened for the assembly of a small number of Deere tractors from Waterloo as a part of a major training program for workers who would later be involved in the manufacturing plant. This pilot project soon had tractors rolling off its assembly line; one of them achieved considerable publicity when President Pérez rode it during the presidential election campaign in 1978. Within months, substantial numbers were coming off the assembly line, some 3,200 in total from the pilot plant.

But Fanatracto had enemies, and they began insinuating that the company was somehow cheating on the original agreement—just assembling tractors rather than manufacturing them. The company might have survived these charges had not two unexpected developments occurred. First, there was a substantial economic downturn beginning in 1979, bringing outcries in the farming communities about the high price of the domestic tractor as against what the farmers might pay were foreign tractors to be allowed in the country. The second event was a change in administration, with the election of President Luis Herrera Campins in 1979. Herrera had a substantially different economic philosophy from Pérez—he maintained that domestic demand was large enough to absorb both local tractors and substantial numbers of imported tractors, and he began allowing the latter to come into the country. In turn, Herrera refused to follow through on the customs duty protection promised by the previous administration (the minister of development had approved a 25 percent duty in principle, but it was never implemented). By late 1980, Fanatracto was losing so much money because of high interest charges on the construction project and the substantial losses on each unit of production that it was forced to close the assembly operation. The Herrera cabinet members concluded that the project should be terminated because its high-priced tractors "were not economically justified." When protests were made about the closing, Development Minister Quijada snapped that the only local content in Fanatracto's tractors was "the air in the tires." Other critics pointed out that the awarding of a factory to one company was monopolistic and that high Venezuelan manufacturing costs would never allow the tractor price to be competitive. In truth, the project could have avoided what became a chaotic tractor situation: more than thirty tractor models were imported from all over the world, with practically none backed by service and parts organizations. With Fanatracto's equity arranged so that dealers and groups of tractor users could have had ownership interest in the project, some of the prior difficulties in Venezuela of narrowly based, monopolistic companies might have been avoided.

By 1980, the project was moribund and the assembly operation abandoned. The new factory building, just completed, was put up for sale (with little likelihood that there would be a purchaser). A major article in the influential commercial magazine, *Business Venezuela*, concluded that while the project was "hailed during its short life as one of the most practical and best-planned projects initiated under the government of Pérez . . . it suggests to future investors—both local and foreign—that Venezuelan governments are incapable of recognizing and supporting sound projects inherited from their predecessors, and that a 'firm' commitment made by one group of politicians in power can be meaningless when another group takes office." The Deere executives who put five years of their time in a project that simply disappeared probably would second this judgment.[5]

CONTRASTS: IRAN, TURKEY, BRAZIL, AUSTRALIA

Another victim of political instability was Deere's operation in Iran, the change in fortune in this case caused by the Iranian hostage crisis of 1979–1980. In the late 1960s the Iranian government had launched a major agricultural development effort, and, after trying imported tractors from Romania with poor results, had decided that it wanted its own combine manufacturing plant in the country. Deere was contacted, and subsequent negotiations resulted in a proposal for a factory at the town of Arak (Sultanabad), south and west of Teheran. A joint-venture company was formed for the partial manufacture, assembly, and sale of Deere combines, with the intention that tractors and cultivating equipment be added at a later date. Deere's interest was to be 21 percent; a government corporation, the Industrial Development and Renovation Organization (IDRO) took 49 percent, with Deere's distributor in Iran and another private holder taking the remainder.

Arak was a backward area that the government hoped to industrialize; its first step was to erect a massive factory building—a "monument," said one of the Deere executives after seeing it. Though Deere combines did come off the assembly line in small numbers in the early and mid-1970s, the difficulties of manufacture in Arak and the lack of government-promised

Exhibit 14-2. Billboard advertising greater productivity with John Deere machinery, Teheran, Iran, 1976. *JD Journal*

financing of retail sales resulted in high losses. By 1974 Deere stepped back from major responsibility in management. (Deere had up to this point provided the managing director; now this person was to be furnished by the government.) The company continued to provide components to the operation from its German combine factory at Zweibrücken. The capital of the Iranian company was increased in 1976, and Deere decided to maintain its 21 percent ownership (the government's holdings being 51 percent). When relations worsened between Iran and the United States in the hostage crisis, all Deere business was suspended. The plant continued to be maintained, operating under the aegis of a workers council. After the hostage crisis was resolved, the Iranian company once again became an important customer for combines and combine components.

Internal political upheavals also caused difficulties for Deere in Turkey. From before World War II, one of Deere's most stable Mideast dealers had been the Cukurova group in Turkey, which sold both Deere and Caterpillar equipment. In 1966 the Turkish government threatened to close its borders to imported equipment, so the Cukurova group formed a new company for the assembly of Deere combines (with components shipped out of Zweibrücken); they called this new organization Cukurova Makina Imalat ve Ticaret A. S. (Cumitas). Deere invested $200,000 in the organization, approximately 10.6 percent of the stock. By 1969 the company began assembling combines at a factory in Tarsus. By the early 1970s its production was about 400 units a year; as it was the only combine factory in the country, it commanded an overwhelming market share. In 1975 the Turkish government also authorized domestic assembly of tractors and Cumitas expanded its operation accordingly. The economic and political situation in Turkey in the late 1970s was tenuous and, as no financing was forthcoming through the government, Deere agreed to a substantial loan in 1977. In 1979 the political situation in Turkey forced Deere to sell its shares in Cumitas (at a nominal sum) in order to retain the link with the company through the sale of combine parts from Zweibrücken and tractor parts from Mannheim.[6]

Brazil and Australia stood in contrast to these instances of intrusive nationalism. From the mid-1970s, Deere had been searching for a way to break into the Brazilian tractor market. In 1976 contact was made with a combine manufacturer, Schneider, Logemann & Cia., Ltd., in the southern state of Rio Grande do Sul, for a joint venture for tractor manufacturing. The Logemann family was already manufacturing large numbers of combines, having adopted a number of the Deere features. After discussing tractor possibilities, Deere and Logemann decided to expand the combine operations instead. A new company was formed, with Deere taking 20 percent of the capital stock (for a cash purchase price of about $6.6 million). Deere technology was provided to the new organization, with the quality of the Logemann combines upgraded in the process, and production began with high hopes. By the early 1980s the venture had not yet provided a way for Deere to gain a larger foothold in tractors, however.

The Australian operations of Deere, once viewed as a pocket market venture, suggest the pejorative implications of the term were probably unwarranted for two important reasons. First, though Australia did originally meet one of the criteria for being a "pocket"—it produced higher priced agricultural machinery that could not compete internationally—the differential had almost disappeared by the early 1980s. Second, the implication that a pocket market is a small market was becoming patently untrue in Australia. Its natural resources and climate make Australia one of the most important agricultural producers for the rest of the world. For its farm machinery potential, Australia certainly had to be classed in the 1980s as a promising market.

Deere had had a modest distributor network in Australia for many years, with several dealers and a central office in Sydney, but these were combined with one of its vigorous Australian competitors and the only domestically owned tractor manufacturer, Chamberlain Holdings, Ltd., in 1970. Out of this came a new company known as Chamberlain-John Deere, Pty. Ltd., operated as a wholly owned subsidiary of Chamberlain Holdings, Ltd. Deere acquired shares in the latter in a staged arrangement that gave Deere a 49 percent interest by 1976, with the Chamberlain shareholders maintaining a 51 percent holding.

The two lines then needed to be meshed. Deere provided the new company needed expertise in engineering, manufacturing, and marketing, as well as a whole range of Deere products. The larger Deere tractors were

Exhibit 14-3. The 50,000th Chamberlain tractor on display, Australia, 1981. *Deere Archives*

imported under Chamberlain aegis, with substantial assembly operations on them being done in Australia from parts shipped from all over the Deere system. Indeed, Australia became perhaps the most apt example of worldwide Deere sourcing—eventually pasture toppers were imported from Mexico; tractors from Mannheim; combines from Zweibrücken and East Moline; balers, mowers, and conditioners from the French operations; engines from Saran; and small tractors from Yanmar, Japan.

In addition, Chamberlain continued to make its own agricultural and industrial tractors, which had become very popular in Australia. Chamberlain maintained its own color—it was yellow, though not quite the shade of Deere's industrial equipment—and its tractor was marketed as a Chamberlain tractor, not as a Deere tractor. Market share in tractors rose to more than 21 percent, making Chamberlain the largest agricultural tractor company in the country (in 1980, Massey-Ferguson had about 15 percent of the market, International Harvester 13 percent, and Ford about 12 percent).

Australia has been an efficacious model for its foreign collaborators—it has pursued a much different approach than have Mexico, Venezuela, and Argentina in encouraging local industry. Rather than establishing tariff barriers, the government chose to use the bounty system. Thus, in 1981, the local automobile industry was given a 57.5 percent bounty, trucks 35 percent, shoes 34 percent, garments generally 51 percent. In the tractor industry the two companies with outside links, Deere and International Harvester, had been given bounties that had begun at higher percentages but had been progressively reduced to 15 percent in 1981 and to 11.8 percent in 1982. Peter Griffiths, the managing director and chief executive officer of Chamberlain, commented on this: "You can see that it is a lightly protected industry, which is another way of saying that it is efficient." In effect, the quality of the Australian operation had improved so measurably over the 1970s that, even with the lower volumes that were still in effect in that country, it was becoming fully competitive with outside machinery. The implications of this tor Deere in terms of worldwide sourcing and marketing are very important.[7]

IDEOLOGICAL ENCOUNTERS IN SOUTH AFRICA

Deere has operated in a few countries whose political and economic philosophies and practices raise issues that go beyond problems of nationalism per se. A country such as South Africa, with its apartheid racial policy, poses difficulties for foreign companies, both in their operations and their public relations. By the early 1970s the presence of American businesses in South Africa was being criticized increasingly in the United States. Elwood Curtis wrote Frank McGuire in late 1971: "The South African question is getting very hot in the United States right now, particularly since the so-called 'Black Caucus' of all black Congressmen has interested itself in the questions. . . . We do not believe the Black Caucus . . . will be particularly impressed if we say Deere & Company is not involved because we own only 50 percent. . . . We need to be prepared to answer questions and meet accusations."

Deere, not particularly alert to such issues in the 1960s, had allowed its wage rates for unskilled employees to fall below that of the Johannesburg area. In 1971, the top hourly rate in Deere's plant at Nigel—which, because of South African government job restrictions, was used only for white employees—was higher than that at any other of the Deere plants outside of the United States and Canada. At the same time, the top rate for common labor was lower than anywhere in the world except Argentina. (The spread between the common-labor rate and the toolmakers rate in Nigel was 643 percent; the spread in the company's plant in Mexico was 285 percent, in Argentina 178 percent, in Germany 154 percent, and in the United States 148 percent.)

These figures galvanized the Deere board to action, and over the next few years the pattern began to change. By early 1973 McGuire wrote Hewitt: "There is a level of higher paid work referred to as 'European work,' which traditionally was not for the blacks. This has been broken and will continue to be encroached on by steady pressure and without Government approval, but with tolerance if they are not embarrassed by the process." McGuire continued: "We should . . . be at the forefront in our declared policy of equality and fairness equal to all. Wherever we do business around the world we should be out in front. . . . We should not stay out of places where the policies are not today to the level of man's dream of full social justice or run away if a turn backward is experienced. A presence there is a positive influence, even though small, and an important factor contributing toward achieving the goal of equal opportunity.... The force needed is a steady one of encouragement to make more progress."

Realigning South African employee relations practices to fulfill McGuire's goals was not easily carried out, for not only was there government pettifoggery, but the South African employees of the company were reluctant to change. The Moline personnel in charge of employee benefits and wages had isolated a number of discrepancies between practices in the South African operations and the rest of the company. When this was pointed out, the South African managing director wrote back to Moline: "I doubt that there is anyone in South Africa who is familiar enough with Deere & Company wage administration policies and procedures to effectively apply them or make a valid comparison—or adapt them to the peculiarities of the South African labour situation.... I question whether there is anyone in Moline sufficiently well acquainted with the current labour situation in South Africa to effectively analyze the current wage policies and procedures as they are applied to our local operation." He was overruled by Moline.

By April 1974 the board in Moline discussed a major policy statement on its South African operations, elaborating on the company's stepped-up efforts for training black employees and calling attention to the uniformity of benefits applied to all employees at the operation. In doing this, the board registered some caveats: "Deere South Africa does not engage in political activity. Its relations with the Government of South Africa consist basically of complying with that country's laws, including the payment of taxes.... Deere South Africa is required to observe the laws of that country." The statement concluded: "Deere & Company also believes its presence in South Africa is beneficial to its stockholders and to its employees in South Africa, non-white as well as white.... The Company also willingly accepts the social responsibility for being a good citizen wherever it operates, including South Africa."

Deere moved on a wide front over the remainder of the 1970s in bringing up wage rates for black employees in the plant, making promotions available, extending employee benefits, and so forth. By the late 1970s, Deere was paying one of the highest wage rates for black employees in the entire country; however, labor shortages in the skilled occupations (still almost exclusively the domain of the white employee) caused an extremely short supply in that period, and those wages, too, skyrocketed, so that the differential still stood at about two to one. Though government regulations still "required" segregated washrooms and lunchrooms, Deere decided in the late 1970s to eliminate separate facilities. On the Nigel grounds a new facility was constructed for a combination of cafeteria and washroom, the whole to be available to any employee, white or black. The historic unwillingness of the white employees to mix still held firm, however; no white employee used the washroom or lunchroom facilities. Only at major employee functions—service award ceremonies, and so on—were the white employees willing to sit in the same room with the black employees.

The company reevaluated its policy statement on "employment practices in South Africa" in early 1977, and this time the wording was more pointed: "Current racial laws, practices and attitudes in South Africa are in conflict with fundamental views of justice and equality held in this country and in many other nations of the world." The policy statement reaffirmed Deere's belief that "our continued presence in South Africa is constructive and proper." The statement detailed Deere's progress in upgrading wage rates and employee conditions and asserted "that our presence there is helping to improve human conditions for non-whites." While "being present in a country involves, we believe, an obligation for us to obey the law, to be progressive and constructive citizens, and to be sensitive to local institutions," and though the government's constraints "have deep historical roots that have been gradually reinforced by legislation, institutions, customs and time . . . it does not make them more consistent with principles of human dignity or with practices of a fully functioning free market economy. . . . To the extent that South African laws, customs and practices are inconsistent with the principle of equality of opportunity, they are at odds with a fundamental commitment of our Company."

The record in the late 1970s and early '80s documents widely divergent attitudes about the question of foreign investment in South Africa, from both black leaders in South Africa and from representatives of many groups, organizations, and governments outside the country, with the debate particularly acrimonious in the United States. A number of advocacy groups in the United States (particularly various church groups) espoused strong positions, pro and con, on this issue. Some argued for a militant approach, urging American business corporations to close operations in South Africa and leave the country. Others advocated a more evolutionary approach, urging that pressure be put on the South African operations of American companies to upgrade employment conditions and educational and economic opportunities for their employees in South Africa and to become advocates and leaders of social change. A third group opted for a more detached "live and let live" approach—in effect, staying in step with whatever decisions are made by the South African government itself.

Deere became an advocate of the second approach. It decided to stay in South Africa and be a constructive force for change. Deere has always been frank in stating its intention to stay in South Africa; the board documents in 1973 and 1977 made this clear. Thus Deere would not satisfy those who advocated full-scale withdrawal. Deere's advanced employee relations concepts have become widely known in South Africa, however, and no advocacy group pressure has been brought to bear on the company (as it has for so many of Deere's colleague American companies).

This approach by Deere (and a number of other United States companies) has been measurably enhanced since 1977 with the development of the so-called "Sullivan Principles," promulgated by the Reverend Leon Sullivan, minister of the Zion Baptist Church in Philadelphia, Pennsylvania. In 1970 Sullivan had joined the board of directors of the General Motors Corporation, becoming the first black leader to hold such a post in a major American corporation. Sullivan had concluded that those United States corporations that were measurably exerting pressure on the South African government for change should be favorably acknowledged and that those who were standing pat should not be allowed to cover up their position. Sullivan first elaborated a set of principles concerning nonsegregation of the races in eating, comfort, and work facilities; equal and fair employment practices; equal pay for comparable work; training programs to prepare blacks and other non-whites for supervisory and technical jobs; more opportunities for blacks and other non-whites in management and supervision; and improvements in employees' lives outside of the work environment. United States companies were then asked to subscribe publicly to these principles, with Sullivan setting up a monitoring process (using the management consulting firm, Arthur D. Little, Inc.) to ensure that the companies were living up to their rhetoric. By 1979, several dozen United States companies (including Deere) had become "signatory companies," with their results tabulated and made public by the Sullivan group.

The process has not been without problems; the Sullivan group has had difficulty in financing the extensive monitoring that seemed to be required, and a number of black leaders in the United States (as well as in South Africa) have taken exception to the concepts that Sullivan espoused. Other organizations have also carried through similar monitoring efforts; in particular, the Investor Responsibility Research Center (IRRC), a consortium of universities and foundations, has developed its own independent reporting system.

Sullivan ranked Deere in the highest category ("making good progress") after setting up its monitoring system. (There were two other categories, "making acceptable progress" and "needs to become more active.") An IRRC monitoring report of May 1981 affirmed "John Deere's progress to date in establishing generous and equitable pay practices and assisting the local African community" and called the company "one of the more progressive employers in South Africa," all this done "without apparent external pressure." The IRRC corroborated that Deere was paying some of the highest wages of any manufacturer in the country, and it called attention to the company's programs to improve training and advancement opportunities for blacks. (In 1980 Deere had placed a black employee and a non-white employee in the apprenticeship program; the move was only grudgingly accepted by the white employees, who refused to let the two apprentices use the whites' tools, thus forcing the company to purchase tools for them.)

The IRRC also commented on Deere's difficulties in desegregating its facilities. "John Deere has had desegregated facilities in principle, but not in fact. Whites are boycotting the only locker room at the Company as well as the cafeteria set aside for hourly employees." The company had continued to maintain three dining rooms, one in the administration building for its marketing personnel and two cafeterias, one for salaried and one for hourly employees. It was the latter that had been boycotted by the white employees. The IRRC commented, "In South Africa, where most hourly paid workers are black and most salaried employees are white, the companies that have successfully desegregated dining areas are generally those that have built a single cafeteria to serve all employees, both hourly paid and salaried."

In March 1975 the company introduced an African liaison committee, its membership constituted from the black employees. In 1981 a multiracial workers council replaced the previous committee. The subjects discussed by both committees were mostly peripheral issues—bus fares, leave times, and so on. Despite Deere's consistent efforts to make the groups viable entities, the black employees were timid about pushing their views in the face of white employee disapprobation. Consequently, the company decided that it was necessary to conduct a "white-employee reorientation program," and in this the company's principles on the promotion of non-white recognition and integration in the Deere organization have been given special emphasis.

The company also attempted to improve employee conditions outside company environs. A noteworthy contribution was made to the education of black children. Almost all of the Deere employees lived in one area, Duduza, a typical segregated black township, located a few miles outside of the town of Nigel (about a mile from the company plant). The township belied its Zulu name, meaning "Land of Milk and Honey," for most of the homes were small concrete block buildings with metal roofs, few of which had their own indoor plumbing or water supply. Duduza provided schooling only through the eighth grade; the closest high school was in the town of Springs, more than a dozen miles away. Because of this distance, most of the children did not go beyond grade school. Some 180 of the Duduza families had an employee at Deere, a minute fraction of the community's total population, estimated at approximately 25,000. Consequently, the company's options were limited. If it chose to help only its own employees in the town, it would arouse resentment among the others. A suggestion had been made, for example, that Deere provide schoolbooks for children of its own employees, but the company's black employee leaders urged rejection of the idea just for this reason.

After much thought, the company decided in 1980 to build a high school for the township. Through the John Deere Foundation, it contributed $420,000 (R 395,731), enough to build a fully accredited general and technical high school that could accommodate 700 students. Teachers for the program were to be supplied by the South African Department of Education and Training; Deere itself would also sponsor additional evening

IDEOLOGICAL ENCOUNTERS IN SOUTH AFRICA

adult-education courses. The Urban Foundation, a South African foundation dedicated to improving the quality of life for South Africa's urban blacks, was asked to supervise the completion of the school, to be named the M. O. Sebone Senior Secondary School (in honor of one of Nigel's pioneering black educators). The school opened its doors in early 1981.

Yet, despite improvements in wages and working conditions, the company's black employees see their overall situation in the country as only marginally improved, their housing and living conditions in the township of Duduza as still minimal, their opportunity to participate in the broader economy as very limited. When the black employees talk of their Deere jobs, they express enthusiasm and hope; when they refer to their total lives, they mirror discouragement and bitterness. As one company black employee put it, "We hope for the best—and pray for the worst."

The decade of the 1970s was a propitious one for Deere in South Africa, despite the continuation of racial unrest in the country. By 1980 the company had more than 17 percent of the tractor market in the country; Massey-Ferguson and Ford each had about 22 percent, with Fiat at 20. Thus, four

Exhibit 14-4. Opening of the M. O. Sebona Senior Secondary School, Duduza, South Africa, 1981. *Deere Archives*

leading companies split most of the South African market. By the late 1970s the South African government again had become concerned about the need for the indigenous manufacture of diesel engines and was able to contract with Perkins Engine and Daimler Benz to allow their engines (the former in sizes up to 125 horsepower, the latter for larger sizes) to be made in a South African–owned factory. The government then mandated that all tractor companies in the country buy from this engine company. Massey-Ferguson's long-time link with the Perkins group made the problem of transition simple for them—they were already using Perkins motors. But Deere, Ford, and Fiat were required to redesign their tractors to adapt to the Perkins configuration. Deere estimated that some 175 parts needed to be redesigned in order to accomplish the changeover; the total cost of this retooling alone was more than $1 million. Once again, the issue of local content had intruded into decision making. Nonetheless, the financial results of Deere's South African operations improved considerably in the 1970s, as a result of a sharp increase in sales. Every year of the decade was profitable except 1979 (Appendix exhibit 28).

This record, while promising, was still well below the company's profitability in the United States over the same period. Many advocacy groups have stereotyped discussions about operating results, implying that United States companies were profiting hugely in South Africa. Deere's figures on relative returns on investment would belie this accusation. It is noteworthy that the company's best performance years have been the most recent ones, after the movement of its black wage rates to the top of the country's scale.[8]

A LITTLE TRADE WITH THE SOVIET UNION

The huge flow of Deere equipment to the Soviet Union in the late 1920s and early '30s under Josef Stalin's first Five Year Plan had dried up—only a trickle of Deere machinery and parts had gone to the Soviets before and after World War II. Not until 1959 was there a substantial new contact between Deere and the Soviet Union, and it came in an offbeat way—a visit by Premier Nikita S. Khrushchev to the United States and to Deere. That trip produced many striking images, not the least of which was Khrushchev's trip to Iowa. Khrushchev was "corn-conscious," said the *Wall Street Journal*, for his major experiments in corn production in the Soviet Union had been attacked as ineffective. It was natural that he would visit a corn farm. The hybrid seed-corn operation of Roswell Garst was chosen (Garst having earlier supplied substantial quantities of both seed corn and hybrid technology for the Soviet Union). Of more importance to Deere was Khrushchev's visit to the company's plant in Des Moines.

The company prepared assiduously for the visit, even producing a series of English-Russian publications explaining the company's products. Hewitt

Exhibit 14-5. Nikita S. Khrushchev, Premier of the Soviet Union, visits Deere's Des Moines factory, 1959.

himself was on hand to conduct Khrushchev through the factory. "We are competing with you on both farm machinery and corn," Khrushchev told Hewitt. As Khrushchev observed the highly automated Deere factory, he bragged: "We'll beat you, we'll be more automated." The premier's visit was marked by a Soviet order for $100,000 worth of Deere equipment—tractors, combines, harvesters, corn pickers—all in small quantities. The Soviets' intent in purchasing this machinery became clearer when Khrushchev was queried about farm machinery at a major informal dinner in New York City a few days later, when he answered off-the-cuff questions from a number of American executives (including Hewitt). The dinner repartee was reported more or less verbatim in *Izvestia*. The question was put to Khrushchev from the audience: "We are interested in business with you, but why do Soviet organizations purchase only sample machines?" Khrushchev replied, "I will speak to you frankly. If John Dear [sic] wishes to sell us tractors and agricultural equipment, this is not realistic inasmuch as we have our own production of agricultural equipment. We purchase, and may purchase tractors and other agricultural equipment in, for example, ten or one hundred samples. This is done for comparison with our machinery so as to see which is better. But why are you dissatisfied with us? How many tractors or combines do you sell to one farm? Will Garst purchase 1,000 tractors from you? Not a single farmer will buy even a dozen machines. He will buy one or two, and we will possibly buy 100. This dispute about samples is not a realistic thing.

CHAPTER 14

I will say frankly that we are not interested in buying tractors, combines, airplanes, or missiles.... We are interested in chemical equipment, equipment for machinery building, and oil refining plants. As far as tractors generally, we can sell them to you. If you wish, we will even sell you one tractor."

Khrushchev's testy response underlined the difficulties and frustrations of dealing with the Soviets in farm equipment. On the one hand, the Soviet Union had a burgeoning farm machinery business of its own—indeed, it was the largest producer of tractors in the world. (Even in 1960, it made more units than any other country; by 1975 it was producing about 13 percent more than the combined output of the United States, the United Kingdom, and West Germany, three of the leading producers in the non-Communist world.) In horsepower, though, the lead was far less—the Soviet tractors were lower powered and not as modern in their technology. Thus, as Khrushchev's words indicated, though it was not likely that the Soviets would buy a large amount of production-run United States farm machinery, they were interested in upgrading the quality and horsepower of their own machines, and therefore were potential customers for licensing arrangements, perhaps even co-production efforts.

Little happened between Deere and the Soviets in the 1960s, but in late 1972, shortly after President Richard Nixon's visit with Premier Leonid Brezhnev in the Soviet Union and the development of the protocols for US–Soviet trade, a Deere delegation visited the Soviet Union. The Soviets decided to purchase a small number of several different kinds of equipment, again to "test these machines under their own conditions." Along with this, the Soviets insisted on extensive technological backup.

By this time, Hewitt had become a board member of the US–USSR Trade and Economic Council. Over the remainder of the 1970s, Hewitt made a number of trips to the Soviet Union, with many additional visits by other senior executives of the company. Extended, often convoluted, discussions were carried on—almost a "cat and mouse game," as the Deere executives tried to fathom Soviet intentions. There was extensive testing of cotton-picking machinery, with Soviet promises of licensing arrangements in regard to cotton-harvesting technology. This never materialized. The Soviets were also interested in a high-horsepower, low-weight tractor and persuaded Deere to build a prototype, the XR-50, an adaptation of the 125 PTO horsepower Model 4430 with the weight of the Model 4020. Once again a licensing arrangement was discussed, but not concluded. Thus, total sales to the Soviet Union over the twenty-plus years after the Khrushchev visit were painfully small. Balanced against the amount of management time expended cultivating the Russians, the "cost-benefit" equation tilted heavily to the negative. As one Deere executive put it, "If you take all the gross retail sales to Russia—not just our profit—it wouldn't be enough to cover our total travel and subsistence expenses for all those trips to Moscow."[9]

MAINLAND CHINA: DEERE'S "FRIENDSHIP FARM" EXPERIMENT

The tremendous promise of technology for increasing food output, as well as alleviating the drudgery and inefficiency of small farmers, has accentuated the pace of mechanization in agriculture all over the world. Many observers argue that the only way to stem the exodus from the rural areas of the younger, more innovative people is to provide them with the tools that will enable them to modernize their operations. On the other hand, laboring people have long had a deep-seated, almost primitive fear of being displaced by machines. As a result, there is burgeoning interest around the world in matching the proper pace of development with intermediate-level mechanization—and using what has come to be called "appropriate" technology. There are places in the developing world where higher mechanization has produced rural unemployment; there are also places where the opposite has happened, where mechanization has been labor-enhancing. Firm generalization on this complicated question must be linked to specific case situations.

It is in this context of an evolving debate about the appropriate scale of mechanization in the developing world that Deere efforts with the People's Republic of China take on particular meaning. When President Richard Nixon visited that country in 1972, important protocols resulted that not only greatly eased political tensions between the United States and China but also established the groundwork for economic relations between the two countries. In late 1973 the first fruits of this economic link were realized, with a visit to Mainland China by a high-level business delegation, under the auspices of a newly formed private United States trade group, the National Council for United States–China Trade. It was a small group of ten, mostly chief executive officers of major United States corporations. The leader was Donald C. Burnham, chairman of the Westinghouse Electric Corporation; he was also chairman of the National Council. Deere's Hewitt was in the party; he was vice chairman of the National Council. Secretary of State Henry Kissinger, in his sendoff to the party, commented: "Your mission is an historic one. It is the first visit to Peking by a broadly represented American business delegation in twenty-four years."

This first contact quickly led to a wide range of initiatives by both countries. China had established its own counterpart organization, the China Council for the Promotion of International Trade (CCPIT). The latter group sent its own senior delegation to the United States in early 1974 for a three-week visit that included Washington, DC, New York City, Chicago, and Minneapolis, Houston, and San Francisco; there was also a major conference in Washington later that year. In the period 1973–1975, China sent ten separate groups to the United States, and there were six United States commercial delegations to China in the same period. The

National Council functioned during this period as a catalyst for individual contacts between separate United States companies and appropriate units in China (generally central government ministries).

In June 1975 Hewitt succeeded Burnham as chairman of the National Council for a two-year term. This was a particularly important period for the council, for its earlier emphasis on imports from China now was to be broadened to serve the export focus of the American companies. Information about China was in short supply, and the council served as a central clearinghouse for updated information. But the council could not enmesh itself in individual projects; it had to serve primarily as a backup for initiatives by individual firms. By the end of Hewitt's tenure as council chairman, this important transition had taken place, and the pace and scale of United States–China trade in both directions had heightened significantly.

The People's Republic of China, with its one billion–plus inhabitants, has enormous food needs. Agriculture is a critical sector, its scale necessarily very large indeed. The more than two decades under Chairman Mao Zedong had seen many twists and turns in agricultural policy, as Mao and his opponents debated central planning versus decentralized operation. The interaction between political and economic dogmas was accentuated in agriculture as the country moved through the traumatic land reform period in the late 1940s and '50s and then through the stages of collectivization and evolution of the communes in the mid- and late 1950s.

Through all this period there had been great debates over the extent of mechanization in agriculture, both in manufacture (should tractors, for example, be built in large central factories, or put together in various versions by individual communes?), and in usage (should there, for example, be centralized machine tractor stations, or should the individual tractors be assigned to the communes, perhaps even to production brigades within them?). During the Great Leap Forward (1958–1960) there was overwhelming emphasis on "backyard" production in a decentralized mode. Then there was a turn back to centralization in the early 1960s after the failure of the Great Leap Forward, and the machine tractor stations again were utilized on a more centralized basis. This was the period in which Mao's lieutenant, Liu Shaoqi was in the ascendancy, utilizing some of the kinds of Soviet management systems that had been brought to the country in the extensive Soviet–Chinese interchanges. When Mao terminated relations with the Soviets in 1966 and soon began the abortive Cultural Revolution, the pattern of agricultural development and agricultural mechanization again turned away from a centralized frame toward more emphasis on the individual communes. In the mid-1970s, though, in the wake of the wreckage of the Cultural Revolution, there was a shift once more toward a more centrally organized approach to many issues, agricultural mechanization being one. As the United States and the People's Republic of China began their historic new initiative after 1973, China was particularly interested in upgrading the quality and productivity of its agricultural sector, and in the process it became interested in the best of the world's newer technologies in agricultural mechanization. Thus there was a natural interest on the part of Chinese officials in Deere.

Among the Chinese delegations to the United States was one focused on agriculture. This group made an extensive visit to Deere's operations in Moline and Waterloo (as well as a corn and cattle farm in Walcott, Iowa). The company had planned this visit carefully, with all publications printed both in Chinese and English, and with simultaneous translations available at all stages of the visit. The Chinese visitors came away with a broadened understanding of American agriculture, as well as substantial interest in what Deere itself could offer their country. This soon led to one of the most striking projects in China in this period, the "Friendship Farm" collaboration between Deere and China near Jiamusi (Chiamussu), in China's northeast province of Heilongjiang.

Historically, the dominant pattern of Chinese agriculture has been the small farm. Before the Communist revolution these had been privately held plots, with a strong pattern of landlordism. After early land reforms, the plots became state owned, generally run by individual production teams (the smallest administrative unit under the commune). Thus the small farm survived. The famous agricultural operations in the village of Tachai in Shanxi province, where the peasants had literally put together small plot farms with their bare hands from a barren, hostile mountain region, exemplified agriculture in the country under Mao. (He himself coined the pervasive slogan, "In agriculture, learn from Tachai.")

Nevertheless, there were large-scale units in certain parts of the country, generally organized separately from the communes as "state farms." Most of them were located along the northern and eastern borders of the country, in areas of low population and extensive expanses of undeveloped land. From early in the history of the People's Republic, most state farms were centrally managed (though the production brigade-production team basis of organization still was preserved out in the field). Some of these farms were enormous in size, ranging upward from 60,000 hectares. Clearly, in units of this size, agricultural mechanization of substantial amount was needed. In the early days a number of Soviet tractors were used; later the People's Republic itself developed its own wheeled and crawler tractors, ranging up to about 75 horsepower. A Soviet-designed three-bottom plow was the standard tillage instrument for these larger-sized tractors. Low horsepower and the relative ineffectiveness of a number of these machines made Chinese officials in the mid-1970s particularly interested in upgrading the quality and size of the agricultural equipment for state farms.

They turned to Deere for their first project. There was a 1,000-hectare farm in the northern part of Heilongjiang province, just a few miles below the Soviet border, in an area known as the "Great Northern Wilds,"

CHAPTER 14

Exhibit 14-6. William Hewitt on an Iowa farm with delegation from the People's Republic of China, 1975. *Deere Archives*

formerly an endless stretch of almost uninhabited wasteland. It was called "Friendship Farm," a name that had been given to it in the early 1960s, when the Soviets and the Chinese were jointly developing this area. This location was now chosen by the Chinese as their key experimental farm for large-scale equipment. For this, they ordered from Deere the whole gamut of Deere tractors, combines, and tillage equipment. Five tractors were ordered, including two of the largest four-wheel-drive versions; the three combines were of the largest size and the planters, plows, and other equipment were all large. These were purchased for hard currency foreign exchange at regular commercial rates.

In effect, the People's Republic had decided that it would test the whole range of advanced agricultural technology: for example, when the Deere executives asked what accessory options the Chinese wanted on tractors, they replied, "All." Thus the big four-wheel-drive tractors came all the way to the northern reaches of China equipped with air conditioning and tape decks!

This exciting venture opened up an extensive set of relationships between company personnel and their counterparts in China. Several dozen Deere engineers and agricultural technical experts were deputed to the farm over several years, and handpicked Chinese engineers also visited Deere operations in Moline and Waterloo. At one point in October 1979, Hewitt himself made an official visit to the farm; this was not an inconsequential trip, for transportation was difficult in this part of the country and the demonstration of the company's interest by its chairman's visit put special focus on the importance of the project to the United States.

Exhibit 14-7. Local manager of "Friendship Farm," Heilongjiang Province, People's Republic of China, presents ceremonial T-shirts to the William Hewitt family, 1979. *Deere Archives*

The experiment was an eminent success and was soon widely reported in the Chinese press. There followed a large number of visits to the farm by Chinese agriculturalists from all over the country, and it seemed as if Friendship Farm had truly replaced Tachai as the place "to learn from" (to borrow Mao's phrase).

The project was not without its critics in China. In the influential Beijing publication, *Guangming Ribao*, a reporter praised the results as having "shown that the advanced farm equipment and skills have played a positive role in building a modern state farm. It has emancipated people's minds, broadened their vision and made them realize that mechanization is the fundamental way to increase agricultural output. . . . It has battered down management methods that are inappropriate to agricultural mechanization and brought about many changes in farm management." But then the reporter continued: "However, we have recently heard some criticism: first, 'there are mistakes in the calculation of labor productivity of the number 2 team'; second, 'importing farm equipment from the United States results in loss of money'; third, 'the equipment drives people away, producing poor results politically.'" The author explored at some length each one of these "discrepancies"; there were contradictions between "specialized production and social cooperation." Further contradictions were alleged in the organizational structure of the units and, perhaps more importantly, there were said to be contradictions among individuals. Tensions between the assistants and the group leaders were noted; there were arguments over the precise way the equipment was to be operated, concerns about the skills of the operators. There had been difficulties at the farm in maintenance and repair, for the Chinese had a fear of working on such complicated machinery. A Deere fieldman reported once, "The central farm shop mechanic 'dared not repair it.' . . . When asked who is expected to repair the John Deere machines, the answer was 'John Deere.'" This dependency raised the question of just how long the Deere contingent would need to stay on the scene to make the project effective.

The oblique reference to potentially negative employment efforts pointed up China dilemmas in recalibrating the old forms of rural organization that existed under Mao's regime. On the state farms, located mostly in the vast expanses of the north and northwest, large-equipment mechanization made eminent sense. There would be some employment effects in the process, as a smaller number of higher-skilled operators took over. But population concentrations were low in these areas. As the central government turned in the 1980s to new forms of private enterprise in the rural areas, mechanization in the communes of the rest of the country was again being demanded. But in this case it was often for smaller machines (hand tractors, for example) than characteristically had been used earlier on the communes. One tractor factory in Anhui Province was reported to have shut down because its tractors were "too large." Here again, under this "private enterprise" rubric there were negative employment effects; not everyone in the commune was equally able to take advantage of the new opportunities. The enormous population concentrations in many of these rural areas made for a potentially explosive situation, were there to be substantial unemployment.

The *Guangming Ribao* article ended by again praising the results of the actual mechanization itself and the striking increase in productivity. Still, the article and several others in the same period had hinted at what some Chinese officials felt was an incipient elitism among the employees at Friendship Farm. Perhaps the air-conditioned cabs of the tractors and combines were just too far away from existing Chinese practice to make sense to most of the peasants. (As one Chinese driver put it, "You sit all day in the cab—now you want to go home to your wife?") Nevertheless, the substantive success of the project was unquestioned—modern, large-scale agricultural mechanization made real sense in the larger Chinese operations, particularly the state farms.

This first decade of Chinese-United States economic cooperation also had its problems. The Chinese were meticulous, hard bargainers; the wording of one of the Chinese telegrams to Deere illustrates this very well: "On this occasion we would like to invite your attention to the fact that the price you quoted is too high and is not competitive. In particular, your quotation for lump-sum payment is surprisingly high and with no payment terms at all. . . . However, taking into account the friendly relation between us we are still willing to invite you to China to join in the competition in the hope that you will reduce your price to a great extent so as to win the transaction."

A number of companies spent extensive time and money visiting China, only to find that very little had resulted. On the other hand, the People's Republic had initiated industrialization with too euphoric a view of the nation's capacities to absorb the new. The first problem was money—the foreign currency available from China was just not enough to cover all its plans. Thus, over the latter part of the 1970s a number of previously announced projects were cut back and even terminated.

More was involved than just money, though. The economic philosophy of the country was still not settled enough to assure unanimity on what was needed. Tension continued between the old Maoist groups and the new leadership under Vice Premier Deng Xiaoping. The latter, as de facto head of state, was an advocate of a number of modern Western ideas, even countenancing renewal of some private ownership and individual incentive within the structure.

The organization of agriculture became a central part of this debate; questions were raised about how much consolidation to large-scale agriculture should be part of the new plans. A high level of agricultural mechanization fit well at the larger state farms. But whether some of the smaller plots of the central and southern parts of the country should be consolidated in some way would depend on both the amount of money available for the agricultural sector and the prevailing economic philosophy of agricultural development.

After Deere's successful Friendship Farm endeavor (one that continued with input of Deere personnel into the 1980s), the company entered into extensive negotiations with China about collaborating on a tractor factory. One of the country's older, larger tractor factories was located in Loyang, in Honan province. Originally constructed with Soviet help, it continued to make tractors after relations with the Russians soured. The Chinese wanted to upgrade this factory to produce higher powered, diesel-equipped tractors in the 70–110 horsepower range. The numbers needed were quite substantial—40,000 units as a starter. The Chinese also wanted to make larger 125–175 horsepower tractors in another factory, perhaps 5,000 of these, too. Deere and several other manufacturers competed for this project, and the company prepared elaborate printed brochures on the proposal and visited the country for further negotiations. In the end, China backed off.

After the Loyang project fell through, Deere became involved in a smaller project for the development of grasslands in the northeast of the country near the town of Ganchika, in a remote region in the Inner Mongolian Autonomous Region, just short of the Russian border. Some Deere grasslands equipment was sent there, together with a small cadre of Deere technicians. Unfortunately, no licensing agreements were concluded, though the People's Republic did copy some of Deere's hay tools.

Then, in 1981, a major new joint relationship developed involving combines. Deere technology was licensed for use in China's combine factories at Jiamusi, and Kaifeng in Honan province. Plans included the training of a large contingent of Chinese engineers and technicians at Deere's factory at Zweibrücken and concurrent visits by Deere personnel to the Chinese production units. Investment in the project was small, and Deere was to be reimbursed in part by purchase of Chinese-made components, but the company considered it to be another important step in developing a broad commercial and technological relationship with China. Deere was the first United States corporation to sell agricultural technology; this fit well with China's apparent emphasis on buying technology rather than actual equipment.

Total sales to the People's Republic by the company in this first decade were about $15 million, and the personal relationships and official links between the company and Chinese industry pointed to the possibility of more business in the future.[10]

FOREIGN REPRISE

By the end of the decade of the 1970s, there could have been little doubt in anyone's mind that Deere intended to stay in Europe. Major capital expenditures were made during this period—first, assembly operations were expanded at both Mannheim and Zweibrücken, then a major, multistory administrative center was constructed at Mannheim for the Region II offices. By 1979 a 20 percent addition in capacity was accomplished in Spain and a somewhat smaller expansion at the Saran factory in France. That year was also marked by the decision to build a major new factory at Bruchsal, near Mannheim, particularly for the manufacture of protective cabs and other components. There was also a substantial expansion of parts depots—a 430,000-square-foot building at Bruchsal for German operations (for start-up in 1982), as well as smaller depots in Orléans, France, and Milan, Italy. By 1980 a set of new European tractors came off the assembly lines of Mannheim and Getafe, Spain; these were well received and allowed the company to increase its market share.

Unfortunately, by this time worldwide recession had affected European agricultural equipment sales in a major way. Net income from the international operations continued strongly on the positive side for the years 1974–1978, with the $41.3 million of 1976 the high point. (The company as a whole made $242 million that year; the total company had made 7.7 percent on its sales, and overseas had made 5.7 percent.) The European recession in 1979 ended the string of good years. Overseas losses in 1979 were $12.5 million and $54.1 million in 1980. In 1981 the total overseas net income again was positive, just over $1 million. Though the Zweibrücken

Exhibit 14-8. Farm families from Kenya's Masai tribe pose with three of seven John Deere tractors they own, 1979. *JD Journal*

operation in Germany remained modestly profitable in these three years, the Mannheim tractor works had large losses. In France, the engine plant at Saran remained modestly profitable, but the other operations lost money. In addition, there were substantial losses in Argentina, England, and Italy, and disturbing losses in Spain. Fortunately, South Africa, Australia, and Mexico remained in the profit columns to help offset some of these major losses. Nevertheless, as Deere moved ahead in the 1980s, the overseas operations were again a source of concern for company officials.

Endnotes

1. See Frank McGuire, "Proposal for Qualifying Deere & Company to Do Business in Germany," April 25, 1967, DA, AN 80/43/5. The Booz, Allen & Hamilton "Top Organization Study" is in five separate reports: May 11, 1966 (DA, 39247); June 27, 1966 (DA, 39248); September 6, 1966 (DA, 45517); October 24, 1966 (DA, 45518); November 4, 1966 (DA, 39250). Hewitt's remarks on decentralization are in Deere & Company *Minutes*, January 31, 1967. For the analysis of comparative dealer terms in Germany, see Deutz memorandum, "Sales Terms without Discounts," January 27, 1966, in E. Curtis to C. M. Peterson, June 1, 1966, DA, AN 80/4. For discussions of "under-the-table" practices, see A. Earl Lee to E. Curtis, May 24, 1966, DA. See *Mannheimer Morgen* and *Frankfurter Allgemeine Zeitung* (June 15, 1967), and the release of the German Press Agency, Vereinigte Wirtschaftsdienste, for responses to the shareholders' meeting of June 14, 1967. See *Die Zeit,* May 2, 1969, for the quotation concerning "lame deer." See also *Land Technik* 13 (July 1967). McGuire's six alternatives for the German operation are outlined in F. McGuire to W. Hewitt, October 22, 1970, DA, Hewitt ms. 707. The Deutz negotiations are collated in Hewitt ms. 706 and 719 and in DA, AN 80/4. McGuire's remarks on the possible merger are in F. McGuire to W. Hewitt et al., August 4, 1970, DA, Hewitt ms. 706. Comparative information on Deutz and prospects in the marketplace for 1971 are analyzed by Walter Vogel, December 7, 1970, DA, AN 80/4. Positions of the two partners are best seen in the unsigned "Minutes—Deere-Deutz Discussions," October 29, 1970, DA, 719, and Deere &

CHAPTER 14

Company *Minutes*, December 10, 1970, Walter Vogel acting as secretary, DA, AN 80/4. McGuire's views on ownership are in his handwritten memorandum to Hewitt and Curtis, ca. December 1970, DA, 719. For McGuire's notification of the impending Fiat negotiations, see F. McGuire to K-H. Sonne, May 18, 1971; see also Sonne to McGuire, May 26, 1971. Sonne's quotation is from H. J. Schubert to Curtis, August 17, 1971, DA, AN 80/4.

2. For the first formal memorandum on the Fiat project, see Neel Hall to R. W. Boeke, "Notes on a Visit to Fiat, December 15–18, 1970," ca. December 20, 1970, DA, George French ms. See Curtis's memorandum to key personnel on the legal aspects of trading stock, January 13, 1971, DA; his remarks on four-wheel-drive capability are in E. Curtis to W. Hewitt, February 8, 1971, DA, Hewitt ms. 706. The involvement of Henry Dreyfuss is discussed in R. J. Gerstenberger to G. French, April 22, 1971, DA. Market share statistics on the two companies are from Joseph Dain to W. Hewitt, May 3, 1971, DA, Hewitt ms., AN 81/37. See Patrick Neville, "The Prince," *Car and Driver*, February 1971; *Time*, January 17, 1969. Hewitt's first meeting with Giovanni Agnelli is briefly documented in DA, Hewitt ms., AN 81/37. The marketing, manufacturing, and finance task force reports were completed in March and April 1971, DA, Hewitt ms. The "Heads of Agreement" is discussed in Deere & Company *Minutes*, April 27 and May 17, 1971 (the full agreement reprinted in the latter). For the D. H. Glover and George French quotations, see ibid.; for press reactions to the "Heads" proposal, see unsigned memorandum, June 22 1971, DA, Hewitt ms. AN 81/37. For negotiations leading to the final proposal for the "Agreement for Closing" and a Fiat proposal forwarded on December 20, 1971, see DA, Hewitt ms., AN 81/37. George French's misgivings about the project are enumerated in G. French to W. Hewitt, January 7, 1972. Final discussions of the project are in Deere & Company *Minutes*, January 11 and 12, 1972; the press release announcing the termination is dated January 19, 1972. For additional material on the Fiat project, see DA, French ms., AN 80/43; DA, Maguire ms., AN 80/43; DA, Hall ms., AN 80/14, 2176. See *Wall Street Journal*, December 29, 1971; William J. Tagtmeier's report and Hewitt's public reply to shareholders are reprinted in Deere & Company *Minutes*, January 13, 1972. Neel Hall's remarks on competitor dealer raiding are in N. Hall to W. Hewitt, February 18, 1972. See *Forbes*, December 1, 1971. DA, Hewitt ms. 727. See also *Le Figaro*, September 20, 1972. Curtis's remarks on the proposed joint venture with Fiat in 1973 are in E. Curtis to Bruno Beccaria, vice general director of Fiat, July 24, 1973, DA, AN 80/4.

3. For more recent history of Deere's Mexican operation, see Harvard Business School, "John Deere, S.A. (Mexico)," case 4-580-043 (November 1979). The issue of Mexicanization is covered in DA, AN 80/54/2.

4. Tax effects of the Argentine sale are discussed in Deere & Company Annual Report, 1981. See R. Hanson to J. A. Martinez de Hoz, May 19, 1980. The termination of the Argentine subsidiary was voted in the board meeting of July 29, 1980; the sale of Cindelmet was authorized at the meeting of November 20, 1980.

5. For an excellent analysis of the short-lived Venezuelan tractor factory, see Joseph A. Mann Jr., "Fanatracto: Plowed Under," *Business Venezuela*, November–December 1980.

6. For the proposal for Iran, see Deere & Company *Minutes*, July 9 and October 11, 1976; ibid., February 23, 1977. For the original contact with Iran, see ibid., January 30, 1968; for the project proposal, see ibid., May 13 and July 16, 24, and 30, 1968. The reduction in Deere's ownership is discussed in ibid., January 30, 1973; the impairment of capital in ibid., October 9, 1974; the additional investment of Deere in ibid., February 25, 1976, May 19, 1976, and February 17, 1977; the shift to components in ibid., July 25, 1979. For the early history of the operation in Turkey, see ibid., June 17, 1966, and May 5, 1971. For the loan and swap arrangements with Cumitas, see ibid., July 6 and September 26, 1972. For additional capital investment in the company, see ibid., September 12, 1973, and October 9, 1975. For the tractor manufacturing proposal, see ibid., July 6, 1977; for the sale of Deere shares, see ibid., May 3, 1979.

7. The Brazilian tractor joint-venture proposal is first discussed in Deere & Company *Minutes*, July 27, 1976; the final project is described in David H. Stowe Jr. to R. A. Hanson and L. N. Hall, November 28, 1977, DA, Hewitt ms. 732. The joint venture is authorized in Deere & Company *Minutes*, December 1, 1977. See also ibid., July 14 and 25, 1978. For the purchase of equity in Chamberlain Holdings Ltd. authorized by the board, see ibid., August 18, 1970; as the purchases were to be made over a six-year period, the amount authorized was specified as a range of $2.2 to $3.4 million. For a discussion of the earlier history of Chamberlain,

see J. K. Horwood's remarks in "Dialogue for the Eighties—Presentation to Institutional Investors and Financial Analysts," November 21, 1980, Chamberlain Holdings Ltd.; for the quotation from Peter Griffith's report, see ibid.

8. For the evolving situation in the United States in regard to US business performance in South Africa, see E. Curtis to F. McGuire, October 14, 1971; the wage comparisons in 1971 are from Frank M. Dickey's trip report, November 18–December 10, 1971. McGuire's remarks on "European work" are from F. McGuire to W. Hewitt, E. Curtis, and R. A. Hanson, April 23, 1973. Moline's concern about the application of worldwide employee relations standards is in R. Barry Cronin to Fred N. Gilchrist, October 18, 1974; the managing director's reply is in F. N. Gilchrist to R. B. Cronin, September 24, 1974. See "Summary Report of Deere & Company in South Africa," April 30, 1974, DA, Hewitt ms. 708, and the updated version, "Deere & Company Policy Statement on Employment Practices in South Africa," February 4, 1977, DA, Hewitt ms. 729. There is extensive documentation concerning the "Sullivan Principles" and related efforts in monitoring by the Investor Responsibility Research Center; see particularly DA, Hewitt ms. 729. The Investor Responsibility Research Center report on Deere is dated May 1981; it discusses the workers council, the company's training programs, and its wage and benefit systems. The "white employee reorientation program" is discussed in Gustav A. Pohl to Walter Vogel, April 17, 1979; the issue of giving schoolbooks to Deere employees is in R. L. Anderson to L. N. Hall and R. W. Weeks, September 5, 1979. A succinct description of the township of Duduza can be found in a memorandum of Harry G. Hoyt Jr., June 9, 1977. The literature on US corporate involvement in South Africa is voluminous; see, for example, Desaix Myers, III, *Business and Labor in South Africa* (Washington: Investor Responsibility Research Center, Inc., May 1979) and *US Business in South Africa: The Economic, Political and Moral Issues* (Bloomington: Indiana University Press, 1980).

9. For Nikita Khrushchev's visit to Deere's Des Moines factory, see *Wall Street Journal*, September 23, 1959. For Khrushchev's quotation from *Izvestia*, see "Vodka and Caviar," privately printed ms., ca. September 24, 1959, DA, Hewitt ms. 714. Comparative tractor statistics are from Earl M. Rubenking, "The Soviet Tractor Industry: Progress and Problems," in Joint Economic Committee, *Soviet Economy in a New Perspective*, 94th Congress, 2nd Session (October 14, 1976). For the visit of a Deere delegation to Moscow in 1972, see Deere & Company *Minutes*, October 31, 1972; see also DA, Hewitt ms. 719. The company joined the US-Soviet Trade and Economic Council in 1974. The licensing arrangement for the cotton harvesting technology is discussed in Deere & Company *Minutes*, October 9, 1977, and March 8, 1978; for the tractor licensing, see ibid., May, 16, September 5, and December 20, 1979, and October 22, 1980. The Russian tractor is discussed in Robert F. Miller, *One Hundred Thousand Tractors: The MTS and the Development of Controls in Soviet Agriculture* (Cambridge, MA: Harvard University Press, 1970); Lazar Volin, *A Century of Russian Agriculture: From Alexander II to Khrushchev* (Cambridge, MA: Harvard University Press, 1970).

10. For documentation of the early National Council for United States–China Trade, see DA, Hewitt, ms. 716, 727, 730. The formation of the council was widely reported in the press; see, for example, *Business Week*, March 31, 1973. See Henry Kissinger to Donald Burnham, October 27, 1973, DA, Hewitt ms. 727. For a discussion of dilemmas facing the council in its transition period, see Working Paper No. 2, August 11, 1975, DA. For an overall discussion of the development of agricultural mechanization policies in the Mao Zedong period of the People's Republic of China, see Bernard Stavis, *The Politics of Agricultural Mechanization in China* (Ithaca: Cornell University Press, 1978). See *Guangming Ribao*, July 29, 1979; see also "A Test-Case of Farm Mechanization," *China Reconstructs*, December 1979. For the quotations on repairs, see G. R. White to R. A. Thompson, May 31, 1980. For the quotation on the Anhui tractor factory, see *New York Times*, April 10, 1983. For the telegram from the Chinese, dated May 23, 1981, see DA, Hall ms.

CHAPTER 15

CHALLENGES OF THE 1980s

Deere has gone around and actually to the detriment of their balance sheet given support to their dealers. They're the only people who are going to end up with a healthy dealer network. 10 to 12 percent of the farm machinery dealers in 1982 went out of business, but the number of Deere dealers that went out of business was miniscule, and they were immediately replaced by people with stronger capitalization. So not only are they the low cost producer, they're going to be the only people around with dealers.

Wall Street Transcript, *1983*

During the first three years of the 1980s, the bright promise for agriculture seen by most experts at the beginning of the decade quickly faded. The American farmer had produced bumper crops in 1980 and '81, and the carryover of grains from these years brought severe downward pressures on agricultural prices just as a general recession hit the United States and the world. These events seriously affected the agricultural equipment industry.

◀ Computer-aided design, John Deere Technical Center, 1983.

WORLD AGRICULTURAL MARKETS

World agriculture went through a profound change in the 1970s. Bad weather—following a diversion of agricultural resources into farm animals—forced the Soviet Union to enter the world grain markets with massive purchases beginning in 1972. Large United States grain exports to that country ensued, and when the crisis in the world's oil markets followed the next year, with its massive transfers of capital and effects on foreign exchange rates, commodities became an even more attractive investment. The resulting upsurge in agricultural prices, with the year 1973 particularly a boom year, enhanced agriculture's prosperity dramatically in a short period of time. It remained strong for most of the rest of the 1970s, to the benefit of the net exporting countries such as the United States. The persistent fear in the 1960s of unwanted surpluses now was replaced by hopes of a long-term balancing of supply and demand. As the 1980s opened, however, the world business downturn that had begun a year or so earlier worsened. Inflationary pressures subsided in the United States in 1981 and 1982, and prices to farmers declined, after bumper crops in those two years. Agriculture was used in a new role as a political weapon by the United States after the Soviet Union invaded Afghanistan in 1980. A grain embargo was imposed, and it had two important results. First, other exporting nations stepped into the breach to fulfill Soviet needs and enter the general world markets more aggressively. Argentina and Brazil, in particular, now became major players in world agriculture. Second, many experts felt that the United States had damaged its credibility as a reliable world supplier and they worried that the United States might be returning to a position of residual supplier.

Projections of world population growth rates in the 1980s, when compared to available food supplies, still appeared to put United States agriculture in a commanding position over the decade. But for this scenario to come true, significant productivity gains would be necessary. These would not come as readily as they had in the past, for several negative factors now intruded:

(1) it would likely cost more for United States agriculturalists to bid additional land and other resources from competing sectors of the economy;
(2) costs of producing would be higher and yields lower on less fertile, fragile land that would be brought into farming;
(3) costs of fertilizer and other inputs, which had been rising rapidly, would likely continue doing so, adding to the expense of increasing yields;
(4) potential gains in yields from current technologies had already been widely attained, so further gains were estimated to be increasingly expensive; and

(5) greater production might entail increased environmental costs from chemical pollution of rivers and streams and increased soil erosion.

CHANGES IN US AGRICULTURAL PRACTICES

These important developments put pressure on the agricultural machinery manufacturers for heightened productivity in their new models—fuel-efficient equipment that could adapt to the many new agronomic practices now demanded by the times. "Conservation farming" seemed destined to be a watchword of the decade—countering erosion threats to topsoil, heightening the regeneration of soil fertility, and monitoring off-farm side effects, such as the depletion of water tables, silting, and chemical runoff from soil erosion. Likely the future would see fewer trips over the field, with more combined operations, reduced tillage, and an overall concern for soil and water conservation practices. Fall tillage would probably be reduced, with rougher surfaces remaining and more residue on top.

Row-crop, monoculture patterns of production potentially could be threatening to soil conservation, though they are often the most financially rewarding. High demand for food had persuaded a great many farmers to move away from traditional rotational practices that included soil-retaining pastures in favor of continuous cropping of corn and other row crops. While American farmers are not likely to turn away from their traditional crops—wheat, corn, and soybeans—renewed sensitivity about soil conservation will bring many back to a more crop-rotating pattern in the future.

Even with single-crop row crops, much can be accomplished by a more sophisticated application of modern reduced tillage practices. Many names are used to describe them—ecofallow, mulch tillage, till planting, less till, minimum tillage, no-till, zero tillage. In essence, all cut down work done to the soil. Rather than conventional tillage—plowing, plus two to five secondary tillage steps, including preplow operations such as disking to knock down weeds and cut crop residues—steps short of full tillage can be practiced. At first, weeds were a perennial bane, but modern herbicides now have mitigated this problem. Selective herbicides, particularly the introduction of 2, 4D in the late 1940s, proved effective on many of the broadleaf weeds found particularly in corn. Since that time, more than one hundred other herbicides have been developed, making practical the many variants of reduced tillage. Under "minimum tillage," there might be only one tillage operation before planting; often chisel plowing or mulch tillage would be used in place of a moldboard plow. Zero tillage or no-till involves planting into an essentially unprepared seedbed; no prior tillage would be done except for a narrow strip in front of the planter that might

CHAPTER 15

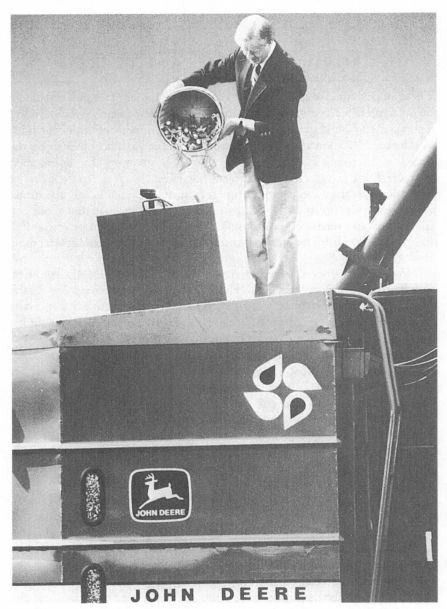

Exhibit 15-1. President Jimmy Carter ceremoniously dumps a basket of corn cobs into Deere's experimental biomass converter during his tour of the company's Moline facilities, 1979. *JD Journal*

be cut and loosened somewhat by a coulter or some other device at the moment of planting.

The key to the extent of reduced tillage would lie particularly in the draining of the soil. With good drainage, a reduced or no-tillage system could emerge as best; in poorly drained soil, no-tillage methods lowered yield significantly in many cases, as much as 30 percent lower for corn.

The moldboard plow has garnered a particularly bad press in the "reduced tillage" debate. The attack has been considerably more pervasive than in the days when Edward H. Faulkner wrote his little book *Plowman's Folly* in the early 1940s. (He had an unbending antipathy toward the moldboard plow, believing that it carried the surface tilth too far into the ground and that in the process nutrients necessary for that first inch or two of the surface soil were lost.) Now more than one conservation purist imputes to the moldboard most or all of the ills of soil erosion. But this oversimplifies. Today's sophisticated farmers understand the value of the right tool for the right situation. Most continue to purchase moldboards, along with the newer chisels and disks, the former for those times involving a complete turning for weed and pest control and the latter for incorporation of fertilizers and herbicides and for reduced tillage needs. As field speeds have increased, tillage machinery engineers have developed new moldboard configurations and adjustable width bottoms, so that dirt will not be thrown as far (and thus not be so susceptible to wind or water erosion). The farmer has now developed a set of rotation tillage practices in addition to crop rotation patterns, testimony to the vastly increased complexity of today's agronomic decisions on the farm.

Reduced tillage does allow some savings in energy. In the case of no-till, upwards of 75 percent of tillage energy can be eliminated, but part of the savings is canceled out by the cost of the energy to produce the chemicals for weed and insect control. Further, the energy costs for subsequent operations, including harvesting and drying, are not altered, so the overall savings are modest (perhaps in the neighborhood of 4 percent). Thus the primary rationale for reduced tillage lies particularly in its erosion-prevention potential, the savings from which may be incalculable for the country over the long run.[1]

SHIFTING MARKET SHARES AND RECESSIONARY FALLOUT

The recession of the early 1980s had profound effects on the agricultural machinery industry. In the fast-paced growth of agriculture in the 1970s, farmers took opportunities to leverage themselves with debt for both land and equipment, given what seemed to be a halcyon future. In the process, the farm machinery industry did well. Deere, International Harvester, and Massey-Ferguson continued as giants of the industry, with Allis-Chalmers,

Ford, J. I. Case, and New Holland taking significant roles, too. Two companies captured an increasing share in the four-wheel-drive tractor market—the Vancouver, BC, firm of Versatile Manufacturing Company, and the Fargo, North Dakota, firm of Steiger Tractor Company (the latter with a minority holding by International Harvester from 1975). The mid-1970s also brought a new player on the North American scene with the entry from Japan of Kubota, Ltd.

When the farmers' hopes were dashed in the abrupt turnaround, one of the inevitable results was a "make do" mentality on equipment purchases. In the face of a deadly combination of high interest rates and falling farm prices, many potential agricultural machinery purchasers just opted out. The farm equipment industry was left in disarray. Business sagged precipitously, and just about every farm equipment manufacturer, large or small, was faced with cutbacks in schedules, stretchouts in capital spending plans, and, frequently, reductions in staff.

Some manufacturers fared relatively much worse than others. Massey-Ferguson experienced its own set of problems even before the onslaught of the downturn; in 1978 it lost a startling $262 million. After a small profit in 1979, the company again lost heavily in 1980, more than $225 million. The firm was saved from bankruptcy only by a massive financial transaction involving a group of some 250 banks and insurance companies from around the world, together with the Canadian and British governments. Even this $730 million (CDN) rescue package could not stem the tide of red ink for Massey-Ferguson; the company lost $194.8 million in 1981 and losses continued in 1982.

International Harvester had only modest profits in 1976–1978, a much better performance in 1979. In 1980 the combination of the business sag, an abortive strike, and certain internal management missteps brought a colossal loss of $397 million. In December of that year Harvester, too, had need for a rescue plan, in this case involving some 200 banks and insurance companies, with a total package of $4.15 billion. 1981 brought a further loss of $393 million; in May 1982 Archie R. McCardell, the chief executive officer from 1977, resigned. The continuing saga of bad news at these two business giants is well known; their difficulties were not just an industry story, for the general press throughout the country widely chronicled the two firms' efforts to forestall bankruptcy.

But there were also many other changes upward and downward in the fortunes of individual companies. Market shares were indeed shifting in a way hardly imaginable in prior years, and some famous names were almost gone altogether from the scene. The White Motor Corporation (the successor combination of Minneapolis–Moline and Oliver) went bankrupt in 1980; its farm equipment division was sold to the TIC Investment Corporation, a Dallas firm. Operations for White continued, but at a low volume with substantial losses. Others, and this included preeminently

Deere, picked up market share and relative strength in this period, even while they themselves struggled to maintain equilibrium in the face of very depressing sales outlooks.

Appendix exhibits 29–34 picture in a dramatic way the market share position of Deere in agricultural equipment. Appendix exhibits 29 and 30 graphically show Deere's strong position in the North American market; the first records the company's nearly 30 percent piece of all farm equipment sales, the second eloquently documents its leadership in individual products, where it is the major supplier for six of the ten listed categories.

The company's reach worldwide is pictured in Appendix exhibit 31; its 17 percent market share is the only one in double figures for 1980. Overseas sales are elaborated in Appendix exhibits 32, 33, and 34, both by total sales and country-by-country for the major markets, for both tractors and combines. In overseas sales alone, Massey-Ferguson leads, with Deere second in 1980 (from a third place in 1975, fourth in 1970). The overseas markets are much more fragmented; national boundaries yield major differences, with some countries over-full in numbers of competitors. Deere's overseas success has been greater in combines than in tractors, and there are only a few markets in which the company has a "substantial" market share of 20 percent or more—Spain (tractors, combines), Australia (tractors), and Brazil (combines). Deere's progress has been painfully slow in the two key markets of France and Germany.

Given the desperate condition of some of the larger agricultural machinery manufacturers in the 1980–1982 recession, patterns of disruptive "fire-sale" pricing dominated the retail scene, and one could not predict with any certainty how market shares were moving. What had been gradual trends in the 1970s might prove to be more dramatic in the 1980s. It was not beyond the realm of possibility that one or more major companies would either cease manufacturing altogether or at least reduce its product lines, no longer attempting to remain "full line."[2]

DEERE'S NON-AGRICULTURAL DIVISIONS

Deere's strong showing in farm machinery in the 1970s and early '80s must be seen additionally in the context of the entire company; i.e., one must include the two major efforts of manufacturing diversification—industrial and construction equipment, and consumer products. The declining position of the former from 1980 was sobering; the latter's growth was positive, but modest.

The industrial equipment division was a very large endeavor in its own right, reaching almost $1 billion in sales in 1979 (more than 20 percent of the company total that year). It had cumulated modest profits during the

1970s, with the highest figure some $97 million in 1979 (17.7 percent of Deere's operating profit that year). The recession hit this industry very hard, and sales in 1981 dropped to just over $780 million, with a $38 million loss. Appendix exhibit 35 illustrates the division's market share in its key product lines for 1979–1981.

The dominant competitor in industrial and construction equipment sales was Caterpillar, whose dollar volume was about nine times that of Deere's industrial division. Second in the industry was Komatsu, the Japanese producer, with three times the dollar volume of Deere. Third was Case, with about twice that of Deere. Next was the German producer, IBH Holding AG, half again as large as Deere; International Harvester was slightly smaller than Deere. Thus Deere was in fifth place, running considerably behind the industry leader.

In a bit of euphoria in 1978, after its strong showing that year, the industrial equipment division stated its goal to be "number three in the industry by 1990, and number two by 2000." These long-term objectives were scaled down in the strategic plans developed in early 1982, in the light of the sobering statistics of the division in 1981. Lowered profitability had reduced the availability of capital for the planned product expansion into the large-machine arena. A shakeout had been expected in Europe, where there were a number of financially troubled companies. However, IBH, an aggressive company, had been picking up significant numbers of these companies; IBH's approach in this period was to produce as close to capacity as possible and price almost at cost to move what was made. This made for fierce price competition in the European market.

Competition from Japan had also been rigorous. A Boston consulting-group study for Deere in 1980 pointed to a 25–33 percent cost disadvantage in laying goods down in a typical Far Eastern market (the Far Eastern markets representing about 30 percent of industry sales). Komatsu, the key Japanese competitor, had been able to market in the United States at very competitive prices; the American companies had been surprised at the ease with which they established a dealer organization.

A number of positive factors about Deere's efforts mitigated some of the pessimism. The industrial equipment division had excellent manufacturing facilities, having expanded its operations significantly with another large plant in Davenport, Iowa. There was a remarkably close relationship to farm equipment in terms of parts and components commonalities and manufacturing similarities (less of a commonality in marketing). Aggressive company-wide efforts to develop sophisticated manufacturing operations had found application in the division's factories, though the depleted workforce at the two key plants at Dubuque and Davenport in 1982 made some of those plans difficult to effect. Plans for new back-hoe models, small crawlers, and a line of excavators were underway. The division continued to exploit its major position as supplier to the forestry industry and in the small and medium segments of the earthmoving markets.

Exhibit 15-2. Top, John Deere was the official supplier of industrial equipment for the 1980 Olympic Winter Games at Lake Placid, NY, 1980; bottom, Davenport Works employees with the JD 890 excavator. *JD Journal*

There were still important questions remaining about the extent of the line (by 1983 it had competed only in the small and medium sizes) and about the focus of the division in terms of industries served. Plans called for a more selective approach, both for customers and in terms of geography. Efforts in Europe would be kept at par, a push would be made to reach competitiveness in the Far Eastern markets (perhaps with the help of another Japanese company), and the opportunities in Mexico and several other Latin American countries, in Canada, and in several of the Middle Eastern and African countries (particularly South Africa) would be pursued aggressively.

While there was stark pessimism in the construction industry in the early 1980s, and little optimism in any of the industries using utility vehicles, over the longer term the industrial and construction opportunities seemed quite substantial. Growth rates could likely have been expected to exceed those of the farm equipment business, according to a number of analysts. The long-term position of the industrial equipment division in the total of Deere activities seemed assured.

The consumer products division also was a substantial contributor to the company's revenues, accounting for more than one-third of $1 billion in sales each year in the years 1979–1981; it had been profitable over a number of years, accounting for more than 11 percent of the company's operating profit in 1979 (this dropped to below 8 percent for both 1980 and 1981). It was in a different posture than the other two divisions, however. First, the parts commonalities, manufacturing linkages, and so forth were far less substantial. Though the division produced the tractors in the lower end of the line, they were lawn and garden tractors, of a much different manufacturing configuration and product characteristic than the larger tractors. The other major consumer products also were unique to this division—snowmobiles, chain saws, and so on.

A substantial percentage of the consumer products marketing was done through some 1,000 independent consumer products dealers; other marketing was done through about 2,000 of the John Deere agricultural dealers (sales split about evenly between these two groups). The most substantial growth in the division was in commercial mowing products, the least in snowmobiles. The Horicon, Wisconsin, plant, where the consumer products manufacturing was done, also had been involved at several stages in the development and manufacturing of the compact utility class of tractors. Total market share was more difficult to determine here; by 1982 Deere had about 25 percent of the lawn and garden business, but considerably less than that in snowmobiles.

The closely related John Deere merchandise unit marketed a wide array of consumer products, ranging from a line of ski clothing to high-pressure washers, welding machines, space heaters, outside cooking grills, shop tools, and so forth. Most were purchased from outside contractors, with marketing through the Deere organization (and again made available not only through consumer products division dealers, but also through agricultural and industrial equipment dealers). Interestingly in light of its venture in bicycles in the 1890s, the company returned again to the bicycle in the mid-1970s, selling a Taiwan-made vehicle under the John Deere trade name. The bicycles were not well enough made to stand as good representatives of the John Deere quality image, and they were discontinued after just a few years of sales.

The verdict was still out about whether diversification in manufacturing had strengthened Deere's balance sheet and profit and loss statement. A simple financial analysis would not tell the whole story, however, for one must look behind raw contribution figures for other less readily identifiable interactions. There had been real synergy between the farm and industrial divisions in the commonalities of parts and manufacturing; though there had not been the same degree of commonality between consumer products and the other two, the widespread involvement of Deere's agricultural dealers in consumer-product marketing gave both divisions some added strength.

One can readily hypothesize that agriculture does not move in lock step with construction, nor with consumer sales. Yet there was little evidence by 1983 to support the notion that either the industrial or consumer products divisions could be a significant contracyclical influence on the agricultural division. Thus Deere's manufacturing diversification efforts had the primary effect of deepening corporate resources, and less that of spreading economic risks for the company.

Exhibit 15-3. New line of lawn and garden tractors, 1983. *Deere Archives*

Deere's "financial services" operating units—the credit and insurance companies—were both profitable in the recession of 1980–1982, countering in a significant way the losses in industrial and agricultural equipment. (The retail finance subsidiaries earned $57.7 million in 1981, the insurance companies an additional $27.2 million.) Thus they were a financial offset to the decline in the equipment business. But, the financial service entities were just that—they were not a contracyclical manufacturing balance. Further, both groups were interdependent to a significant degree with equipment—neither was, in 1983, a fully arm's-length, stand-alone organization. Still, both legitimately could be considered viable efforts at diversification, and, differing with industrial and consumer products, did have the effect of spreading some of the company's risk.

There was no doubt, though, that up to 1983 the agricultural segment of Deere's business was dominant, particularly its North American agricultural sales. There were difficulties from excessively high dealer and farmer receivables, due both to farmer caution and excessive price cutting by some of the financially strapped competitors, but Deere's hold on market share in the North American farm equipment business strengthened markedly in the 1980–1982 recession period. This was the area that most outsiders looked to in assessing Deere's strength; the statement of an analyst at the annual review of the industry conducted in 1982 by *The Wall Street Transcript* seemed to mirror the prevailing view about the company: "The rich are going to get richer. That's John Deere. Their Chairman, Bill Hewitt, said it best at their meeting in Moline last summer, when he said that this was the chance that occurs once in every three generations. Deere has one strategic advantage over everyone else. It's called access to capital, and their willingness to use capital for strategic purposes. Their enormous capital investments, their strong dealer network, the fierce brand loyalty among farmers—all suggest that Deere is in a very favored position to sharply increase its market share over the next couple of years."[3]

DEERE INCREASES ITS MARKET SHARE

The first ingredient of "using capital for strategic purposes" is to understand the underlying trends in the market, and Deere seems to have done this very well indeed. A major *New York Times* article comparing Deere and International Harvester concluded: "Deere correctly saw the post-World War II trend toward fewer and bigger farms. From this came its strategy: continual reinvestment in product and manufacturing innovations, close attention to costs and quality, and lavish support for its dealer network. As a result, Deere is in the enviable position of being the low-cost producer and the high-quality provider."

The forty-five-year-long decline in the number of US farms was reversed in 1981—there were 8,000 net new units in the 1981 farm count for that year. Officials of the US Department of Agriculture (USDA) estimated, however, that most of these were in the small-farm, part-time category likely reflected in the movement away from commercial farming.

The commercial potential of American agriculture seems oriented ever more strongly toward the large farm (defined by the USDA as those with gross annual sales of $100,000 or more). By 1981, the largest one percent of the farms in the United States was producing about one-fourth of the nation's food; by the year 2000, according to the USDA, the largest million farms will operate almost all of the nation's farmland. Three-fourths of the farmland will be in the hands of the top 200,000 operations; indeed, the largest 50,000 will produce about two-thirds of all farm output.

Deere analysts consistently emphasized the central importance of this group to the agricultural machinery industry. In the census of agriculture of 1978, those farms with annual sales of $40,000 to $499,999—some 560,000 units—accounted for almost 60 percent of farm products sold and in the process took more than 70 percent of the industry's machinery sales. As the *New York Times* analyst put it, "Deere had the strategy and foresight to see where the market was going . . . and became the market leader in large equipment, which carries the highest margins."[4]

DEERE TRACTOR STRATEGY

By almost any standard, the "large equipment" that is the most important is the tractor. An understanding of Deere's moves on this key product is essential for assessing the company's present strength.

The number of on-farm tractors in the United States had peaked at 4,787,000 in 1965; by 1980 the number had fallen to 4,324,000. Still, throughout this period, the total horsepower continued to rise, with the 248-million-horsepower total of 1980 the highest to that date. Thus there were fewer but larger, higher-horsepowered tractors in the country.

Deere responded to this by introducing seven additional tractor sizes above 100 horsepower in the 1970s, giving the company in 1982 a total of thirteen models in 40 horsepower (PTO) and above. The Generation II tractors introduced in 1972 (the 30 series), particularly strengthened the company in the medium-sized (40–150 horsepower) row-crop tractors; the configurations of the thirteen 40 series tractors introduced in 1977 were essentially the same, but with more internal "guts," as the parallel increase in the width and size of tillage and planting equipment put ever-increasing demands on the tractor.

The number of these medium-sized row-crop tractors sold by Deere was very large. A total of 74,580 of Model 4430, at 126 horsepower (PTO), was sold in 1972–1977; the 4440 (at 130 horsepower) sold at the same pace in

1978-80. Their companions in the medium-sized range also sold equally well. By 1981, Deere's market share of United States row-crop tractors was more than 45 percent and in Canada was almost 31 percent.

But it was in the new 50 series, introduced in 1982, that the link between innovation and new field demands became so striking. So important was the breakthrough technology of this new line that the company marketing group chose to rival the historic "Day in Dallas" of 1960, when its first four-cylinder tractors were introduced. Over a week-long period in New Orleans in July 1982, dealers were flown in from all over the world to witness, in the spectacular setting of the Superdome, the unveiling of ten new row-crop tractors. All had increased horsepower (from five to ten horsepower, depending on the model); all were available in mechanical front-wheel-drive versions. The most revolutionary new ideas were reserved for the five machines in the 100–190 horsepower range. Each had a new fifteen-speed power shift transmission that gave the operator extremely efficient operation, with an estimated 7 percent increase in fuel economy. Seven of the fifteen forward speeds were in the field working range (with choices of speeds from 3 to 7.5 miles per hour), four more were in the PTO range (below 3), and four were in the faster transport range.

A second feature of these five models—a true innovation, patented by the company—was a clever configuration for the axle of the mechanical front-wheel drive, called CASTER/ACTION. One of the problems in mechanical front-wheel-drive adaptations for row-crop tractors had been its increase in the turning radius. When farmers come to the end of a row they demand minimum lost time in the turning swing, before returning to the next row for productive work. With the CASTER/ACTION principle, the front wheels cant as they turn, dipping and giving the effect of turning under a bit, just enough to allow a substantially shortened turning radius. Introduction of the new 50 series tractors was in process in 1983 and field acceptance by farmers remained to be determined. The models' potential seemed outstanding.

Deere's market share in the top-of-the-line, four-wheel-drive tractors was more modest. The aggressive inroads on this market by the two small companies, Versatile and Steiger (each with horsepower sizes much larger than Deere), coupled with some reliability problems with one of Deere's earlier models (in the early 1970s) resulted in a modest market share for the company. (It had risen to almost 18 percent by 1980, but fell back again in 1981 to just over 14 percent.) The eventual success of the 50 series would have much to say on this; the introduction later in the decade of a larger horsepower top-of-the-line, four-wheel-drive tractor to complement the three members of the 8000 series (the 8450 at 185 PTO horsepower, the 8650 at 235, and the 8850 at 300) would help to determine whether Deere could increase its market share in this behemothian size.

Exhibit 15-4. Top, 50 Series tractor, showing the patented CASTER/ACTION principle; bottom, Minimum turning radius of a 50 Series CASTER/ACTION tractor, 1983. *Deere Archives*

THE SMALL TRACTOR: THE DEERE-YANMAR COLLABORATION

At the lower end of tractor size, in the utility range, several interesting developments in the 1970s profoundly affected Deere strategy and planning. By the mid-1970s, the issue of the small tractor had become one of the company's most puzzling dilemmas. In 1970 the company was still producing a 30-horsepower tractor. But by 1976, the model line gradually had been shifted upward to the point at which the smallest tractor was the Model 2040 at 40 horsepower. Below this were only the seven lawn and garden tractors, from 7.5 to 14.5 horsepower, manufactured in the Horicon, Wisconsin, plant and marketed through the consumer products division of the company. The lawn and garden tractors were just that—well-made, lighter-weight machines for the homeowner to take care of his small- or medium-sized yard and garden. In no sense were these machines intended to be full-scale utility service agricultural tractors—their size was too small, their construction not rugged enough for such purposes.

In that gap left between the homeowner and the larger commercial farmer were substantial numbers of people who needed—and would buy—a tractor. Part-time farmers, for example, had increased during the early 1970s—and had done well in income, both in their on- and off-farm efforts. Heretofore, these "sun-downers" (as company marketing people dubbed them) would likely have bought a used, outdated tractor, in the older, lower horsepower sizes (the venerable two-cylinder "Poppin' Johnnies" of the pre-1960 John Deere line being great favorites). Now these people could afford a new tractor, built to the size and specifications they demanded for part-time utility work. These new operators might not necessarily have in mind much commercial farming, as a USDA analyst recently pointed out: "To some extent, the small farm comeback may represent the transition of resources out of agriculture, as full-time commercial enterprises are transferred into the hands of others with only secondary interests in farming. . . . People with fairly sizable non-farm incomes can pay prices for land that mid-size farmers can't match. So they can bid farm resources (such as land) away from all except very well-heeled farmers." Still, whether these people farmed on any substantial scale, they nevertheless represented a potent market for agricultural machinery.

For a while a few American manufacturers continued to make full-scale utility tractors in the range of 30 horsepower and below—Ford's 30-horsepower model was one of the most popular—but this intermediate horsepower range (13–30 horsepower) was increasingly coopted by the Japanese. Already the Japanese were the preeminent producers of the two-wheel, walk-behind tractor used all over Asia; now they perfected the four-wheel, under 30-horsepower tractor in many sizes. Moreover,

THE SMALL TRACTOR: THE DEERE-YANMAR COLLABORATION

Exhibit 15-5. Three horsepower sizes of the Yanmar-built Deere tractor: left, 33 horsepower; center, 27 horsepower; right, 22 horsepower.

they did this with economy, for the most difficult problem in the smaller horsepower tractor was to keep its comparative price low (its retail price per pound and per horsepower). The numbers of units needed in the range of 20–50 horsepower, worldwide, were large indeed. Deere economists estimated in 1974 that of the total free-world tractor sales in 1975 of approximately 670,000, more than 214,000 would be in that range. Perhaps 40,000 of them would be sold in the United States, substantial numbers in Western Europe, and even greater amounts in Asia, with Japan alone taking some 26,000.

Deere reacted in the early 1970s to the fact that the gap in its line was damaging, and the company began an experimental program at the product engineering center and at the Horicon plant to design a low horsepower tractor (somewhere under 30) that would be competitive with the Japanese machines. By the mid-1970s, the Japanese were selling into the American market in substantial quantities in these utility sizes; by 1977, Japanese shipments into the United States of under 40-horsepower tractors exceeded the numbers sold by all the United States and European manufacturers. Unfortunately, Deere's experimental model—the XR80—turned out to be too expensive for the marketplace. R. W. Boeke, who headed Deere's manufacturing at the time, put the problem succinctly: "Unless we decide to make major capital investments based on marginal returns, it appears likely that we would continue to eliminate lower hp. tractors, and essentially abdicate the smaller hp. sizes on a gradual basis. This could result in direct benefit to Japanese manufacturers." Boeke saw the company's Waterloo plant as making the "large-sized" tractors, the Mannheim plant as the source of "middle-sized" tractors, and the Dubuque plant, which had formerly produced some of the mid-size tractors, turning its production largely to industrial tractors. Spain, Mexico, Argentina, and Australia all likely would remain "pocket markets," none being low-cost manufacturers, and thus they would be able to export competitively only if subsidies were given by their local governments (a step that had been taken only by Spain).

Concurrent with the reduction in production of smaller tractors by the United States producers, the Japanese were aggressively expanding their horsepower sizes from the 15–25 horsepower range up to the 40–50 horsepower range. Several American manufacturers saw opportunities for joining with the Japanese—Ford had a link with Ishi (Ishikawajima-Shibaura), and, later, Massey-Ferguson with Satoh and White with Iseki. Deere had been marketing its Japanese-bound agricultural equipment through Yanmar, a major Japanese diesel engine manufacturer, but earlier had rejected Yanmar's collateral proposal for a possible manufacturing and/or marketing collaboration for small tractors in the North American market.

It seemed wise for Deere to reconsider, for the company now needed a source for tractors in the 20–40 horsepower size (which it did not want to produce itself), and additionally desired a source for the 40–60 horsepower sizes that would be more competitive for the North American market than the company's German-built tractors (the Deutsche mark–dollar exchange rate at that time produced a relatively high dollar figure when German goods were imported).

Deere's fresh start with Yanmar was just in time, too, for Deere soon found out that Yanmar already had been talking with International Harvester. As the Deere-Yanmar sales relationship had been a good one, Yanmar now decided to close out their negotiations with Harvester; the latter then acquired a link with another Japanese company, Kubota. Out of the ensuing discussions between Deere and Yanmar came an important Deere initiative—the introduction of certain sizes of Yanmar-built tractors into the United States as full members of the John Deere line. The tractor would be manufactured in Japan by the Yanmar Diesel Engine Company, Ltd., the manufacturing arm of the Yanmar group. Yanmar Diesel built industrial and marine diesel engines, outboard engines, fuel injection pumps, construction equipment, and agricultural implements and tractors. There already were seven models of the last in the Yanmar line, from 11 PTO horsepower to just over 26; diesel engines were manufactured in a range from 3 to 3,000 horsepower. Yanmar was the second largest agricultural tractor manufacturer in Japan—in 1975 it produced some 38,000 units to Kubota's 80,000 (of a total production of approximately 204,000).

Exhibit 15-6. Deere-Yanmar "summit meeting" in Japan: left to right, Takeo Yoshikawa, executive managing director, Yanmar; Robert Hanson; William Hewitt; Tadao Yamaoka, president, Yanmar. *Deere Archives*

The Yanmar Agricultural Equipment Company was the marketing organization not only for Deere in Japan, but for all the agricultural equipment manufactured by Yanmar Diesel. Under the new arrangement, this Yanmar arm would still be Deere's distributor in Japan; in turn, certain jointly designed and Yanmar-produced tractors would be sold in North America (and selected countries elsewhere) by Deere—with the Deere name—under a sales agreement with Yanmar. Deere would have no equity ownership in either one of these companies—Yanmar was a family company and did not desire such an arrangement.

Positioned in the middle of all of this as the linchpin was a third entity, a new joint-venture engineering company, owned 50 percent each by Deere and Yanmar, to carry out the development of designs and specifications for agricultural tractors in the range of 22–60 horsepower. Some of the tractors that came out of this effort would be sold under the joint-venture marketing agreement, but both the parent companies also would have the right to use the joint-venture group's designs and specifications that ensued for manufacturing their own tractors. Almost all of the detail design work would be done by Yanmar engineers; Deere would be primarily responsible for setting specifications and for testing the final result. In sum, Deere would elaborate the specifications needed to make a Yanmar tractor under Deere standards, but the joint-venture company would do the design work itself. Initially, the plan contemplated that some Deere engineers would be resident in Japan, working with their Japanese counterparts. In practice, this was not done.

The project was consummated in June 1977, following a moment of unease when the Federal Trade Commission took a look at the project, querying Deere about why it did not build its own vehicles in these sizes. The economics of the situation were clear, though, and the FTC chose not to intrude into the arrangement. This question answered, the project went forward apace. By late 1978, the joint-venture company had produced designs for two new tractors, first a 22 PTO horsepower, then a 27 PTO horsepower. By 1979 the actual tractors were in place in the United States as part of the Deere line (becoming models 850 and 950).

Meanwhile, Yanmar Diesel moved ahead by itself to modify two of its existing models, a 14.5 horsepower and an 18 horsepower, both built to Deere specifications. These, too, entered the Deere line, in this case, though, not under the initial joint-venture marketing arrangement, but in a more arm's-length marketing contract between Deere and Yanmar Diesel. The latter also continued to build its own line of tractors for the United States market, first selling through Yanmar dealers and later in collaboration with Mitsui dealers.

Design and development work by the joint-venture engineering group also went forward immediately for two larger models, at 33 horsepower and 40 horsepower. These models also joined the Deere line. The initial agreement provided that Deere was to be the sole distributor of all Yanmar tractors in the 27–60 horsepower range; two larger sizes, a 50 horsepower and a 60 horsepower, came into the line in 1983, marketed under the joint-venture arrangement.

Were Yanmar Diesel to become interested in building tractors beyond 60 horsepower and marketing them on their own in the rest of the world or even in the United States—a move contemplated by some Japanese tractor manufacturers—they would be free to proceed on their own, for the Deere-Yanmar marketing arrangements did not provide for any relationship beyond 60 horsepower.

THE SMALL TRACTOR: THE DEERE-YANMAR COLLABORATION

By 1983, the Yanmar-made tractors under the Deere name had primarily been marketed in the United States. At the start of the arrangement a complicated territorial agreement was made. Yanmar took most of Asia (its traditional marketing area) but also Brazil and the Soviet Union, as well as the key Middle East countries and several important countries on the northern and western perimeters of Africa. Deere was given North America, the rest of South America, Europe, and Africa, as well as a number of countries in the Middle East, certain countries in South Asia, and Australia and New Zealand.

Deere had to concern itself with a complicated interface in this Yanmar relationship—the Yanmar-made tractors against its own models, manufactured by the company itself. There appeared to be only modest problems at the lower end of the scale—the two smaller tractors, the 14.5 horsepower and the 18 horsepower, sometimes compete with the Horicon lawn and garden tractors for large-lawn and park jobs, but the utility tractors are really quite different from the lawn and garden models. Similarly, in the 18–40 horsepower range, Deere chose not to produce any models itself—the Japanese volume was so dominant that a United States manufacturer was not likely to be able to challenge their unit costs.

At the upper end, though, there were some significant relationships raised by 1983. The joint-venture-designed model 1250, introduced in 1982, is a 40-horsepower version, and the Deere-manufactured model 2040, built in Mannheim, is also a 40-horsepower tractor (though its counterpart, model 2150, introduced in late 1982, is a 45-horsepower model). The rubric was established that the Yanmar model was the "lean and trim" version, the Mannheim a "deluxe." The price of the latter at first was substantially higher than the Japanese tractor, but in the early 1980s the unfavorable Mannheim differential began to narrow as the dollar strengthened against the Deutsche mark, and the Mannheim tractors at the interface of horsepower size could then be landed in the United States at a price much closer to Yanmar than at the start of the latter's relationship with Deere. The Yanmar machines became less "lean" (though presumably equally "trim"). Thus, this interface problem would always be plagued by the bugaboo of exchange fluctuations, among the dollar, the Deutsche mark, and the yen.

The Deere-Yanmar relationship proved to be a promising one. The quality of the design work of the joint venture was outstanding, and the manufacturing capabilities of Yanmar Diesel were certainly close to those of Deere. Through 1981 the Yanmar tractors had not yet been particularly profitable for Deere, given the substantial up-front development costs. The arrangement had the effect of keeping other aggressive Japanese manufacturers (e.g., Kubota) out of Deere dealer lots. But, more than that, the collaboration was one of goodwill and mutual give-and-take; the relations between the two companies' chief executive officers were a particularly strong force, Hewitt and Tadao Yamaoka (the head of Yanmar) visiting each year for what they came to call fondly "the summit meeting." The choice of Yanmar as a major joint-venture partner has been a felicitous one.[5]

GLOBAL TRACTOR STRATEGY

The Yanmar relationship remained just part of the larger issue of the required global tractor strategy for Deere in the 1980s. Deere's strategic plans that extended out over the decade were proprietary both for competitive relationships with Deere's always-aggressive sister companies and also for the sensitive internal changes that would result from them. Factory assignments would be changed, marketing structures would be shifted—the future always has portents for those whose lives are affected. Certain key questions likely will form the basis for Deere's strategy.

For the tractor product lines, how many total models can be successfully sustained? By 1987, the worldwide plans call for some thirty-three separate model numbers across twenty-nine horsepower ratings. Can production runs on each of these models be large enough to exceed break-evens for each, or is this an over-proliferation of models? Product definition appears well structured at the very low and very high ends of the horsepower spectrum; there is less clear production definition in the middle range. Horsepower ratings are close, sometimes identical. Can a strategy of "lean and trim" models (Yanmar) and "deluxe" (European) versions work together? How high in the horsepower range should tractors sourced from outside Deere & Company proper (Yanmar, others) be allowed to go in the Deere line?

For tractor design strategy, how can a tripartite manufacturing responsibility (the Japanese joint-venture company for the lowest horsepower, the low and medium horsepower in Europe, and the high horsepower in the United States) be successfully meshed? What will bring maximum parts commonality and design integrity? Can Deere & Company itself maintain requisite control over major componentry?

In tractor sourcing, how many tractors should come from outside Deere & Company? At what level in outside sourcing does excess capacity appear in Deere factories, and in which ones? Should any whole-goods factories be transformed to componentry operations? What are the longer-run potentials of the factories in Turkey and Iran? What will be the effects of Spain's entry into the Common Market on the sourcing role of Deere's operation there? Are there other countries with promising sourcing potential, for example, the People's Republic of China or Brazil?

For the tractor strategy, two principal markets are envisioned—North America and Europe. Can all other areas of the world be fully served by models developed for one of these two markets? If the market focus in North America is indeed clear-cut, can the same be said for Europe? Can "pocket markets" be sustained over time, with high-cost inefficiencies built in? Overall, how can volatile currency rate fluctuations and political risks of the world best be understood and mitigated?[6]

DILEMMAS IN COMBINES AND TILLAGE AND HAYING EQUIPMENT

The Yanmar relationship particularly focused company attention on the question of worldwide tractor strategy, probably the most important of the questions facing Deere in the 1980s. Nevertheless, tractors are inextricably linked with the other segments of the company's lines—combines, tillage equipment, and so on.

The Deere combines were particularly strong competitors in the late 1970s and early '80s; they commanded more than 40 percent of the United States market in 1981. In this case, the company adopted a conservative strategy, continuing with the conventional combine, not building the new "rotary" configuration adopted first by New Holland and then International Harvester in the 1970s. The rotary had some inherent advantages—instead of a reciprocal motion in the conventional configuration, the rotary had a simpler rotary motion. The rotary took more power than did the conventional for a given size, but its construction was more compact. The rotary sometimes did not work well in wet or tough straw and was particularly ineffective when there was a bad season of rain or a heavy weed content. Deere had been testing its own version of a rotary combine since 1958 but at several stages had chosen not to adopt the new concept. In the early 1980s, it was decided that "while John Deere will continue long-term basic research with the goal of achieving a true breakthrough in threshing and separating concepts, our combines will continue to feature our current concepts." The dealers at the New Orleans show in 1982 were explicitly told: "There is no rotary combine on John Deere's horizon." Throughout most of the recent period of the rotary, Deere has picked up market share over its competitors, testimony to the validity of this important decision.

With farmers' demands for combined operations—for example, using a mower, a conditioner, and a windrower in one combination—new, more adaptable tillage and hay-conditioning equipment were obviously needed. Field speed had continued to edge up in the mid- and late 1970s, yet the farmer wanted further gains in total efficiency and productivity and greater savings of his time. This could only be done by wider equipment. Soon there were many eight- and twelve-row, and later twenty-four row, tillage and planting instruments. The soybean crop in the United States more than doubled (in bushels) from 1970 to 1980, and this brought new innovations especially designed for soybean harvesting—for example, the flexible combine platform and the use of a row-crop head (the latter a Deere exclusive in those years).

Especially with planters, the tension for the farmer greatly increased with increased width—it became even more crucial that each of the many rows be working correctly. In the mid-1970s, Deere engineers developed a remarkable new planting concept in the Max-Emerge planter. The engineers began by asking, "What does the seed want?" First, it needs positive contact with firm soil at a precisely set depth. Most planters at this time cut a large trench; before the seed could be placed, much loose dirt fell in the trench and the seed dropped on this loose mass. Further, the seed, once there, germinates best if the soil to fill the trench is firmly packed around the seed, not compacted heavily right over the top. The Max-Emerge planter uses an ingenious combination of a narrow "V" cut and a pair of slanted wheels to drop the seed at a precise depth in the bottom of the "V," touching firm soil, then pressing dirt around the seed with the wheels, but not compacting downward. The planting rate of each row is monitored electronically, so that the operator in his cab knows exactly how each row is planting. More recently, concern was put to the issue of not just monitoring the planting itself, but correcting any faults automatically. Deere's Max-Emerge, a patented concept, dominated the field in the early 1980s; the company's planter market share rose as high as 75 percent in one of those years.

Innovations were also made by the industry in baling—the round baler became very popular in the 1970s. The "hydrapush" manure spreader was also well received, particularly in the colder climates, where the more traditional conveyor chain often froze or broke. Deere was a follower for the baler but pioneered with the new spreader.

The broader questions about worldwide strategy applied as well to combines. With combines being made both in East Moline and Zweibrücken and being assembled in several other parts of the world, essentially the same kinds of questions posed earlier about the tractor surfaced once again. Other equipment—hay tools, cotton pickers, tillage instruments—also is made and/or assembled in several locations, and questions of multiple, integrated sourcing apply there, too. The company's consumer products sourcing was limited in the main to North America, but sales had become international. In construction equipment, major sourcing and marketing questions remained to be resolved, questions of almost as much complexity as those relating to agricultural tractors.

THE NURTURING OF INNOVATION

These puzzles concerning worldwide product strategy relate fundamentally to the company's posture in respect to manufacturing, and the research and development efforts that underlie it. A number of nagging questions had to be resolved in the 1970s for both research and development and production.

Exhibit 15-7. Introductory advertisement for the Max-Emerge planter, 1974. *Deere Archives*

Seldom in the history of the company had there ever been any question about the inherent quality of Deere's engineers. Over the years, the very hallmarks of "John Deere" were its sturdily built, well-engineered products. Questions of coordination in engineering did arise, however, particularly in the research and development dimension. At root was that longstanding enigma of the company: How much decentralization?

From the 1910s and the days of Joseph Dain's experimental work with tractors, and later in the 1920s and '30s with Theo Brown's "experimental department," there had been a small amount of centrally coordinated research and development. But most of the key work was done by engineers back at each factory, working on their own particular machines and jealously guarding their "turf." In 1944 Charles Stone shifted his portfolio from vice president of manufacturing to vice president of product development and two central departments—a product research department and a product development department—were established; this placed some of the basic product research at the company level.

In the 1960s, when Frank T. McGuire was made vice president–research, a further move took place: a new engineering research department (incorporating the old product development department) was established, Gordon H. Millar becoming its director of research. These moves signaled to everyone in the company that the research arm was now more centralized, though McGuire still felt it necessary in his memorandum announcing the role of Millar to state: "The role of research is to complement and not compete with our current product engineering efforts which are the responsibility of the factory organization." McGuire described the effort to the Deere board as "non-competitive and compatible with the factory product engineering effort." The annual research expenditure at this time was budgeted at $1.8 million, representing just 0.3 percent of the approximate $600 million sales level.

When McGuire left the company temporarily in mid-1965, though, there was a malaise in the central engineering research group, further compounded in early 1969 when Millar became assistant general manager at the Waterloo Tractor Works. Shortly after, all of the engineering group responsible for engines and transmissions had also been relocated there. The bulletin explaining this move was frank in stating that "this reorganization represents a basic change from the structure that has had considerable emphasis on our research." Now the function of the "technical center" (as it was renamed) was to explore computer techniques relating to engineering design, to provide a liaison with and appraisal of university experimental work, and to develop "advanced concepts of new products that do not have a direct relationship to our present product line." The technical center work was to be constrained, for the bulletin continued: "By the very nature of our decentralized organization, the engineering activities of the various factory Product Engineering Departments have a tendency to overlap or duplicate engineering work." The message was that such duplication should be eliminated—the factory engineering groups themselves should do the primary development work.

There was considerable unease about this downgrading, and later in that same year the board made a decision that was to have profound effect on the entire research posture of the company. Dr. Fredrick C. Lindvall, a world renowned engineering educator at the California Institute of Technology, was brought to Deere as vice president, engineering. Lindvall, a Moline native, knew a number of the people he was to work with; Hewitt, at this time a Cal Tech board trustee, had become particularly well acquainted with him. Lindvall's charge was to assist especially Deere & Company's principal officers and their staffs in planning "the technological development of the Company."

It took Lindvall only a few months to assess the situation, and in May 1970 he presented a major memorandum to the board. It was blunt in its criticism of the proprietary biases of the individual factory engineering groups. There, he felt: "New product ideas naturally tend to conform to a particular plant's experience and competence. Competitive pressure demands that substantial product engineering effort be concentrated on renovative changes and product line increments which sustain today's business. Cost reduction studies on a large volume item may present a more attractive profit potential than a new item with an uncertain market. Yet such efforts may not assure a longer range future nor anticipate 'the next generation' of demand." The general managers of the major factories had substantial discretionary money involved in this, and extensive engineering and experimental machine testing might have gone on before a "product review" brought the new product development to the attention of the general company officers. "A high level of security can thus be maintained at the factory in the early stages of a development if this appears to be necessary or desirable. However, a lack of communication among product engineers at the several factories may lead to duplication of effort, development of incompatible products, or failure to exploit a good idea which has Company-wide potential." Lindvall felt that the "present informal process through which new products are developed" lacked specific checkpoints and product reviews and that the factory manager, in the absence of clearly expressed views or recommendations from outside his own territory, "is obliged to make his own decision in an atmosphere of uncertainty in which 'silence gives consent.'"

Lindvall recommended "a formal procedure," in which clear statements of objectives would be elaborated, with definite check or review points at which the design or the prototype could be measured against these objectives, all this to be done with good documentation in order that management might later evaluate the development process. Lindvall acknowledged that the particular factories probably had such a procedure already, but that there needed to be "the benefit of inputs and review by personnel of other factories and Deere & Company . . . early enough in the design and evaluation study." Lindvall quoted at length from Patrick Haggerty, the chief executive officer of Texas Instruments, who seemed to be speaking directly to Deere in an article in the magazine *Innovation*. Haggerty pointed out that many companies in the process of developing a high degree of decentralization had found themselves "in danger of becoming no more than the sum total of the decentralized parts, loosely governed at the corporate level, primarily from a financial point of view." Haggerty continued: "At that point, the biggest job the corporation can handle has to be related to the biggest job that one, or at the most a few of the decentralized working units working together can handle. . . . The span of responsibility practically always exceeds the span of authority. Each manager has an authority which extends only to his decentralized unit, but a responsibility which extends across the corporation. Without some countervailing force, even when good innovative managers do develop in a decentralized organization, their innovations are ordinarily restricted to the entity for which they have responsibility or only narrowly and obviously beyond it." Thus, Haggerty felt: "Innovative efforts, even when they exist, tend to come out to fit the size of the decentralized units.... There are few or none of what I call breakthrough strategies; that is, strategic and innovative courses of action which, if they succeed, will impact the whole corporation in a major way."

Lindvall's memorandum was a bombshell, for it challenged as no other effort earlier had done the vested interests of the individual factory managers and their engineering groups. The board moved quickly to implement Lindvall's suggestions. In August 1970, two new committees were established: the product engineering council and the product engineering technical committee. The bulletin explaining these two new arms paid homage to the past: "In keeping with the policy of decentralized operations . . . product engineering strategy, problems and staff work . . . have been and continue to be based firmly on the assignment of direct responsibility for product strategy and product engineering activities of each John Deere factory to the general manager of that factory." Then the bulletin continued: "Without either changing or limiting this responsibility, product engineering effectiveness can be strengthened." Now there were to be "means for corporate coordination" and "means for organized communication."[7]

Lindvall, then at retirement age, stayed with the company only until August 1972. At this point, Gordon Millar was brought back into the role for which he had originally been hired—vice president, engineering. This time the organizational groundwork was laid for a far more effective effort by general-company research and development work. Frank McGuire had begun a practice of bringing in senior outsiders, engineers on the cutting edge of technological developments around the country; now this was again an accepted pattern (although not always setting well with all of the company's entrenched engineering groups in the factories). The payoff for this move was great; the decade of the 1970s witnessed great strides in Deere & Company research and development efforts, moves that were fundamental to the ability of the company to become so strong in the industry. In the period 1972–1976, total expenditures on capital improvements was some $561 million; research and development expenditures added an additional $427 million. Thus the total of the two came to almost $1 billion. In the second half of the decade, 1977–1981, the capital improvement amount rose to $1.3 billion, with the research and development expenditure at $951 million, a total for the two well over $2.2 billion.

Each year *Business Week* keeps an "R&D Scoreboard" for the major companies in most of American industry. Deere's expenditures on research and development consistently have been at the top of the agricultural machinery industry, indeed exceeding the others by quite a margin (Appendix exhibit 36).

Such growth indices as the *Business Week* research and development figures are a bit misleading, though, for it matters a great deal just how the mix of these expenditures is being used by a given company. There is a subtle, hard-to-define equation between development work put to existing products as against new products. Particularly for the latter, the question of the degree of innovation sought is also a preoccupation of research and development managers. There is a threat to the profitability of such endeavors if excessive risk is taken or, contrariwise, if not enough risk is taken—a "me too" posture. In exhibit 15-8, both high-risk and moderate-risk activities are shown with a confidence band that broadens with the increased degree of innovation, or with increasing portions of the research and development budget invested in innovative concepts. Both profit opportunities and exposure to loss of profits increase with an increasing degree of innovation.

Research and development costs are further affected by the degree of vertical integration present in a given organization and by the volumes that can be attained with given product lines within this structure. Too much vertical integration sometimes can diminish profitability because of excessive investment needs in research and development, so that high figures on the *Business Week* tabulations may sometimes be misleading. Similarly, not enough internal vertical integration within a company can reduce profits, even though research and development costs drop significantly in such a mode. This balancing act can be seen clearly in another organizational battle that took place in the 1970s at Deere, centering on the original equipment manufacturing (OEM) group.

SHOULD DEERE SELL ITS TECHNOLOGY TO COMPETITORS?

Deere has often been a supplier to other companies of various pieces of Deere componentry (engines, hydraulics, transmissions, castings, etc.). If the company has substantial in-house manufacturing capacity, why not spread overhead over a larger volume, selling the amounts in excess of needs to other manufacturers? For some components this notion is quite straightforward, for others the idea has a threatening dimension. For example, will the selling of engines give away to competitors important proprietary technological benefits? Could Deere's research and development costs be recovered under such an arrangement, or would the company just be "giving the product away"?

Robert J. Carlson, senior vice president for North American farm equipment and consumer products in 1974, put the negative side bluntly, in regard to a proposal for selling a particular pump as an OEM product: "I believe

Exhibit 15-8. *The Relation between Innovation and Profitability*

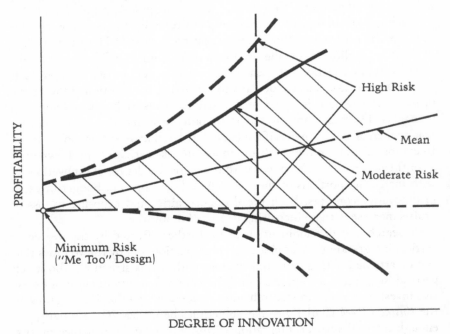

Source: R. W. Boeke to Robert A. Hanson, August 7, 1980.

the pump to be more unique than an engine, a Sound Gard body, or a corn head. . . . Our hydraulic system is the most advanced feature of our tractor line—in short, the area where we have the greatest visible advantage. . . . The pump is the heart of the system . . . the problems inherent with the service of hydraulics are difficult ones. A competitor would help himself greatly if he could point to the industry leader, in this case Deere, and say, 'Ours is the same, don't worry about problems.'"

The key question is whether a technological advantage is being dissipated; in the pump case, Gordon Millar, Deere's vice president of engineering, differed with Carlson: "I don't sense that industry is beating our doors down . . . on the basis of the pump's intrinsic capabilities. The real reason people want to buy our pumps is because they are well made and priced competitively. With regard to the pump itself there is nothing intrinsic in its design or manufacture that cannot be copied by competition. . . . I appreciate the position that Marketing has taken . . . but I personally think the risk is greater in not expanding our manufacturing base than it is in giving away technology."

The process works the other way, too. It might well be that for a particular component an outside manufacturer could supply a better and/or less expensive component than Deere itself could make. Each time, though, it erodes some of Deere's traditional vertical integration. Selling one's own machines in the hurly-burly of the competitive world sharpens one's own competitiveness on price and product development. If a company just buys componentry, its internal capability tends to dissipate; as Millar pointed out, "Not going to be in the OEM business does not preclude the requirement for Deere making substantial investment at some future date to remain competitive in the diesel engine marketplace, even if we are our own exclusive customer.... We don't lull ourselves into the trap of thinking that just because we choose not to be in the OEM diesel engine business that this somehow precludes spending sizable sums of money to remain competitive ... not just look at the status quo as keeping us competitive forever."

Collaborative arrangements for selling diesel engines to other manufacturers—in either noncompetitive business, or even for trucks or farm equipment—certainly are possibilities. With the enormous capital costs necessary for redesigning a particular diesel engine (for even the redesign and relocation of production of one series of engines, the 300 series, Deere spent an estimated $175 million), such sales would help to recover some research and development costs. A strong OEM business often has a tendency to drive, rather than follow, internal component manufacturing. A higher degree of standardization is likely needed—for example, more compatible interface connections to other makes. Some special in-house engineering may have to go in the process. A components division of a company can be greatly strengthened by a strong OEM business, however, both by virtue of increased volume and by a more competitively oriented product development.[8]

FURTHER IMPLICATIONS OF CENTRALIZATION

In 1972 and 1973 tractor sales shot up much beyond expectations; forecasts not only predicted higher sales but also indicated rapid increase in the proportion of over-80-horsepower tractors. It seemed incumbent to expand tractor manufacturing facilities.

Engines would be the bottleneck in doing this. Potentials for outside sourcing were considered impractical, so a decision was made to develop a new engine facility and a new foundry. Everyone agreed that the foundry's location should be Waterloo. For the engine factory, however, there was sharp controversy in management about its location. Most of the Waterloo manufacturing group preferred to expand in Waterloo; indeed, Robert J. Carlson, the senior vice president in charge, wanted to have the existing downtown facility expanded by a high-rise building (there being no room to expand horizontally). Others, particularly Delno W. Brown, the senior vice president of the industrial division, advocated moving the engine works away from Waterloo altogether, arguing that the Waterloo organization was getting too large "both as a facility and also in relation to the available labor market."

But this was only part of Brown's argument, for he and others became at this time champions of a far-reaching notion, namely, that the engine decision was the propitious moment to decide upon the establishment of a full-fledged "components division." Brown faulted once again the dominance of the factories (especially Waterloo) in a letter to Hewitt: "The increased interfactory business we are experiencing has exposed a major problem—of priorities in the design and manufacture of interfactory components. A factory manager's success or failure is based on his performance in relation to his own products. This means that components for other factories may be given a lot of 'lip service' but very little priority when there is a conflict with the factory's own product."

The John Deere Engine Works indeed was built, the final decision on location, Waterloo. On a 191-acre site more than one-third of a mile outside of the city's limits, a 21-acre engine manufacturing operation was put under roof. A wholly new electric foundry also began operations in Waterloo in 1972 with a capacity that soon rose to some 90,000 tons annually (with a 1978 addition heightening this capacity to a total of 236,000 tons). In 1975, as the new Engine Works opened at its new site in Waterloo, land was acquired just outside Waterloo at the so-called "Northeast Site" for the construction of a new tractor factory. Over a period of five years, a new, 2-million-square-foot factory building was brought into being, opening in May 1981. This was the star in the raiment of Deere's capital expansion program of the 1970s, a remarkable award-winning plant that dramatized more than any other single effort the Deere capital expansion program.

Meanwhile, the notion of a components division was kept alive. The product engineering center became the focal point for some of the tensions between the decentralized factory personnel, particularly at the Waterloo Tractor Works, and those Moliners of a more centralized, coordinating persuasion as advocated a few years earlier by Lindvall. By 1975, when the Engine Works came on line as a separate factory, the product engineering

Exhibit 15-9. Interior, Deere's Waterloo Engine Works, 1981. *Deere Archives*

center had developed a separate engine group and the center's reporting relationship was changed to the director of manufacturing, to whom also reported the Tractor Engine Works; this was the first time that the Waterloo Tractor Works did not have its product engineering center unit reporting directly to it. The practical effect of this for the remainder of the 1970s was to concentrate corporate component engineering effort, as well as the design and development of agricultural tractors worldwide, at the Waterloo location. Though some of the overseas operations—Mannheim, Saran, Australia—continued to play a modest role in engineering, the pinnacle was in Waterloo. Similarly, in industrial and construction equipment, the worldwide engineering effort was concentrated at the Dubuque engineering entity. Combine engineering, though, continued to be fully decentralized as in past practice, with the East Moline engineering group taking care of North American combines and the Zweibrücken group designing the European combines. Gordon Millar, the vice president of engineering for the company, worried that the company was not taking advantage of a worldwide product concept for combines: "We have enormous internal difficulties rationalizing a worldwide engineering effort with our concept of decentralization. Breathe the phrase 'central engineering' and you can shoot down in one fell swoop design control, engineering coordination and anything that would relinquish from the factory structure any local control of product design. It may be that we can afford the premium costs and some of the attendant beliefs in our combine engineering effort of the total decentralized approach.... I doubt it."

In the summer of 1979, Senior Vice President Robert J. Carlson resigned from Deere to accept a position at United Technologies, Inc. At the same time, the retirements of Senior Vice Presidents A. B. Lundahl and Delno Brown brought the opportunity for appointment of new senior management in these posts. Boyd C. Bartlett became senior vice president of farm equipment and consumer products–United States and Canada. Charles A. Hanson became senior vice president of corporate staff, Joseph Dain became senior vice president of corporate planning and was succeeded, on his retirement a year later, by Thomas A. Gildehaus. L. N. Hall remained senior vice president of the overseas division–farm equipment and consumer products, and Clifford L. Peterson continued as senior vice president of the finance division. Robert Boeke, who had been named a senior vice president in 1979, headed the new components division. The major sections of the new components and design manufacturing division initially consisted of two components manufacturing units (the Tractor Works and the Engine Works at Waterloo), as well as the production engineering center, the John Deere foundry, and parts distribution.

The establishment of the components division as an equal in reporting relationship to the other three major operating divisions of the company (agriculture and consumer products, industrial equipment, and overseas) clearly signaled a new dimension of Deere manufacturing strategy. The historic focus of the company on whole-goods design and manufacturing now was to be balanced by a separate concept of worldwide components design and manufacturing.

In the process, the long-sensitive issue of the product engineering center seemed to have been resolved—it was given a reporting relationship directly to the senior vice president of the components division. However, things appear never to be completely settled in the age-old tug-of-war between centralization and decentralization. Late in 1980 operational responsibility for the Tractor Works was taken from components and given back to the farm equipment and consumer products division. Then in August 1982, all two hundred or so of the engineers working on the agricultural tractor vehicles (i.e., everything excluding engines, transmissions, and final drives) were taken away from components, now to report to the general manager of the Tractor Works (they continued to physically reside in the product engineering center, though). How this would work out should components decide to build, for example, an engine that could be sold outside of the company, too, and the tractor division engineers plan a vehicle with special features only available to John Deere, remained to be seen.

In 1982 a task force on a "global strategy tractor plan" looked once again at Deere's overall strategy to be the "world's leading manufacturer of agricultural-type tractors" and elaborated explicitly on the component division's mission: *"A componentized product strategy* appears best suited.... High production volumes will insure the *lower unit cost* of all critical components. Tractors can then be assembled of extremely cost-effective components, insuring the best cost/price ratios and the *highest profits* in the industry for the total line." The task force identified four specific areas for concentration: (1) providing a family of modular nonstructural power-shift transmissions for all corporate requirements above 90 PTO horsepower; (2) becoming cost effective in the 40 to 80 PTO horsepower tractors by developing a "skinny machine," with a particular focus on volume, rather than a primary emphasis on design of many different special-need products; (3) instituting a feasibility study toward the notion of providing sourcing for tractors in the 80 to 130 horsepower group, from either North American or European sources, with an upward extension of the present utility line in Europe and a downward extension in North America of the row-crop line; and (4) maximizing the commonality between the grounds-care and compact utility tractors in the 10 to 39 PTO horsepower group.[9]

THE "AUTOMATIC FACTORY"

A Deere engineer at a manufacturing strategy conference in 1980, hearing a comment from a fellow participant about a ray of light for Deere's European manufacturing profitability, commented: "That ray of light might turn out

Exhibit 15-10. Aerial view, Deere's Waterloo Tractor Works, 1982. *The Furrow*

to be the headlight of the Japanese Express Train." The preoccupation in American industry with the "Japanese challenge" was pervasive at that time.

These issues really centered on the factory, for it was in manufacturing that the Japanese had acquired such a comparative advantage. (As a matter of fact, there was considerable evidence that American marketers were significantly more adept than their Japanese counterparts, at least at that time.) In the factory, two key components dominated: (1) the configuration of the factory itself, its equipment, logistics, and technical backup; and (2) the people who manned it, their work relations, compensation, and so forth. The new Waterloo Tractor Works was at the cutting edge in regard to both.

The term "automatic factory" was used in a number of write-ups of the Waterloo factory in outside publications. The term tends to conjure up an almost frightening image of an enormous complex of steel and electronics, with robots all over the place at the work stations sending parts here and there and assembling products all automatically, guided by giant computers in some inaccessible, antiseptic room in the bowels of the plant. While such a Frankensteinian vision of the automatic factory seems overly apocalpytic, there are critical questions at issue in regard to labor's views of the new developments in manufacturing technology at Deere and other companies.

The computer does indeed lie at the heart of the Tractor Works. Even before the factory floor, the computer is critical in the design function; the widely used acronym, CAD-CAM (computer-aided design, computer-aided manufacture) is used as a generic term for a family of sophisticated computer-generated aids to design and manufacturing engineers. In the creation of a particular part, the engineer can use an electronic pen in a geometric modeling system to draw the part directly into the computer, detail it, add dimensional lines, and even reshape it. Three-dimensional visual capability permits the part to be rotated and viewed from any direction on the computer screen. A set of thirty-six standard dimensional descriptions—"round," "hexagonal," and so on—are already in the computer, so that the engineer can instantly call up a particular combination of them to picture a part before his eyes.

The real power of the system, though, is in the next step, for the computer can transform the designer's concept of a part—his mental data set—into an electronic data set that becomes a precise definition of the part, machine-readable in the computer. Thus this data base cumulates past practice; having all of this information in the computer gives the engineer tremendous power to be able to identify whether a part has already been made somewhere in the system before. Instead of engineers at different locations and at different times redesigning the same part, perhaps with slight variations, the process becomes far more standardized. Where perhaps nine or a dozen different variations of a particular part might have been in the system before, now only one or two may be needed.

It is still necessary for design engineers to take advantage of this opportunity for standardization, a process not always easy to accomplish. James F. Lardner, Deere's vice president of manufacturing development, put the dilemma well: "Component planners and design engineers need to develop a constructively skeptical attitude about the degree of uniqueness a component must have for a given end product application, while end product designers must recognize the risk inherent to creating a design for which a unique variant of a standard component must be improvised hurriedly to meet scheduled production dates."

The process of what Deere calls group technology (GT) has already markedly decreased the number of Deere parts in the total system, a remarkable savings when one considers that even in 1983 there were several hundred thousand parts in total in the company (with some 70,000 already in the GT database).

The cost accounting specialist, incidentally, became a handmaiden in this effort. Formerly, an engineer could redesign a part for a minor savings in material or labor time and get kudos for the presumed lower cost. Yet in the process he was more than likely proliferating the total number of parts, a result far more expensive than the savings that ensued from the "new" part. Thus an ostensible savings of cost by redesign of a new part could only become effective if the part was used in realistic amounts. The Deere accounting group has applied a revised concept of "volume costing," in which the actual volume of a particular part used helps to define its cost. (Standard cost systems are often notoriously insensitive to volume, and sometimes produce "answers" contrary to experience and common sense.)

On the factory floor, too, new concepts (often abetted by the computer) have really come into their own. Of the many Japanese innovations in manufacturing, one of the most potent is their application of the "just in time" concept (what the professional literature calls the Kanban system, named from a Japanese word for a production order card). To put it simply, the theoretical idea would be to have just one part at every work station at the moment that that particular part was needed for machining or assembly. There would never be any downtime because the part was not there in time, nor would there be any excess parts inventory at any prior stage. In practical terms, this concept can never be fully reached—one needs part storage, work-in-process, and so forth. Yet the potential for savings in inventory in process by tight control of materials management is very important indeed. Conventional wisdom in manufacturing would hold that inasmuch as one should never be caught short of materials, a substantial buffer should be on hand at all times. One management consultant estimated that the Japanese automobile industry was running on $800 million in inventory to produce 11 million cars; the United States companies, he stated, needed $8.5 billion to do the same. Deere needed only to remember its overbuying debate of 1920 to be reminded of the dangers of such thinking.

But concentrating on inventory savings alone misses the real story of Kanban. In truth, the Japanese made these steps primarily to eliminate manufacturing inefficiencies—to take away the artificial cushion of a large parts and inventory backup, a prop for ineffectual managers to "cover their tracks." "The Japanese consider inventories a cover-up for production problems and waste," a Deere executive commented. "They lowered inventories to reveal production deficiencies—once revealed, the problems could be solved." The Japanese coupled this with an innovative use of sequential continuous manufacturing, in which each individual work station is reduced to the ultimate in simplicity but many are linked together in a "cell" arrangement that reduces the time that a part or component stays "in process." (Deere found, for example, that certain components taking only 3-1/2 hours in actual process time were taking five weeks to get through the system; in effect, the components were sitting idle in bins 95 percent of the time.) Such a cell arrangement also cuts down the many jurisdictions that components move through in more conventional systems; the savings in overhead are substantial.

It is in the Waterloo Tractor Works that Deere has progressed farthest in these new manufacturing ideas. The Tractor Works superficially resembles many other manufacturing plants of its type—its conveyors, machine tools, welders, even its programmable controllers, all are familiar and well proven. "Nonetheless," lyrically reported one of the trade journals, "the TW shines as a supernova in the galaxy of today's US manufacturing plants." What makes the Deere factory unique and award winning is its integrated weaving together of many, many advanced manufacturing concepts. At the heart is the concept of computer-aided manufacturing (CAM). A host computer system gives orders to ten minicomputer systems scattered through the plant; the latter carry most of the burden of managing parts ordering, material flow, storage and retrieval of work-in-process, synchronization, tracking, routing, verification, and logging. Tied to the minicomputers are some 130 computer terminals, so that foremen and production managers can instantly determine where everything is within their systems. An innovative mechanical integration of conveyors, tow lines, monorails, automated storage and retrieval systems, and cranes free the operative for more advanced technical responsibilities.

The potential savings in in-plant transportation and machine usage can be astounding. Deere's president, Robert Hanson, at the annual meeting of the company in 1982, recalled that manufacturing executives had found that whereas the manufacturing of fifty-nine parts had earlier required twenty-three machine tools and twenty-eight different routes through the factory, by the use of another planning concept just being implemented in the company (computer-aided process planning [CAPP], along with the group technology underpinning) these same fifty-nine parts were estimated to be made by only nine machines and fourteen standard routes.

More than a dozen robots were in operation at various Deere factories in the early 1980s, the bulk of them at the Tractor Works, the Dubuque Works, and the Harvester Works. Their uses ranged from the automatic spray painting of individual parts and complete assemblies to loading parts in heat-treat operations and welding. Studies were also underway concerning possible assembly operations by robots. The use of the robot contributed to lower product unit cost through the reduction of a number of labor-intensive jobs. Further, as a company publication noted, "Robots have been found to demonstrate as high as 95–99% uptime and to provide consistently high quality work. In most cases, the robots are doing jobs that can be hazardous or highly repetitive for workers." The report's authors continued: "Justification of some robots is difficult because certain benefits are difficult to quantify (e.g., scrap, repair, quality, consistency, safety, health, etc.). It is almost a 'Catch 22' situation since the robot may need to be in place to gather the data needed to justify the robot."

The acceptance of robotics by Deere's employees has been remarkably good. There is, of course, a primitive, almost instinctive fear of machines taking over from man. As one committeeman in the United Automobile Workers local at the Tractor Works put it, "When people in the plant see these little robots moving down the aisles, they worry about their jobs . . . they think, 'The next robot could take mine.'" Yet, as the president of the Waterloo UAW local put it, "We're concerned, but the UAW has never opposed new technology."

Deere management has been effective in working with employees in planning and understanding the system and in using a participative style of management for group support. A pioneer program at the Waterloo Engine Works for "employee participation" was a significant success, and there were plans to implement it at other plants in the future. The company, incidentally, eschewed such buzz words as "quality circles" in describing its plan and attempted to stay away from Japanese comparisons; as a company report put it, "An encouraging point is that their success is not purely attributable to the Japanese culture. . . . The Japanese style is working very well in many parts of the world—using Australian, British and American workers."

A United States management consultant visited Deere's Tractor Works and was quoted later in the press as saying: "If you offered the Japanese that plant at a bargain price, they would not take it." (He was particularly critical of the company's high-rise storage units.) But such comments miss the point; as Robert Hanson put it in a keynote speech on Japan before a conference of business leaders in 1982: "The United States is not Japan. Not everything that works there would work here. We're a pluralistic society, not a racially and culturally identical one. Job mobility—a labor force willing and able to move, following new and changing opportunities, is a tradition here; almost the very opposite is the case in Japan. Our unions are generally craft unions, not company-wide unions, and relationships between companies and unions here are at arm's length, unlike the usual situation in Japan. Most companies here have outside directors on their boards, and this is an increasing trend. Few outsiders sit on the boards of Japanese companies. Women have far fewer opportunities in Japanese industry than they do in our own country. And the relationship between large Japanese companies and the Japanese Ministry of International Trade and Industry is a very cozy one. A similar arrangement here might be considered collusion rather than cooperation. So although I think there is much to admire in the Japanese way of doing business and certain areas where we may be able to adapt some of their procedures, I would not suggest that we mimic them."[10]

All of this serves to emphasize that the employee relations implications of the "automatic factory" are critically important to the system's effectiveness.

"GIVE BACKS" IN INDUSTRIAL RELATIONS

Deere, Caterpillar, and International Harvester all had contracts with the United Automobile Workers that were due to terminate in 1979. The situation for these negotiations was complicated first by a set of developments in Caterpillar's UAW local. There a new leadership had gained control, a militant group that soon took Caterpillar out on a strike that was to last some eighty days. The international office of the union had given only desultory support to the calling of the strike and found the acrimonious approach by the local to be counterproductive. This made the situation for the rest of the farm equipment and industrial products companies tenuous and settlements difficult to reach.

At International Harvester, the inflexible approach taken by both its chief executive officer and the UAW complicated an already difficult situation. Harvester's chief executive officer, Archie McCardell, had come to that firm in 1977 with a reputation as a "cost-cutting tiger." He did indeed pare down International Harvester's excess costs, and the corporation showed earnings of $360 million in 1979. It was just at this point that negotiations with the UAW were opening for a new contract, and McCardell took the opportunity to press the union for "give backs" in certain expensive features of the previous contract. The UAW was reputed at that time to have had on hand a $90 million strike fund and, as one commentator put it, "a burning need to prove its virility." The clash between McCardell and the union was traumatic. In the first three months of the strike the company lost some $222 million, at end of the following quarter they had lost another $350 million. By the time the employees got back to work, in mid-1980, they found that the recession had decimated the heavy truck market and International Harvester's sales had plummeted. After that, with high interest rates intruding, things went from bad to worse, leading to the difficulties pictured earlier in the chapter.[11]

Deere, meanwhile, had also taken a strike, but it had lasted only twenty-one days. The final settlement provided for modest wage increases, but also contained provisions for tightening the sick-benefit program.

The key issue in the International Harvester strike, the demand by the company for "give-backs," became a dominating one in American industry during the recessionary period of the early 1980s. What had always been assumed as sacrosanct—union-negotiated gains—began to be taken back through negotiation as early as 1977, when the Chrysler Corporation asked its union and nonunion employees to pitch in, along with lenders, suppliers, and dealers, to keep the company going. The pattern of these concessions spread through the automobile industry, with both General Motors and Ford negotiating downward structuring of compensation rates. Worker sacrifices had traditionally accompanied economic recessions, but in the past most of these sacrifices had been in the form of layoffs. Now wage freezes and wage cuts had been advanced as an alternative to such layoffs.

Lurking in the background of this extremely sensitive issue was the bugaboo of Japanese competition. The great promise of the "automatic factory" gave new hope that the apparently vast competitive advantage of the Japanese against comparable American companies might be mitigated. But there was another key competitive imbalance between Japan and the United States, that of employee wage rates. Some unions became skeptical that employers were crying "give backs" like "wolf," alleging noncompetitive wage rates without being willing to back up the comparisons with explicit figures. Nevertheless, there was a widespread belief, from all quarters of

American industry, that American wage rates were indeed substantially higher than those in Japan—perhaps by as much as $8.00 per hour.

Precise comparisons are very difficult to make because Japanese industrial relations patterns are quite different from those in the United States. On the one hand, job security in terms of the euphemistic "lifetime employment" of the larger Japanese firms is a plus for their workers. But the smaller firms generally do not have this policy, and the total "safety net" of benefits for the Japanese worker, both in small firms and large, is much below that of the United States. There seems no doubt, though, that when the lower Japanese wage rates are combined with their higher productivity, there is a significant total wage cost advantage for the Japanese, a potent competitive burden for American companies attempting to compete in world markets.

The farm machinery industry, Deere included, initiated its next round of contract negotiations in late 1982, as the three-year contracts of 1979 terminated. With skyrocketing losses at so many companies in the industry, there undoubtedly would be pressure from companies, almost across the board, for wage concessions of one sort or another. It seemed clear from preliminary indications in the public press that the UAW would seek to hold on to, and extend, wage and working-conditions levels in the more successful companies in the industry, knowing that they might face some heavy pressure in the more desperate companies for concessions that may need to be given just to hold on to the companies' jobs. Deere, as one of the more successful companies in the industry at the time, rightfully was concerned that its wage and working-conditions patterns might be thrown comparatively out of line with those of its competitors in the process of the year's bargaining.

FINANCIAL PERFORMANCE

The company has prided itself on its growth and profitability since the 1970s and its financial conservatism in attaining some very attractive performance figures. The overall story is quickly shown in Appendix exhibit 37. The dominance of Deere's profitability in the farm machinery industry, and its second-place position to Caterpillar in the industrial equipment industry, are clearly shown for the same time period in Appendix exhibit 38.

For a number of years, Deere's sales figures put it among the one hundred largest industrial companies in the United States (these rankings provided each year by *Fortune* magazine). These firms provide an excellent standard of performance against which Deere can be measured, for they are the companies with which Deere competes for people and capital. It is revealing of Deere's excellent performance in the years 1972–1981 and its weak performance in 1982 to look at the company's rankings on some other key indices for these same one hundred largest industrials, again provided by *Fortune* (Appendix exhibit 39).

By a number of financial criteria, the agricultural machinery industry is an unusual business. Most of its larger companies are involved not only in manufacturing but in retail financing. Deere's retail finance subsidiaries—John Deere Credit Company in the United States, John Deere Finance Limited in Canada, and John Deere Leasing Company (US)—total well over $2 billion in terms of total assets. Their function is to purchase and finance retail notes from Deere's sales branches and industrial equipment sales regions; these originate from retail installment sales by John Deere dealers of new Deere products and used equipment.

These companies are essentially freestanding entities, not consolidated into the parent company financial statements and carried on the latter's balance sheet as investments, producing income (or losses). The parent company has had an agreement with the credit companies to take any action necessary to maintain the credit companies' ratio of earnings to fixed charges for each of its fiscal years through 1982 at not less than 1.5 to 1 (after 1982, at 1.25 to 1). The Deere & Company annual reports have made it clear that "the agreement is not intended to make Deere & Company responsible for payment of the obligations of John Deere Credit Company." Nevertheless, the link between the parent company and its credit subsidiaries is less than arm's length.

The relationship between the insurance subsidiaries and the parent company is similar to that of the John Deere Credit Company—freestanding in just about all of their operations, yet with certain corporate ties, such as marketing linkages, and so forth. Plans were afoot in the early 1980s to break this umbilical cord altogether, setting up a multi-line insurance company, completely freestanding, that could be measured not only against Deere corporate goals but against other such insurance companies. A holding company would likely be set up that would consolidate all of the subsidiary boards of directors and bring about a complete separation of the marketing function from the company's equipment marketing. There is no intrinsic reason why the credit companies cannot be managed in at least a quasi-independent mode—indeed, outsiders such as regulatory authorities, customers, and dealers tend to press for this. It would be more complex a process than for insurance, though, given the historically stronger captive relationship and marketing considerations.[12]

Another important financial parameter of the agricultural equipment industry is the presence of long selling terms. Deere, like most of the farm machinery manufacturers, sells to its dealers, not directly to the customer. In other words, the company takes its profits from the dealer. In turn, the company extends terms to these dealers some fifteen or more months out into the future, depending on when the particular product involved is shipped or when it is finally sold to a customer. In effect, goods are shipped on open account, under provisions codified in the Uniform Commercial Code. When the product is sold to a customer, or when the terms expire, the dealer is obligated to pay off 90 percent of the amount due. The remaining 10 percent is then due at the beginning of the second month following the month in which the 90 percent was due. The dealers are not obligated to pay interest during the entire term's period, except on tractors, for which interest is charged after the first nine months.

Often it takes a substantial length of time to sell capital goods such as tractors—a hiatus of many months on the dealer's lot is not uncommon. Some of this is due to the typical seasonal swing in sales, but often a substantial cyclical effect exacerbates this inventory problem. Excessive inventories are a perennial worry of dealers anywhere, in any business; in the case of the farm machinery business, it is equally a worry of the manufacturer, by the nature of this lengthy credit exposure. Thus there is a great need for control of dealer receivables. In some years Deere's receivables at year-end (the key point for measurement) ran as high as 60 percent of the sales for the same year; a "normal" ratio might be around 40 percent, and in the late 1970s the ratio was as low as 35 percent. In 1981, the figure was about 44 percent; in 1982 it had drifted up again to more than 58 percent. (Several of Deere's competitors had much higher inventory percentages in these two years, and this led to much distress selling in the markets as these companies attempted to convert product to cash.)

PREOCCUPATION WITH THE RECESSION

This chronicle concludes at a time of uncertainty—a serious worldwide recession, compounded for a number of industries in the United States by a vigorous foreign challenge. The effects have been felt at Deere in a significant way. Earnings sagged from the excellent profitability records of the period 1972–1979. In 1981 indefinite furloughs of several thousand line employees of the company were instituted. The longstanding bonuses paid salaried and managerial employees were temporarily suspended in 1982 and wages were frozen for this group (except for the cost-of-living adjustment). The company's contribution to the employees stock purchase plan was temporarily halted and rigorous cost-control measures were implemented throughout the organization. There is no doubt that the recession of the early 1980s had a significant detrimental effect on the agricultural machinery industry (for example, the crisis situation in Massey-Ferguson and International Harvester); while Deere came through the difficulties in better shape than just about all other members of the industry, it still faced major short-term concerns.

It is this modifier, "short-term," that is the conundrum for American industry and for Deere—and for this book. The reader deserves a realistic view of how seriously the recession has affected Deere, yet the recession is not over, and the nature of "short-term" not yet clear. The following discussion is based primarily upon the situation pictured in the annual report of 1982, when it was determined that the essentially conservative posture of the company had stood it in good stead in the three difficult years of 1980–1982. Deere had never chosen to leverage itself with large holdings of debt, a pattern that had been followed by some of the other larger companies of the industry. The percent of short- and long-term borrowings to equity had been in the 40 percent range in 1977 through 1979, had risen sharply to more than 67 percent in 1980, had dropped to about 55 percent in 1981, but had risen again in 1982 to more than 77 percent. Long-term debt, which had been at 31 percent in 1977 and had risen to more than 36 percent in 1978, dropped to just under 28 percent in 1981 but increased to more than 37 percent in 1982. Inventories remained remarkably stable in the three years 1979–1981, but decreased by $111 million in 1982.

The short-term debt was a more serious concern. Total current liabilities stood at $1,465 million in 1979, were up to $2,304 million in 1981, and reached $2,395 million in 1982. The shorter-term sales situation was where the real concerns lay. Trade receivables, almost all of which were with Deere's dealers, stood at $1,401 million in 1979, rose sharply in 1980 to $2,093 million, and were up to $2,374 million in 1981 and $2,661 million in 1982. By the end of fiscal 1981, the company's retail finance subsidiaries had an additional $2.3 billion in receivables (after selling some $572 million of their retail notes to financial institutions in 1980, and following this with another sale of $833 million in 1981); an additional $1,035 million was sold in 1982, and the total dropped to just under $2.1 billion. In total, trade receivables and retail finance receivables cumulated to a sobering figure of $4.7 billion at the end of 1981, holding at approximately the same figure held for 1982.

Rapid fluctuations in interest rates complicate this in a major way. The retail notes, for example, can be outstanding for an average of some three years, up to five in some cases. Back three years, prior to 1982, the charges to farmer customers were in the neighborhood of 13–14 percent; in 1982 the company was paying as high as 20–21 percent for its own short-term needs, but this percent had dropped to about 9.47 by the end of 1982.

Clifford L. Peterson, the senior vice president of finance in 1981, commented to a management group about the difficult situation in that year: "We have financed *incremental* assets at an average cost of 17 percent so far this year. For the last 12 months Deere has earned 8.2 percent pretax, preinterest return on assets. This means that we have lost money on every additional asset added so far this year—whether it was dealer inventory, factory inventory, parts inventory or many of our new fixed assets. . . . The 'engine of leverage' is running in reverse."

A significant complication was the influence of waived finance charge programs, instituted by the company through its dealers in 1980, 1981, and 1982 to stimulate retail sales. The concept of "waiving the interest" as a sales incentive device had long historical roots, but generally was applied only in the low-selling season, typically November through March. The pattern in the recent period of year-round waiving—that is, in-season waiving—was a more troubling development.

The effect of this waiving on the parent company was substantial; the annual report of 1982 explained: "The retail finance subsidiaries received compensation from Deere & Company approximately equal to the normal net finance income on retail notes for periods during which finance charges are waived." Another way of saying this, of course, is that the parent company is subsidizing the credit companies (and therefore the customers), undoubtedly a necessary program to keep dealer strength in the face of a serious situation.

It is not just these waived finance charges that the parent company has had to absorb; in 1981, the company was required to provide a $74 million support payment to the United States retail finance subsidiaries to satisfy the bankers' restrictions on loans. Hewitt explained this in his report at the annual meeting of 1981: "We had to divert $74 million from other projects in order to support our retail finance subsidiary in 1981. But we believe that was good business practice because year after year we depend largely on the same group of loyal customers. We are not in the banking business, but our customers had a special need for credit assistance during this difficult period of high interest rates." Hewitt continued with a description of a new approach to consumer finance: "We are offering a new variable-rate finance plan to help our customers. This plan enables them to avoid being locked into the payment of high interest costs when rates come down in a later period."

Deere's access to the capital markets has been excellent over the past decade. The company's credit rating has moved back and forth between A and AA at various times during this period; in 1982 both the parent company and the credit companies were able to maintain an AA rating. In 1983, however, these ratings were reduced to A due to difficult economic conditions that also affected many other industrial firms' ratings. It seems clear, though, that the capital markets continue to respect Deere's inherent strengths and its longer-term future. Deere's credibility in the equity markets was excellent throughout this period; the longstanding pattern of regular dividends brought some imposing returns for equity owners. For example, 100 shares purchased in 1955 for $3,462 would have grown to 824 shares (through stock dividends and splits) by the end of 1981 (worth $25,995) and $14,771 would have been paid in dividends. Clearly, stockholders have fared well.

As recently as 1981, the company increased its equity position by a public sale through underwriters; in January of that year 4 million shares of common stock were sold, with the company receiving net proceeds of $166 million, the purchasers paying something in excess of $41 per share.

Having noted here a plethora of puzzling financial problems in the recessionary period of 1982, it is well to refer to the quality of the company's products, as attested to by its strong and increasing market share almost across the whole range of products. When the recession subsides, and the farmer returns to the market, Deere's products would seem to be eminently strong. The truly significant amounts of funds that the company has invested in itself—capital improvements of the nature of the Waterloo Tractors Works, sophisticated computer-based analytical underpinnings such as the CAD/CAM system, and research and development—all bode very well indeed for the ability of the company to maintain and improve its already strong line of products. Management itself, at an important transition point in 1983, has been widely recognized as having depth and quality.

Finally, the marketing organization needs always to be remembered. If there is any single arm of the company that has had preeminent historical strength, going back to the very days of John and Charles Deere, it is the marketing structure that came into its own through the sophisticated branch-house system and the development of a corps of John Deere dealers that has had no equal anywhere in the world. As one industry analyst put it in *The Wall Street Transcript* review of the industry in March 1983: "Deere has gone around and actually to the detriment of their balance sheet given support to their dealers. They're the only people who are going to end up with a healthy dealer network. 10 to 12 percent of the farm machinery dealers in 1982 went out of business, but the number of Deere dealers that went out of business was miniscule, and they were immediately replaced by people with stronger capitalization. So not only are they the low cost producer, they're going to be the only people around with dealers."

"John Deere" has always been close to the farmers, with its personnel attuned to the farmers' beliefs and ways and with a keen sense of the need for loyalty from customers. Other companies have built up loyal customers, too, and Deere will not be without competitors. The fact remains, though, that Deere has a superb marketing organization, positioned very well for the longer run.

Endnotes

1. For United States agriculture in the 1970s, see J. B. Penn, "Economic Developments in US Agriculture During the 1970s," in D. Gale Johnson, ed., *Food and Agricultural Policy for the 1980s* (Washington: American Enterprise Institute for Public Policy Research, 1981). Recent world population trends are discussed in Rafael M. Salas, *The State of World Population, 1982* (New York: United Nations Fund for Population Activities, June 1982). Negative factors in agricultural productivity are discussed in United States Department of Agriculture, *Farmline*, June 1981; for changes in farm size and structure, see William Lin, George Coffman, and J. B. Penn, *US Farm Numbers, Sizes, and Related Structural Dimensions: Projections to the Year 2000*, Technical Bulletin 1625 (Washington: Department of Agriculture, Economics, Statistics, and Cooperatives Service, July 1980); Luther Tweeten, "Prospective Changes in US Agricultural Structure," in Johnson, ed., *Food and Agricultural Policy for the 1980s*. For the moldboard plow controversy of 1943, see Edward H. Faulkner, *Plowman's Folly* (Norman: University of Oklahoma Press, 1943). The conservation tillage development is discussed in United States Department of Agriculture, Office of Planning and Evaluation, *Minimum Tillage: A Preliminary Technology Assessment* (Washington:

Government Printing Office, 1975); Glover B. Triplett, Jr., and David M. Van Doren Jr., "Agriculture Without Tillage," *Scientific American*, January 1977; Ronald E. Phillips et al., "No-Tillage Agriculture," *Science*, June 6, 1980. For a review of soil erosion problems, see Lester R. Brown, "World Population Growth, Soil Erosion, and Food Security," *Science*, November 27, 1981. Agronomic aspects are discussed in S. H. Phillips and H. M. Young Jr., *No-Tillage Farming* (Milwaukee: Reiman Associates, 1973). The issue of on-farm energy use compared to food energy produced is analyzed in David Pimental et al., "Food Production and the Energy Crisis," *Science*, November 2, 1973, 443–48; John S. Steinhart and Carol E. Steinhart, "Energy Use in the US Food System," *Science*, April 19, 1974, 307–15.

2. For an analysis of Massey-Ferguson's problems, see Peter Cook, *Massey at the Brink*, 245–73. See also Harvard Business School, "Massey-Ferguson, Ltd.," 4-582-082 (1982). References to International Harvester's travails are extensive; a particularly good summary is "International Harvester's Last Chance," *Fortune*, April 19, 1982. See also "Harvester's Bitter Harvest," *Dun's Business Month*, January 1982; "International Harvester: Can It Survive When the Banks Move In?" *Business Week*, June 22, 1981. White Motor Corporation's difficulties are discussed in "What the Banks Did to White Motor," *Business Week*, June 22, 1981; the sale of its United States operations to the TIC Investment Corp. is reported in *Wall Street Journal*, November 21, 1980; the sale of its Canadian operations to TIC is discussed in *Toronto Globe & Mail*, May 6, 1982.

3. For the industrial equipment division analysis, see "Strategic Plan Update, 1983–1987," March 1982, DA; the Boston consulting group's study is discussed on V: 10–14. See also Brian R. Alm, "A History of the John Deere Industrial Equipment Division," 1982. For consumer products division analyses, see "Strategic Plans, 1983–1987," March 26, 1982, DA. The quotation on the "rich getting richer" is from Eli S. Lustgarten, "Farm Equipment Industry," *The Wall Street Transcript*, February 1, 1982.

4. For Deere's recognition of the "large-farm" trend, see *New York Times*, August 22, 1982. For an analysis of the issue, see Luther Tweeten, "Prospective Changes in US Agricultural Structure," in Johnson, *Food and Agricultural Policy for the 1980s*, 115–16. Some of the inconsistencies of the "Jeffersonian ideal" are elaborated on in William Flinn and Frederick Buttel, "Sociological Aspects of Farm Size," *American Journal of Agricultural Economics* 62 (1980): 946. For "most efficient" farm usage, see Thomas A. Miller, Gordon E. Rodewald, and Robert G. McElroy, *Economies of Size in US Field Crop Farming*, US Department of Agriculture, Agricultural Economic Report 472 (July 1981). Further analysis of farm size is in Harald R. Jensen, Thomas C. Hatch, and David H. Harrington, *Economic Well-Being of Farms: Third Annual Report to Congress on the Status of Family Farms*, US Department of Agriculture, Agricultural Economic Report 469 (July 1981); see also *New York Times*, March 2, 1982. The Deere memorandum on the $40,000–449,000 farm sales group in Roy Harrington, "Farm Equipment Markets Versus Farm Cash Receipts," June 18, 1982, DA.

5. The two Yanmar organizations are described in H. Peter Keller to H. W. Becherer, May 25, 1976, DA, L. N. Hall ms.; the Yanmar agricultural tractor line is analyzed in T. J. Wilkinson to L. N. Hall, July 30, 1976, DA, L. N. Hall ms. An analysis of Deere's XR80 experimental tractor is in R. W. Boeke, "Memo-Planning for Yanmar Visit, June, 1976," May 12, 1976, DA, L. N. Hall ms. For doubts about the profitability of the XR80, see Robert A. Hanson to L. N. Hall, June 25, 1976, DA, L. N. Hall ms.; doubts on the Yanmar tractor's adaptability to Europe are in Walter Vogel to L. N. Hall, July 14, 1976. The joint-venture company is described in R. A. Hanson to Susumu Nitta, of Yanmar Diesel, June 23, 1976; the proposal was authorized by the Deere board in its meeting of May 11, 1977. Discussions about the overlap between Yanmar's Model 1250 and Deere's Model 2040 are analyzed in Heinz Laier to L. N. Hall, October 29, 1979, and L. N. Hall to R. W. Boeke, W. Vogel, and B. C. Bartlett, July 13, 1979. For quotation on the Japanese relation to North American agricultural machinery companies, see *Wall Street Journal*, July 25, 1980. For discussion about Yanmar's move into larger models, see H. W. Becherer to L. N. Hall, June 12 and September 21, 1981. Yanmar's Asian sales are discussed in Christopher M. Cowan to H. W. Becherer, February 1, 1978, and in H. W. Becherer to L. N. Hall, April 28, 1978, May 4 and 5, 1978, and October 18, 1970; pricing issues are discussed in H. W. Becherer to L. N. Hall, March 6, 1981.

6. For issues of longer-term, worldwide tractor strategic planning, see R. W. Boeke to L. N. Hall, April 30, 1980; Thomas A. Gildehaus to R. A. Hanson, September 2, 1980; and M. C. Rostvold to B. C. Bartlett (with attached notes of Eugene L. Shotanus), March 9, 1981, DA, Hall ms.

CHAPTER 15

7. For Deere's philosophy of research, see F. McGuire to W. Hewitt, April 27, 1964, DA, Hewitt ms. The redefinition of the research division is described in bulletin B113, the appointment of F. C. Lindvall in B304; the change of name to "technical center" in B290. See the K. W. Anderson memorandum to Lindvall, "Review of Research and Development Activities in Deere & Company General," February 17, 1970; Lindvall's report to Hewitt is dated May 13, 1970, DA, Hewitt ms. 706. The quotations concerning the product engineering council and the product engineering technical committee are from bulletin B338, August 28, 1970. See Lindvall's memorandum to Hewitt and Curtis proposing the committee, June 11, 1970, DA, Hewitt ms. 706.
8. For Robert J. Carlson's remarks on the OEM pump, see R. J. Carlson to W. Hewitt, February 12, 1974; see also Gordon Millar to Joseph Dain, April 7, 1977. The $175 million estimate for the development and relocation of the 300 series engine is in E. R. Curtis to W. Hewitt, March 29, 1979; G. Millar to A. B. Lundahl, February 13, 1974.
9. See Delno W. Brown to William Hewitt, September 14, 1973, DA, Hewitt ms. 706; Gordon H. Millar to Robert A. Hanson, June 23, 1978. The task force report of 1982 is in "Interim Report—Global Strategic Tractor Plan Task Force," May 7, 1982.
10. For the quotation on the "Japanese express train," see "Manufacturing Strategy Conference," vol. 1 (December 1980), vii–20. Deere's group technology system is described in "John Deere Group Technology System and Worldwide Information System for Engineering: A Comparative Overview" (August 1981). See also "Implementing CAD CAM Technology in Industry," a speech by James F. Lardner before the Autofact III Conference & Exposition, November 9–12, 1981; "The Computer and John Deere," *American Machinist*, June 1982. Lardner's quotation is from his speech at the Product Engineers' Meeting, April 20–21, 1981. For an excellent description of the "Kanban" system, see Y. Sugimori et al., "Toyota Production System and Kanban System—Materialization of Just-in-Time and Respect-for-Human System," 4th International Conference on Production Research (Tokyo: February 1981). For the quotation comparing Japanese and United States automobile inventory totals, see J. McElroy, "Making Just-In-Time Production Pay Off," *Chilton's Automotive Industries*, February 1982. Similar figures are reported in "Here Comes Kanban," *Ward's Auto World*, June 1982. For Japanese views of inventory as "cover up," see "Industrial Equipment Division Strategic Plan Update, 1983–87," March 1982, DA, v–11. See also "'Just in Time' at the Tractor Works," *Materials Handling Engineering*, June 1982. For the quotation on robotics, see "Manufacturing Technology Profile," vol. 1, "Industrial Robots," Manufacturing Engineering Department, Deere & Company (August 1981). For the quotations from UAW committeeman and the local union president, see *Chicago Tribune*, May 22, 1982. For the quotation on Japanese culture, see "Industrial Equipment Division Strategic Plan Update," v–13. For R. Harmon's comment about the Japanese view of the Waterloo Tractor Works, see *Waterloo Courier*, ca. May 1981. Hanson's remarks on Japan are from his speech before the ICED Outlook Conference, Bloomingdale, Illinois, June 8, 1982.
11. For the quotations about Archie McArdell and the UAW strike fund at International Harvester, see *Times* (London), April 19, 1981.
12. The Rock River Insurance Company was organized on December 22, 1964; the Fulton Insurance Company, predecessor of the John Deere Insurance Company, was founded in 1929, and its entire assets were purchased by Deere in 1969; the Sierra General Life Insurance Company and the Tahoe Insurance Company were incorporated on July 23, 1973; Security-Gard Holdings, Inc., was incorporated on July 9, 1976. All these companies are wholly owned subsidiaries of Deere & Company.

CONCLUSION

CHAPTER 16

CHARACTER OF A CORPORATION

[Deere was an exception,] unusual in both the intensity with which it is connected to the past and for the ability to turn handsome profits. . . . Deere's sensibilities, nostalgia and respect for the past make it peculiarly adept at finding continuities, at making events spring out of other events, new departures take off from well-worn paths. . . . It is a company which its executives are devoted to and dependent on, as a company. These men are unlike that more modern breed of manager which, management sociologists tell us, is "career-oriented", and which considers one company as useful to him as any other. . . . The Deere men have a more romantic orientation. They believe in their company and not, as most salesmen do, out of expediency. They honestly feel that Deere deserves their fealty and reverence. The view is not exactly fashionable these days. But it has stood the test of time.

Management Today *(England), 1973*

The era of William Hewitt as chief executive officer ended with his retirement from Deere & Company in August 1982 at the mandatory retirement age he himself established for all management when he first became president in 1955. Hewitt's personal reputation was at its zenith. He had become a key representative of American industry, widely recognized all over the world. When *US News and World Report* published its eighth annual

◀ Deere Administrative Center. *Deere Archives*

report—"Who Runs America?"—he was included, the only business executive to be listed in the category of agriculture. His services continued to be called upon after retirement. In October 1982, he accepted an appointment as United States ambassador to Jamaica.

AS SEEN BY OTHERS

In 1982 Deere's reputation was also at a high point, judging from comments appearing in the business and popular press. It had been praised in the early 1980s when American management generally had been faulted for low performance, particularly in the face of challenges from foreign competitors, and for taking too short-run a view. As one analyst wrote in 1982 in the *New York Times:* "The truly great American companies, those most respected and

Exhibit 16-1. William A. Hewitt in his office, 1982. *Photograph © by Arnold Newman*

effective offshore as competitors, are not players of funny money games. Rather, enterprises like Caterpillar, John Deere, Procter & Gamble and Eastman Kodak single-mindedly dedicate themselves to deepening their product lines, improving their product delivery and support systems, manufacturing more efficiently and adapting to evolving world market conditions." In the same year, the *Boston Globe* named Procter & Gamble, Caterpillar, and Deere as three of a half-dozen examples of organizations "that had kept the corporate eye on the far horizon." The *Los Angeles Times* commented that Deere "has avoided the dominant corporate approach today, which relies on a numbers-oriented portfolio strategy aimed at draining so-called 'cash cows' . . . to provide money for operations promising greater returns."

Despite the pace of the modern business corporation, which sometimes has seemed to obscure the longer-term perspective, Deere's "sense of history" remained intact. A British writer commented: "All large corporations have a history. But in most cases that history is an appendix merely, a footnote. They are dominated and ruled by contemporary ethics and values . . . the successful modern corporation is expedient. . . . Like a chameleon it takes on the coloration of each succeeding era. In business, history *is* bunk usually. . . . Probably the majority that revere their past . . . have a profitless present and anticipate a poor future. They are becalmed. . . . History for them is a substitute for action."

This same analyst had concluded that Deere was an exception, "unusual in both the intensity with which it is connected to the past *and* for the ability to turn handsome profits. . . . Deere's sensibilities, nostalgia and respect for the past make it peculiarly adept at finding continuities, at making events spring out of other events, new departures take off from well-worn paths. . . . It is a company which its executives are devoted to and dependent on, as a company. These men are unlike that more modern breed of manager which, management sociologists tell us, is 'career-oriented,' and which considers one company as useful to him as any other. . . . The Deere men have a more romantic orientation. They *believe* in their company and not, as most salesmen do, out of expediency. They honestly feel that Deere *deserves* their fealty and reverence. The view is not exactly fashionable these days. But it has stood the test of time."

Hewitt, representing the fifth generation of the family, felt strongly this sense of history. Early in his tenure as chief executive officer in 1964 he had inaugurated the "Green Bulletin" series, codifying the key goals and policy statements of the company. In his introduction to the series, cast as a letter to all employees, he first commented on the "rapid changes and increasing complexity of the times." He then continued: "It seems particularly important that we should . . . rededicate ourselves to the store of wisdom—tried, tested and proven—accumulated by the men who built and steered the John Deere organization through these many years. It was their guide and remains as their legacy. It must live in our daily work." The British writer remarked about Hewitt's predilection: "Time and again in conversation, Hewitt returns to historical touchstones . . . he has much affection for each mosaic of the past. . . . There is in his voice a warmth, a tenderness for the entire accumulation of human time and energy expended within the boundaries of the Company."

Analysts have commented widely on Deere's comparative strength relative to its domestic competitors in the North American agricultural equipment industry. There will be "fallouts and marriages," as International Harvester's chairman, Donald D. Lennox, put it to the *Wall Street Journal* (in February 1983), as companies seek to reduce costs and cut excess capacity. Deere has been seen to be emerging as the dominant company, the "General Motors" of the industry, as the *Washington Post* put it in early 1983. "They really don't have any weaknesses," the *Washington Post* writer concluded. In the annual *Wall Street Transcript* review of the industry in March 1983, analysts cited Deere's unique position as a low-cost producer and pointed to the company's larger-than-average research and development outlays and major capital expenditures over the past decade as portents of increasing market share when recovery of agriculture comes.

The focus on Deere's preeminence in the North American market does obscure difficulties abroad. Deere's ability to source and to sell worldwide will be critical in the upcoming shakeout that the industry is likely to experience. Foreign competitors will be strong in both agricultural machinery and construction equipment and must be challenged on their own grounds and in their own terms. As the previous chapter elaborated, however, Deere's international operations, though they sometimes have had the patina of success, have often reduced the company's profitability. Renewed losses in Europe stemming from the recession of the 1981–1982 period have counterbalanced successes in other parts of the world.

Nevertheless, Deere's management does appear to have reason to believe its "good press." Despite the short-term worsening of its financial ratios, its conservative financial policies, long a central tenet of Deere management, have stood it in good stead in the recession of the early 1980s. As one financial analyst put it, "The severe deterioration of the balance sheet of many industry participants will not permit them to compete head on against Deere, which is using capital as a strategic weapon." Deere has three well-run profit centers, its manufacturing entities, the marketing units, and its financial organizations (the credit companies and insurance companies). Thus it has been able to use its capital more effectively; at the same time, its competitors have been forced to withdraw potentially productive assets by out-sourcing, spinoffs of divisions, or outright sales of those assets because of short-term demands for cash.

The dealer organization, always a source of strength at Deere, has survived the ravages of the recession to become even stronger. It is the best financed and most stable in the industry. A marketing group of this quality takes many years to build; it seems highly unlikely that any other competitor, domestic or foreign, can hope to match Deere's dealers, at least over the short run.

Deere's employee relations add a further historical strength. Its bargaining with the United Automobile Workers is in marked contrast to that at International Harvester, Allis-Chalmers, and Caterpillar. Since the watershed 1955–1956 negotiations, Deere has had excellent relationships with this union, and the union has been an important barrier to excessive paternalism. In a Boston *Globe* article on the "four companies where quality still counts," a revealing comparison was made between the UAW relationship at Caterpillar and that at Deere. Citing the "long history of difficult labor-management relations at Caterpillar," the *Globe* writer continued: "The Company is a divided society, the salaried workers, who are supposed to have tractor-colored 'yellow blood' on one side, and the hourly union workers on the other." In contrast, at Deere "management speaks a different language," in which "the workers are not divided into white-collar and blue-collar—they just have different responsibilities." Deere has followed a straightforward policy of equal respect for salaried and hourly paid employees alike. The 1980s will undoubtedly put strains on this relationship, as the company drives to reduce the rate of growth of labor costs in the face of aggressive competition from abroad.

Technological and managerial supremacy must be continually re-earned, and Deere executives know this. Further, few if any organizations have no weaknesses, the *Washington Post* notwithstanding. Some of Deere's most important strengths are at the same time potential sources of weakness. The company's persistent difficulties in sorting out conflicting strands on the concept of decentralization is a case in point.

INSTILLING INITIATIVE

Vesting authority in an individual, who at the same time is made personally responsible for the use of that authority, has proven in practice to be a truly liberating influence on people. Since the days of Alexis de Tocqueville's *Democracy in America* (1835) and before, Americans have been characterized as individualists striving to carve out their own destinies. The dominance of the frontier mentality, captured so well by Frederick Jackson Turner, colored the development of the West—and at just the time when John and Charles Deere were contributing to the opening of the prairie. Modern social scientists—for example, the pioneers in the field of organizational behavior, such as Rensis Likert, Douglas McGregor, Warren Bennis, Chris Argyris, and others—have reinterpreted de Tocqueville, Turner, and others to emphasize in contemporary terms the great potency of the concepts of participation and individuality in infusing a sense of both contribution and personal worth in organizational environments.

CHAPTER 16

Decentralization of authority and responsibility has been a dominant ingredient of the Deere character, a recurring theme for at least the last 110 of the company's 145 years. (It emerged in the early 1870s with Charles Deere's novel way of organizing the branch houses.) Through all of these years the company has given full effect to this, not only with management and salaried personnel but also with line employees through the use of a truly actuating incentive plan of compensation. One International Harvester executive was quoted in 1982 as saying: "Deere decided they were going to have a truly decentralized management and, believe me, they *are* decentralized. At Deere, you don't make any decisions in the center of the Company except finance, capital-resource allocation and overall strategy. Everything else is done in those operating groups, with super plant managers making the decisions."

This does make the concept sound deceptively simple, but applying it has always required considerable imagination and good sense. Deere's long experience with decentralization has become not only an article of faith but has provided many operational lessons on just how elusive the precise application of the concept can be. These lessons are still not complete, for there will always be new issues that seem to fit individual initiative and that at the same time demand central coordination and control. Recent efforts to set up a worldwide tractor sourcing plan and the decision to set up a corporate-wide components division both seem to imply more "centralization." At the same time, emphasis on worker participation in decisions and the "quality circle" discussions seem to press toward "decentralization." But the former cases do not cut off the potential for vigorous exercise of individual initiative, and the latter cases do not reject the need for realistic authority to ensure coordination and control.

One cannot gainsay the positive effect that the debate over decentralization has had on the company. In training Deere people for leadership, it has probably been the single most important factor. The delegation of authority to the lowest practical level has instilled a sense of initiative in Deere managers, as well as an ability to respond to local conditions and to changing environments. Paired with this has been management's willingness to understand and accept misjudgments. As Hewitt has put it, "Managers have the right to fail, the same way as in baseball the home run king often is the one who strikes out most often in the season. We cannot, of course, tolerate disastrous failures, but we can tolerate reasonable mistakes." He reiterated the same thought in a *Boston Globe* interview: a well-managed company was one "not out to make instant heroes." Deere has been a supportive—and forgiving—institution over the years.

Whether Deere has been able to inculcate a spirit of entrepreneurship in its present management remains a moot question, not yet answerable in this period of cutback and retrenchment. There is no doubt that the current "shakeout" period will have many entrepreneurial potentials, and Deere commands resources that can take advantage of them to make these operational. Personal skills of entrepreneurship must be present, however. These have been there in the past; they must also be in place for the challenging future.[1]

THE SEARCH FOR EXCELLENCE

Deere has always been known for well-engineered products and superb service. Hewitt chose dramatic ways to emphasize the pursuit of excellence embodied in these achievements. His efforts reflected an enduring interest in style, both in design and as a quality permeating the company in its physical aspects, buildings, products, advertising, and literature, as well as the actions of its people. The electric effect on company personnel of the Saarinen-designed Deere Administrative Center has continued to be felt and was enhanced in 1978 when Deere added a new "West Wing" to the center. It proved to be no ordinary addition, for architect Kevin Roche, Eero Saarinen's successor, created an award-winning facility that was not only visually stunning but that enhanced the work environment for the staff. Roche chose to complement the Saarinen masterpiece, connecting it to his wing with a fourth-level corridor-bridge. Featured was an interior atrium through the center of the building, with the offices on all sides overlooking its gardens. The building won a number of awards, including the *Administrative Management* "Office of the Year" citation, just fifteen years after the administrative center had taken the same award.

The company's and Hewitt's continuing concern for architectural and aesthetic excellence has earned widespread recognition. The German magazine *Der Spiegel* commented: "One of the best known patrons of the new glasshouse vogue is William A. Hewitt, an American entrepreneur whose aim has been for 25 years to make outstanding architectural works a sort of trademark for his Company. This has earned him a place among the ranks of the Medicis and the Bourbons by the American Institute of Architects."

The comparison to the Medicis may be extravagant but it makes an important point. A few corporations have had profound effects on the culture and aesthetic life of communities—Deere in Moline; the Corning Glass Company in Corning, New York; the Cummins Engine Company in Columbus, Indiana. Each has become internationally known for its harmonious relationships to its community, expressed in part by excellent taste in company architecture and art. Critics sometimes have raised the fear of Philistinism in corporate incursions into the arts, but one would be hard-pressed to say this about Deere, Corning, and Cummins, all of which seem to have accomplished these efforts with perception and sensitivity. All three are firms that have been family owned, with more than a bit of paternalism, enlightened to be sure, as their past rationale.

Exhibit 16-2. Interior atrium, West Office Building, Deere Administrative Center, Moline, 1978. *JD Journal*

In the case of Deere, at least, there is much more to the story than the issue of "corporation as patron." Involvement in the arts was meant to be substance as well as style, as a number of analysts have recognized. An article in *Smithsonian* magazine on architect Kevin Roche featured the Deere Center, "which takes its responsibility to provide a good workplace as seriously as aesthetics." Commenting on Roche's West Wing, the author continued: "Their 1978 addition . . . extends the attempt begun a decade earlier at the Ford Foundation to better the nine-to-five life of office workers. It is an expansion on the idea developed there of combining efficiency and enchantment." A book published by the Yale University Press, *Spaces: Dimensions of the Human Landscape*, remarked on the location of the center: "Deere &

Company chose, for consciously formulated symbolic reasons, to build in the Illinois prairie which had brought the company into being. . . . It appears to be a consequence of an unusual and admirable preoccupation not merely with the consumer's image of the company's product but with the self-image of the people who do the producing. . . . When man-made and natural form are combined, consciously or unconsciously, with the symbols that make explicit the culture of the people who use and experience a particular geographical space . . . the sense of place is to the environment what personality is to the person. When a diversity of cultures share a space, environmental forms that unite them create a larger sense of place. . . . Only occasionally do talented people like Saarinen, Sasaki, and Hewitt bring it about all at once. When such designers achieve this, it is not so much because they have the ability to 'create' form as to discover it in the elements that are already there." (Sasaki Associates was the landscape architect for the Center.)

This focus on art and architecture has spilled over to heighten the historic concern for excellence in products, service, and corporate communications. *Fortune* commented: "Hewitt recognized years ago that what Deere needed most of all was an expanded vision of what was possible. His emphasis on perfection and style, symbolized by the Saarinen office building, puzzled many of the Company's older executives, but not a few of them eventually caught on to what he was doing." *Fortune* quoted one of Deere's executives: "Bill made us realize just how good we were . . . before he came along, we were just a bunch of good old country boys. We had a lot of self-respect; we knew we made good farm equipment, but we never really looked at ourselves as in the big league. We knew we could make it in Moline, but we didn't know that we could make it in New York. Bill showed us we could." *Forbes* reiterated this same thought: Hewitt "helped give Deere people the confidence they needed in order to begin thinking in world terms."

THE TRANSITION IN LEADERSHIP

Hewitt and Curtis had been charting an orderly management succession for several years. The Fiat merger proposal, had it gone through in 1972, would have required the selection of a chief executive officer, and four senior executives were considered—Robert Hanson, Robert Boeke, Delno Brown, and Robert J. Carlson. The proposal fell through, but these four men later became the prime candidates for the next step forward. In early February 1975, Hewitt and Curtis (with the counsel of the outside board members) decided to reinstitute the post of executive vice president. The tension surrounding this anticipated move was substantial. On February 10, 1975, in a meeting attended by all the senior officers of the company, Hewitt announced the choice of Robert A. Hanson for the position (and for a place in the chairman's office). Though it was not automatic that Hanson would later become president, his promotion established the intended line of succession. The decision was not without its disappointments for the others, but it was overwhelmingly popular. The personality of Hanson was compatible with the personalities of his two senior colleagues in the chairman's office. Further, Hanson brought a background of general management experience and breadth that was distinctive within the company.

A potential weakness of any functionally organized company is its tendency toward lifetime compartmentalization of its employees. Deere's

Exhibit 16-3. Helicopter puts in place Henry Moore's "Hill Arches," Deere Administrative Center, 1974. *Deere Archives*

tradition of decentralized power had fostered great loyalties and depth of knowledge, but had also tended to breed insularity and parochialism. Many senior managers had risen through promotions in one function, never once having left it. At the time that Hewitt turned the company toward international markets, few of the senior management had ever been abroad for business purposes. Even domestically, it was not typical for a manager from manufacturing to move to marketing at some stage of his career, or vice versa. Indeed, in the early days it was not common for branch personnel to move around among branches.

At the top management level, a company often can use teams of people to achieve the cross-fertilization and interaction necessary for a broad perspective. The long-time relationship between Hewitt and Curtis is an apt example. Hewitt's keen sense of organization, marketing, and external relations blended with the financial acumen and analytical power of Curtis. But many senior management positions require wide-ranging skills and a general management perspective that is not typically developed in a functional system.

One spot in Deere where such perspective is built into the job is the managing directorship of one of the international operations. There are several of these at the middle management level, managing directors of smaller operations, such as South Africa and Mexico. There are major general management posts in Spain, France, and Germany. Finally, the managing directors of the two regions, Region I (Latin America, Australia, and the Far East) and Region II (Europe, Africa, and the Mideast) are called on to exercise broad general management responsibility and perspective.

It was this international career path that shaped Hanson. Boeke, too, had had some international experience. Brown and Carlson largely presided over domestic functions; Brown had headed industrial relations and then taken over industrial equipment, the foundries, and OEM; Carlson worked almost exclusively in marketing.

At the end of the 1970s, several significant senior management appointments were made, each mirroring the importance of general management skills. First, in April 1978, Robert Hanson was elected president of the company, and Curtis moved to the newly created post of vice chairman; Hanson assumed the additional title of chief operating officer in April 1979. A month later, Curtis retired as vice chairman, and Walter Vogel, who had been vice president and managing director of Region II, was given a double-step promotion to the vacant post of executive vice president. This appointment reflected Hewitt's concern that the company think of itself as a truly multinational entity. (Vogel, a West German national, gained almost all of his experience in Europe.) A year later Thomas A. Gildehaus was appointed vice president, and within a few months senior vice president, for the corporate planning portfolio. Gildehaus had come directly to the post from outside the company. He had been an officer in the consulting firm of Temple, Barker, and Sloane, which had worked closely with Deere in long-range planning during the previous seven years. In November 1982 Gildehaus was appointed executive vice president. The appointments of both Vogel and Gildehaus brought strong generalist skills to the top management of the company.

Concern about the dearth of opportunities to move across functional lines has long been matched by concern over recruiting policy. Historically, most managers have been recruited from a small circle. The company selected its new people heavily from plants and branches in the locality. For professional personnel, the colleges and universities in the region have been drawn upon heavily. The company's recruiting power at the Midwestern universities has been impressive, especially for engineers and agriculturally trained personnel. This deliberately local and regional preference in recruiting has given the company enormous strengths stemming from family and friendship ties. Deere became a company with second-, third-, even fourth-generation Deere employees. Deere vice president Robert W. VanSant commented in a *Los Angeles Times* article comparing Deere and International Harvester: "Our customers and our employees talk the same language, share the same commitment to the traditional American work ethic. . . . It's a very nice match from which we derive a lot of strength." An article in *Crain's Chicago Business* in 1982 contrasted International Harvester's downtown Chicago headquarters with Deere's: "One of the good things about Deere is the geographic awareness because most of their executives live on a farm, or near one. . . . Deere's got farm all around it; those guys walk through a history of agriculture before they get to their office. There's a different mentality."

But does this cast too narrow a net? Frank McGuire thought so when he headed engineering in the 1960s. He deliberately plumped for more engineering and technical talent from diverse sources; the Lindvall appointment in the early 1970s epitomized this broader perspective of looking to national figures for talent. Hewitt, right from the start of his assignment as chief executive officer, also stressed this policy: "Hire the best talent from anywhere in the nation—or in the world, for that matter."

It has never been easy for Deere to recruit top personnel, for the small-town atmosphere of Moline and the factory towns of Waterloo, Dubuque, Horicon, and Ottumwa seem provincial and parochial to businessmen who have grown up or worked in large urban centers. The choice of Moline for the technical center and Waterloo for the product engineering center are excellent ones in many ways, but these towns are not strong drawing cards for talent.

Deere has worked hard to make its home office and factory towns more exciting and satisfying places to live. Moline itself has been especially enriched by such cultural efforts as the visiting arts series (exported to most of the other Deere locations, too). The company has made generous contributions to local educational and community efforts, and the John Deere Foundation has supported community projects. Deere also has taken a number of steps to make outsiders aware of the quality of the facilities in Moline, in Waterloo, and in other company areas. Periodically, approximately every third year, the company invites a group of more than one hundred financial analysts to a two-day visit in Moline and its environs (generally with a trip to Waterloo, to show the new Tractor Works). The success of these endeavors was chronicled by the Financial Analysts Federation in 1982, which gave Deere an award for excellence in corporate reporting, shared only by thirty other companies in the country.

But there is also a positive side to the small-town environment. *Fortune* magazine commented in an article in 1981 on the family feeling: "Both Corning and Deere have kept their headquarters in small cities, close to some of their plants so executives can visit informally to get a feeling of factory ambience.... At both these companies, product people count." Some corporate critics might bristle at this statement, thinking that it suggests an overly paternal top management. Yet Deere seems to have escaped most of the negative features of this close relationship. "Deere is paternalistic, needless to say—perhaps more so than most corporations," commented a British management journal, "but that paternalism may be more benign and less ruthless than is commonly the case."

If the word "paternalism" is taken to mean "a system under which management undertakes to supply needs or regulate conduct of employees in matters affecting them as individuals as well as their interpersonal relations," it does not seem applicable to Deere. As the British quotation states, paternalism was historically often condescending, hierarchical, and all too often ruthless. George Pullman, out of professed benevolence, developed a model town for his employees in the 1880s and '90s, but his Orwellian insistence on total control was a caricature of benevolence, a harsh, authoritarian paternalism run amok. A bitter strike resulted in 1894, one of the landmark cases in American labor relations. This was the time of Charles Deere, and his company, too, had a keen sense of concern and beneficence toward its employees. But paternalism was not its base. Perhaps William Butterworth came closer to the image of a father figure in the company, but again in no sense an authoritarian, intrusive leader. In recent years, Charles Wiman and particularly William Hewitt both exemplified professional managership, a philosophy almost antithetical to paternalism, though they retained a strong attachment to the community.

It is still a fact, though, that many knowledgeable people do not quickly recognize the name "Deere & Company" or know that the company is located in Moline, or even that it is a Midwestern firm. What is recognized much more often are the two words "John Deere." The man, John Deere, is an authentic folk hero of America, and his plow has won an honored place among the country's important artifacts. As Americana, the crowning

CHAPTER 16

achievement may have come in the "Peanuts" comic strip reproduced above. The company may well have been unwise in jettisoning the word "John" from the corporate name back in the 1850s. But so long as the company's products carry the well-recognized "John Deere," it's not likely that its sales will be adversely affected.

LOOKING AHEAD

Hanson's accession to the post of president, chairman, and chief executive officer in October 1982 marked the beginning of a new era, which followed a period of unbroken family leadership in the company for 145 years, incredibly with only five chief executive officers. (Burton Peek assumed a stewardship role for two years during World War II, while Charles Wiman was colonel of ordnance in Washington, DC.) Hewitt had paved the way for guiding the transition to professional corporate leadership. Although a member of the Charles Deere Wiman family (by marriage), he came to the company by an independent route, built on professional management training.

In 1973 the first outside director was appointed; by 1975 there were a total of five out of fifteen. Samuel C. Johnson, chief executive officer of S. C. Johnson & Son, Inc., was Deere's first outside director, elected at the annual meeting of 1973. Later that year, William B. Graham, chief executive officer of Baxter Travenol Laboratories, Inc., joined, and the following year Lloyd B. Smith, chief executive officer of A. O. Smith, Inc., was added. In 1975 two more additions, Edward E. Carlson, chairman of UAL, Inc., and United Airlines, and Robben W. Fleming, president of the University of Michigan, brought the complement of outside board members to five. The total board size stayed at fifteen until 1979, when it was increased to seventeen (the outsiders remaining at five). Robert D. Stuart Jr., chief executive officer of the Quaker Oats Company, replaced Fleming in 1979. In 1982, Ms. Juanita Kreps, vice president of Duke University, became the sixth outside director, the first woman to hold this post. (There had been one woman corporate officer, during the tenure of Ms. Elizabeth Denkhoff as corporate secretary, 1971–1982.) With the addition of the five outsiders, a more general-management atmosphere was slowly inculcated (though some slippage remained because it always is hard for an inside director to "wear two hats").

At the time of the appointment of the first outside director, the "family" owned just over 12 percent of the company's outstanding shares (most of them owned by descendants of the Charles Deere family); the remainder was owned widely by institutions and the general public. Until 1974 these family shares were held in trusteeship and therefore represented a single voting unit, the owners showing only desultory interest in the operation of the company. After the shares were dispersed into individual hands, even the unifying element of trusteeship was lost. There was no lack of pride among the heirs of John Deere, but the company had outgrown the family. Deere & Company was no longer family owned. With the election of outsiders to the board, the remaining vestige of a "closely held" corporation had been eliminated.

Deere & Company thus seems to have been spared most of the disabilities that many family-owned companies suffer after more than one or two generations—idiosyncratic and often fragmented leadership, excessive focus on the past, and difficulties in management succession. Each of the five chief executives surmounted a distinctive set of challenges. John Deere's was survival in a frontier environment. Rapid growth and change characterized Charles Deere's world as the company rose to national prominence by the turn of the century. William Butterworth consolidated the organization, in the process adding several new companies and product lines, including tractors and the combine. Charles Wiman carried the company through the Great Depression of the 1930s and World War II, all the time exhorting everyone about the critical role of product development. Hewitt presided over another era of rapid expansion that encompassed the world. He rebuilt the physical plant, enlarged the workforce, and broadened the company's management to take on the global challenge, while he preached quality and style. The company became the largest agricultural machinery producer in the world.

The challenges facing Robert Hanson were daunting, for he began his tenure at a deep trough in the agricultural machinery business cycle. Yet the long-run outlook remained promising. Deere's resources in research and development, its strength in the market, and the quality of its plant and organization allowed the company to come through the recession of the early 1980s poised for growth. Both Charles Wiman and Hewitt often exhorted their colleagues that "Deere's best years are ahead"—and the adage still seems appropriate. Hanson would need to resolve a number of knotty problems. Pressures to cut costs would be very strong—companies would need to be "lean and trim," to paraphrase Deere's slogan for its Yanmar-made tractors, in order to meet "the Japanese challenge" (a term embracing much more than just the actions of Japan itself). Deere's worldwide sourcing issues would be complex in the near future, as would be the related question of the degree of standardization possible in an industry serving a larger clientele with differing needs and geographically dispersed locations.

Could the company remain a predominantly single-industry organization, or would further diversification be necessary? A *Los Angeles Times* analyst commented: "Until recently, such devotion to a single industry has been considered a sign of weakness in U.S. corporate circles." But Hewitt pointed out in the same article: "We have zero interest in getting into a business with which we have no familiarity," and he commented on the "growing recognition of the drawbacks of excessive diversification and merger mania." Hanson might have to confront the issue again, however, since research and development and decentralization are persisting concerns.[2]

Deere has a set of enduring qualities that has served it well indeed over a 145-year span. It has been a preeminent example of a well-run family endeavor, and there is no reason why it should not retain some characteristics of a "family company"—the close ties between corporation and community,

Exhibit 16-4. Robert A. Hanson, chairman and chief executive officer, Deere & Company, 1982.

the employees' sense of identification with the products, and management's concern for the long run. The corporation is now independent of the family, but the "John Deere" legacy is firmly embedded in the organization that "gave to the world the steel plow" and that has provided a model of innovative and responsible American business behavior.

Endnotes

1. "Who Runs America?," *US News and World Report*, May 18, 1981. For the quotation about "funny money games," see John B. Schnapp, "Who for the Pedestal Now?," *New York Times*, July 11, 1982. For the quotation on the "corporate eye on the far horizon," see *Boston Globe*, October 3, 1982; for the quotation on "cash cows," see *Los Angeles Times*, August 8, 1982. For Deere's view of history, see *Management Today* (London), October 1973. For the quotation concerning "fallouts and marriages," see *Wall Street Journal*, February 11, 1983; for the "General Motors of the industry" quotation, see *Washington Post*, March 6, 1983; see also *Wall Street Transcript*, March 28, 1983. For comparison of Deere and Caterpillar labor relations and the comment on "instant heroes," see *Boston Globe*, October 3, 1982. For the quotation on "truly decentralized management," see *Crain's Chicago Business*, November 15, 1982.
2. For the West Wing addition, see *Fortune*, October 9, 1978; the "Office of the Year" citation is from *Administrative Management*, February 1980. For the comment on "the Medicis," see *Der Spiegel*, August 17, 1981. The article on Kevin Roche is in *Smithsonian*, August 1982; for quotation on a "sense of place," see Barrie B. Greenbie, *Spaces: Dimensions of the Human Landscape* (New Haven: Yale University Press, 1981). For the quotation on "big league," see *Fortune*, August 1976; see also, *Forbes*, January 21, 1980. For the "traditional American work ethic" quotation, see *Los Angeles Times*, August 8, 1982; for the comparison of Deere's and International Harvester's offices, see *Crain's Chicago Business*, November 17, 1982. See "Rediscovering the Factory," *Fortune*, July 13, 1981; for the comment on paternalism, see *Management Today* (London), October 1973; for the quotation on single-industry firms, see *Los Angeles Times*, August 8, 1982.

LIST OF APPENDIX EXHIBITS

Exhibits 1–20 appear in John Deere's Company, Volume I
(Hardcover ISBN 978-1-64234-080-8; Softcover ISBN 978-1-64234-163-8))

Exhibit 21	Estimated Horsepower-Hours of Power Developed Annually for Farm Operations, 1925
Exhibit 22	Market Share, Deere and International Harvester 1921 and 1929
Exhibit 23	Costs per Dollar of Sales, Various Firms, 1929 (in percent)
Exhibit 24	Return on Investment in Farm Machinery Business (After Taxes), Selected Companies, Long- and Short-Line, 1927–1929
Exhibit 25	Industry and Deere-Lanz Tractor Sales
Exhibit 26	Domestic and Overseas Sales and Income, Deere & Company, 1965–1970
Exhibit 27	Investment in Overseas Subsidiaries, October 31, 1970
Exhibit 28	Deere's Operation in South Africa, Sales and Profitability, 1964–1980
Exhibit 29	Farm Equipment Sales in North America: Market Share and Growth Rates, 1970–1980
Exhibit 30	North American Market Positions, 1981

LIST OF APPENDIX EXHIBITS

Exhibit 31	Farm Equipment Sales Worldwide: Market Shares and Growth Rates, 1970–1980
Exhibit 32	Farm Equipment Sales Overseas: Market Share and Growth Rates, 1970–1980
Exhibit 33	Deere's Position in Largest Overseas Markets, Tractor Market Share, 1980 FY
Exhibit 34	Deere's Position in Largest Overseas Markets, Combine Market Share, 1980 FY
Exhibit 35	Position of Deere in the North American Industrial and Construction Equipment Industry
Exhibit 36	Research and Development Expenses as Percent of Sales and Dollars Per Employee
Exhibit 37	Deere & Company, 1972–1982
Exhibit 38	Deere Profitability Ratios
Exhibit 39	Deere's Rankings among Fortune Magazine's 100 Largest Industrials, 1972–1982
Exhibit A	Deere Family Links in Deere & Company
Exhibit B	Five Generations of Deere & Company Chief Executive Officers, 1837–1982

APPENDIX EXHIBITS

Appendix Exhibit 21
Estimated Horsepower-Hours of Power Developed Annually for Farm Operations, 1925

Source: C. D. Kinsman, An Appraisal of Power Used on Farms in the United States. *United States Department of Agriculture, Bulletin 1348 (July, 1925) 5*

Appendix Exhibit 22
Market Share, Deere and International Harvester 1921 and 1929

		Deere %	International Harvester %	Total as percent of industry
Mowers (horse or tractor)	1921	14.3	62.4	76.7
	1929	20.7	64.6	85.3
Rakes (sulky, dump)	1921	14.7	51.2	65.9
	1929	20.8	56.2	77.0
Binders (grain and rice)	1921	13.5	73.2	86.7
	1929	25.9	67.9	93.8
Combines	1921	—	85.1	85.1
	1929	6.8	31.8	37.8
Hay loaders	1921	26.3	43.9	70.2
	1929	22.7	53.1	75.8
Corn pickers	1921	—	97.5	97.5
	1929	29.4	48.5	77.9
Sulky plows	1921	12.2	13.0	25.2
	1929	25.2	19.7	44.7
Tractor plows	1921	5.3	15.8	21.1
	1929	26.5	49.1	75.6
Disk harrows	1921	17.9	32.2	50.1
	1929	22.4	41.8	64.2
Corn planters	1921	25.0	32.3	57.3
	1929	41.2	32.0	73.2
Tractors	1921	NA	NA	NA
	1929	21.1	59.9	81.0

Source: Federal Trade Commission, Report on the Agricultural Machinery Industry. *Washington: Government Printing Office, 1938, 150–53*

Appendix Exhibit 23
Costs per Dollar of Sales, Various Firms, 1929 (in percent)

	Manu-facturing	Selling	Transfer	Collection	General Administrative	Bad Debts
Deere	57.90	8.52	2.28	0.20	1.76	0.27
International Harvester	69.04	8.35	0.97	1.24	0.87	3.05
Allis-Chalmers	77.51	8.46[a]	NA	—	2.59	0.46
Case	59.39	23.15	1.07	—	1.07	2.42
Oliver	65.25	22.25[a]	NA	—	2.89	5.07
Minneapolis-Moline	71.38	15.95[b]	—	—	—	5.59
Massey-Harris	76.60	14.61	1.79	1.06	1.79	—
B. F. Avery[c]	60.43	18.50[b]	3.86	—	10.34	—

[a] Includes collection expense.
[b] Segregation of items not made: includes selling, transfer, collection, general administrative, and bad debt expenses.
[c] The Federal Trade Commission considered B. F. Avery to be a long-line company, even though it did not make tractors. (It did have a marketing link for its tractor equipment with Allis-Chalmers.)
Source: PhD thesis, Warren Wright Shearer, "Competition through Merger: An Economic Analysis of the Farm Machinery Industry" (Harvard University, 1951), 236

Appendix Exhibit 24
Return on Investment in Farm Machinery Business (After Taxes), Selected Companies, Long- and Short-Line, 1927–1929

	1927 (%)	1928 (%)	1929 (%)
Long-line companies			
Deere & Co.	20.30	24.65	26.67
International Harvester	12.30	17.06	17.92
J. I. Case	14.22	14.48	7.77
Allis-Chalmers	8.58	9.04	11.90
Oliver Corp	—	—	6.63
Minneapolis-Moline	—	—	7.81
Massey-Harris[a]	0.70	2.07[b]	0.64
B. F. Avery	6.04	9.83	3.98
Short-line companies	—	—	—
More than $3,000,000 investment	1.24[b]	5.04	3.40
$1–3,000,000 investment	9.50	9.99	6.66
$300,000–$1,000,000 investment	2.59	6.80	5.89
Less than $300,000 investment	2.65	3.62	3.12

[a] Includes United States operations only.
[b] Loss.

Source: PhD thesis, Warren Wright Shearer, "Competition through Merger: An Economic Analysis of the Farm Machinery Industry" (Harvard University, 1951), 233

Appendix Exhibit 25
Industry and Deere-Lanz Tractor Sales
(in units)

Country		1955	1960	1965
Western Europe	Industry	299,430	313,520	331,830
	Deere-Lanz	13,528	5,949	11,420
	Percent of Industry	4.5	1.9	3.4
Germany	Industry	98,850	88,680	84,250
	Deere-Lanz	10,848	2,123	4,644
	Percent of Industry	11.0	2.4	5.5
France	Industry	66,190	73,340	71,550
	Deere-Lanz	374	635	2,096
	Percent of Industry	0.6	0.9	2.9
Spain	Industry	5,210	9,760	17,660
	Deere-Lanz	587	2,710	2,630
	Percent of Industry	11.3	27.8	14.9

Source: Market Economics Department, Deere & Company.

Appendix Exhibit 26
**Domestic and Overseas Sales and Income,
Deere & Company, 1965–1970
($000,000)**

Net Sales

Year	United States and Canada	Overseas	Worldwide
1965	741.6	145.0	886.6
1966	924.7	137.4	1,062.1
1967	935.4	151.0	1,086.4
1968	855.5	175.0	1,030.5
1969	856.2	186.8	1,043.0
1970	939.6	198.1	1,137.7

Net Income

Year	United States and Canada	Overseas Before United States Income Tax Credit	Net	Worldwide
1965	74.2	Not Applicable	(18.4)	55.8
1966	96.1	(20.4)	(17.4)	78.7
1967	82.3	(32.8)	(24.7)	57.6
1968	49.7	(19.9)	(7.1)	42.6
1969	59.2	(18.1)	(5.1)	54.1
1970	59.7	(29.5)	(13.7)	46.0

Source: 1970 Comptroller's Report, Deere & Company.

Appendix Exhibit 27
Investment in Overseas Subsidiaries
October 31, 1970
($000,000)

	Investment in Capital Stock and Surplus or Deficit	Total Investment as Shown on Balance Sheet	Total Investment at Cost
Germany	(60.6)	20.9	123.0
France[a]	26.8	39.6	46.8
Spain[a]	9.8	11.1	15.1
Mexico	9.2	10.1	8.4
Argentina	9.0	24.2	39.3
South Africa	(0.8)	3.6	5.4
Australia	1.6	4.5	4.5
JDIL (Export)	13.8	19.5	5.7
Other	1.5	3.8	6.3
TOTAL	10.4	137.3	254.3

[a] *Includes separate manufacturing and sales companies; totals not congruent because of rounding.*
Source: Comptroller's Report, Deere & Company (1970).

Appendix Exhibit 28
Deere's Operation in South Africa
Sales and Profitability, 1964–1980

	Sales ($000,000)	Net Income ($000,000)
1964	4.9	(0.4)
1965	5.1	(0.0)
1966	5.3	(1.1)
1967	10.6	0.1
1968	11.5	0.1
1969	11.3	(0.2)
1970	15.9	0.7
1971	18.2	1.0
1972	15.9	0.7
1973	24.3	1.6
1974	30.5	1.8
1975	63.0	5.7
1976	51.0	3.8
1977	42.0	2.0
1978	44.0	1.2
1979	33.0	(0.2)
1980	93.0	8.8
1981	139.0	10.1

Source: Deere & Company Comptroller's Report; sales figures 1975–1981 include Industrial Division shipments from Belgium.

Appendix Exhibit 29
Farm Equipment Sales in North America:
Market Share and Growth Rates, 1970–1980

Company	Year			Annual Growth Rate		
	1970	1975	1980	1970-75	1975-80	1970-80
Deere	26%	25%	29%	21%	12%	16%
International Harvester	18	17	14	21	5	12
Massey-Ferguson	8	8	7	20	7	13
Fiat	NA	—	3	NA	NM	NA
Hesston	1	2	NA	34	NM	NM
Sperry-New Holland	4	3	5	16	18	17
Ford	5	4	4	18	9	13
J. I. Case	4	6	6	29	8	18
Allis-Chalmers	6	6	5	22	7	14
Versatile	1	1	2	34	16	24
Steiger	—	1	1	91	6	42
Claas	NA	—	—	NA	NM	NA
Deutz	NA	—	1	NA	79	NA
Kubota	NA	NA	NA	NA	NA	NA
Total Industry Sales[2]	$3[b]	$8[b]	$12[b]	22%	9%	15%

NA – Not available; NM – Not meaningful.
[a]*Industry farm equipment dollar sales shown on Appendix Exhibits 29, 31, and 32 differ in definition from the US Department of Commerce farm machinery and lawn and garden equiptment sales. Not included in industry sales on the tables are farm dairy machines and equipment; farm poultry equipment; hog equipment; other barn and barnyard equipment; and irrigation systems. These categories have been excluded because the companies shown produce little or no equipment of these types.*
[b]*$000,000,000*
Source: Annual reports, supplemented by Deere & Company estimates.

Appendix Exhibit 30
North American Market Positions, 1981

	Deere	IH	MF	Hesston	New Holland	Ford	Case	AC	Versatile	Steiger
TRACTORS:										
Under 40 HP	2	4	5	—	—	2	—	5	—	—
40–89 HP	2	3	2	5	—	②	4	5	—	—
Over 90 HP	①	2	5	5	—	5	3	4	—	—
4WD	2	②	5	—	—	5	3	5	2	2
HARVESTING EQUIPMENT:										
Combines[a]	①	2	2	—	4	—	—	3	—	—
Forage Harvesters	②	3	5	3	2	5	—	—	—	—
Cotton Harvesters	①	2	—	—	—	—	—	3	—	—
TILLAGE EQUIPMENT	②	2	5	—	—	5	5	5	—	—
PLANTERS/ DRILLERS	①	2	5	—	—	5	—	4	—	—
HAY EQUIPMENT	2	3	3	3	②	5	—	—	4	—

	Claas	Deutz	Kubota	White	Yanmar	New Idea	Gehl	Satoh	Vermeer
TRACTORS:									
Under 40 HP	—	—	①	5	5	—	—	5	—
40–89 HP	—	5	5	5	—	—	—	—	—
Over 90 HP	—	5	—	5	—	—	—	—	—
4WD	—	—	—	5	—	—	—	—	—
HARVESTING EQUIPMENT:									
Combines[a]	—	—	—	5	—	5	—	—	—
Forage Harvesters	—	—	—	—	—	5	2	—	—
Cotton Harvesters									
TILLAGE EQUIPMENT	—	—	—	3	—	—	—	—	—
PLANTERS/ DRILLERS				4					
HAY EQUIPMENT	—	—	—	5	—	5	5	—	4

KEY:
1 - 30% and over
2 - 15–30%
3 - 10–15%
4 - 5–10%
5 - 0–5%
O - Major supplier.

[a] Combines can be split into two categories: conventional and rotary. Deere does not consider a rotary combine to be a distinct product with applications different from a conventional combine.
Source: Deere Company estimates.

Appendix Exhibit 31
Farm Equipment Sales Worldwide: Market Shares and Growth Rates, 1970–1980

Company	Year			Annual Growth Rate		
	1970	1975	1980	1970-75	1975-80	1970-80
Deere	15%	16%	17%	23%	12%	17%
International Harvester	12	14	9	25	4	14
Massey-Ferguson	10	12	9	24	5	14
Fiat	NA	4	5	NA	19	NA
Hesston	1	1	NA	38	NA	NA
Sperry-New Holland	4	3	4	19	14	17
Ford	7	5	4	14	6	10
J. I. Case	2	4	4	32	10	21
Allis-Chalmers	3	3	3	21	7	14
Versatile	—	1	1	34	20	27
Steiger	—	1	1	94	8	45
Claas	NA	2	2	NA	11	NA
Deutz	NA	2	2	NA	18	NA
Kubota	NA	4	4	NA	10	NA
Total Industry Sales	$6[a]	$16[a]	$27[a]	21%	12%	16%

NA—Not available.
Source: Deere & Company annual reports, estimates.
[a] $000,000,000

Appendix Exhibit 32
Farm Equipment Sales Overseas: Market Share and Growth Rates, 1970–1980

Company	Year			Annual Growth Rate		
	1970	1975	1980	1970-75	1975-80	1970-80
Deere	5%	8%	7%	30%	12%	21%
International Harvester	6	10	6	34	2	17
Massey-Ferguson	12	16	10	26	5	15
Fiat	NA	7	7	NA	13	NA
Hesston	—	—	NA	97	NA	NA
Sperry-New Holland	4	4	3	22	11	16
Ford	8	6	4	13	3	8
J. I. Case	1	2	2	47	15	30
Allis-Chalmers	1	—	—	9	8	8
Versatile	—	—	—	—	—	NM
Steiger	—	—	—	NM	25	NM
Claas	NA	3	3	NA	10	NA
Deutz	NA	3	3	NA	15	NA
Kubota	NA	NA	NA	NA	NA	NA
Total Industry Sales	$3[a]	$8[a]	$15[a]	20%	14%	17%

NA – Not available; NM-1 Not meaningful.
Source: Annual reports supplemented by Deere & Company estimates.
[a] $000,000,000

Appendix Exhibit 33
Deere's Position in Largest Overseas Markets Tractor Market Share, 1980 FY (%)

Country and Market size[a]	Deere	IH	MF	Fiat	Ford	Deutz	Landini	Other	Other
France								Renault	
(58,719) (units)	8	16	13	11	7	7	—	18	—
Italy								SAME	Lamborghini
(53,502)	2	2	4	25	4	—	8	18	7
Germany								Fendt	
(46,159)	9	20	6	4	2	17	—	16	—
Spain								EBRO	Barreiros
(25,132)	31	1	[b]	5	4	3	—	28	8
UK								Case	Leyland
(22,011)	10	12	24	3	25	—	—	10	5
Mexico									
(18,581)	15	21	30	—	33	—	—	—	—
South Africa									
(16,626)	17	5	22	20	22	3	5	—	—
Australia								Kubota	Case
(14,847)	21	15	14	7	12	4	—	5	8

[a]*Registrations, retail sales, or sales to dealers.*
Source: Deere & Company estimates.
[b]*Tractors manufactured by EBRO.*

Appendix Exhibit 34
Deere's Position in Largest Overseas Markets Combine Market Share, 1980 FY (%)

Country and Market size[a]	Deere	IH	MF	New Holland	Claas	Fahr	Other	Other
Brazil							Santa Matilda	Ideal
(5,437) (units)	29[b]	—	17	33	—	—	9	12
France							Braud	Fiat-Someca
(4,685)	12	13	10	24	20	—	7	10
Germany								
(4,467)	18	11	6	8	37	19	—	—
UK								
(2,081)	14	3	24	29	22	—	—	—
Italy							Laverda	Arbos
(1,981)	8	2	6	19	8	—	39	14
Finland							Sampo	
(1,948)	5	—	21	5	—	21	46	—
Australia								
(1,934)	18	32	30	15	—	—	—	—
Spain								
(1,468)	35	1	7	11	16	12	—	—

[a] *Registrations, retail sales, or sales to dealers.*
Source: *Deere & Company estimates.*
[b] *Deere/Schneider–Logemann Corporation collaboration.*

APPENDIX

Appendix Exhibit 35
Position of Deere in the North American Industrial and Construction Equipment Industry
(Percent of Market Share)

Product	Calendar Year 1979	Fiscal Year 1980	Fiscal Year 1981
Industrial wheel tractors, all sizes	17.0	20.7	20.0
Wheel log skidders, all sizes	31.8	28.6	27.7
Crawler dozers, under 160 NEHP	32.3	30.3	30.0
Crawler loaders, total	30.0	32.6	29.4
Four-wheel drive loaders, 80 & under 150 NEHP & 250 & under 300 NEHP	21.1	20.7	19.1
Elevating scrapers, under 18 cu. yd.	31.8	31.5	36.7
Motor graders, under 200 NEHP	24.9	30.4	30.9
Crawler mounted excavators under 40,000 lb. and 85,000–100,000 lb.[a]	32.9	32.5	26.1

[a] *All excavator data are shipments.*
NEHP Net estimated horsepower
Source: Comptroller's Annual Report, Deere & Company, 1981.

Appendix Exhibit 36
Research and Development Expenses as Percent of Sales and Dollars Per Employee 1977–1981

Year	Deere % of Sales	Deere $ per Employee	International Harvester % of Sales	International Harvester $ per Employee	Allis-Chalmers % of Sales	Allis-Chalmers $ per Employee	Caterpillar % of Sales	Caterpillar $ per Employee
1977	3.8	2,346	1.9	1,230	3.4	1,916	3.8	2,835
1978	3.7	2,618	1.8	1,235	not reported		2.3	1,940
1979	3.8	2,877	2.6	2,230	3.0	2,012	2.5	2,131
1980	4.2	3,788	4.0	2,930	2.4	1,709	2.3	2,444
1981	4.4	3,943	3.5	3,731	3.2	2,431	2.5	2,637

Source: "R&D Scoreboard," Business Week.

Appendix Exhibit 37
Deere & Company, 1972–1982

	Sales ($000,000)	Net Income ($000,000)	Net Income Per Share ($)	Book Value Per Share ($)
1972	1,500	112	1.91	13.73
1973	2,003	168	2.88	15.91
1974	2,495	164	2.78	17.90
1975	2,955	171	2.89	19.86
1976	3,134	242	4.04	22.86
1977	3,604	256	4.24	25.87
1978	4,155	265	4.38	28.84
1979	4,933	311	5.12	32.34
1980	5,470	228	3.72	34.13
1981	5,447	251	3.79	36.29
1982	4,608	53	0.78	35.29

Source: Deere & Company Annual Reports.

Appendix Exhibit 38
Deere Profitability Ratios

	A. Net Income as Percent of Net Sales				
	Deere	International Harvester	Massey-Ferguson	Allis-Chalmers	Caterpillar
1972	7.5	2.5	3.6	0.9	7.9
1973	8.4	2.5	3.9	1.4	7.8
1974	6.6	2.4	3.8	1.8	5.6
1975	6.1	2.1	3.8	2.0	8.0
1976	7.7	3.2	4.3	3.9	7.6
1977	7.1	3.4	1.2	4.4	7.6
1978	6.4	2.8	(8.8)	4.3	7.8
1979	6.3	4.4	1.2	4.1	6.5
1980	4.2	(6.3)	(7.2)	2.3	6.6
1981	4.6	(5.6)	(7.4)	(1.4)	6.3
1982	1.1	(38.2)	(20.1)	(12.9)	(2.8)
	B. Net Income as Percent of Beginning Year Total Assets				
1972	6.2	2.5	3.1	0.7	11.3
1973	8.9	2.8	4.4	1.3	12.8
1974	8.1	2.9	4.7	1.8	10.2
1975	7.3	2.3	5.1	2.3	13.6
1976	8.4	3.1	5.0	4.3	11.4
1977	7.7	3.4	1.1	4.3	11.4
1978	6.3	2.7	(7.5)	4.3	13.0
1979	7.1	4.8	1.1	3.8	9.8
1980	4.6	(4.1)	(6.3)	3.2	10.5
1981	3.7	(3.8)	(5.5)	(1.2)	9.5
1982	0.7	(16.2)	(12.7)	NA	(2.5)

Note: Corporate-wide figures for all companies; finance subsidiary assets consolidated.
NA - Not available.
Source: Deere & Company Comptroller's Annual Reports.

Appendix Exhibit 39
**Deere's Rankings among Fortune Magazine's
100 Largest Industrials, 1972–1982**

Year	Sales	Net Income	ROS	ROE	10-Yr. EPS Growth	10-Yr. Total Return
1972	90	45	20	24	20	14
1973	78	40	21	16	16	12
1974	75	47	27	39	49	13
1975	62	38	23	24	17	16
1976	66	37	15	13	21	34
1977	60	36	20	28	6	7
1978	61	41	26	42	3	2
1979	60	51	32	61	10	10
1980	60	64	63	80	28	13
1981	65	62	51	77	41	16
1982	75	85	84	86	84	54

ROS = Net income as percent of sales.
ROE = Net income as percent of stockholders' equity at end of year.
EPS = Earnings per share of common stock.
Total Assumes Return dividends = Return reinvested to a stockholder, including both common stock appreciation at end of year and dividends.
Source: Fortune *magazine, "The Fortune 500."*

APPENDIX

Appendix Exhibit A

Deere Family Links In Deere & Company*

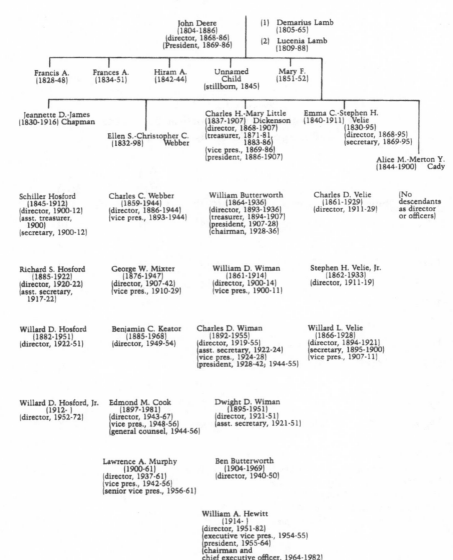

*Included are only officers and directors.

APPENDIX EXHIBITS

Appendix Exhibit B

Five Generations of Deere & Company Chief Executive Officers, 1837–1982

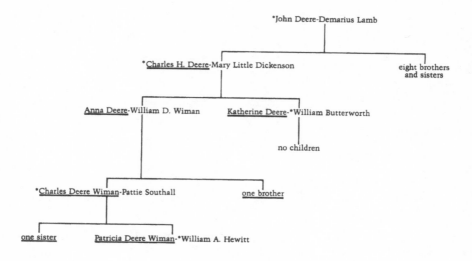

NOTE: Burton F. Peek was president, 1942–1944, when Charles D. Wiman was in the United States Army.

* – *chief executive officer*
— *direct descendant of John Deere*

BIBLIOGRAPHIC NOTES

Taken as a whole, the literature of agriculture is extensive, indeed overwhelming. Certain general studies were meaningful for this work, as were general bibliographic efforts that provide additional detail. Writings specifically relating to the agricultural equipment industry vary widely in quality. While the technical literature is comprehensive, even back into the nineteenth century, the economic and social history of the industry is less well documented. A few key theses from young academics—always a fruitful source for bibliographic depth—and important government hearings in the 1910s, 1920, 1938, and 1948 have added extensive documentation of certain aspects of the industry. Still, there have been few significant business histories of individual firms.

The history of Deere & Company is a seminal one in regard to the Westward Movement. John Deere's emigration from Vermont in 1837 and his settlement in Illinois at the time of the opening of the prairie parallel similar efforts by countless thousands of other Americans in the early and mid-nineteenth century. A number of writings on "the frontier," many relating particularly to the Midwest, are pertinent to the Deere story.

Deere & Company records of the nineteenth century are rich for certain times, poor for others. Just scraps remain from the pre–Civil War days of John Deere, though they fortunately document critical points. John Deere himself was not a record keeper; his son Charles, on the other hand, kept excellent reports both for the firm and for his own personal use. Charles Deere's diary for 1859–1860, for example, reveals many insights, both about the firm and the man. An extensive set of Charles Deere literature for the 1870s and '80s remains, superb material for understanding that hurly-burly period of growth. Records of Deere's efforts to engineer a "plow trust" are extant, as are a number of documents relating to manufacturing that help to build a picture of the technical and labor aspects of the business in that period.

Fortuitously for historians, a trademark court case in the late 1860s—*Henry W. Candee, et al., Appellants v. John Deere, et al., Appellees*, Supreme Court of Illinois (1870)—laid on the record a rich panoply of detail about Deere's business, replete with personal statements by many dozens of individuals relating to a wide range of topics. Also, the company has kept meticulous minutes of every one of its directors' meetings since the inception of incorporation in 1868; the earlier years are particularly evocative, with many dozens of the meetings in the first three decades of the twentieth century recorded almost verbatim. Another fascinating, though often undependable, source about the mid-nineteenth century is the diary of Robert Tate, sometime partner of John Deere ("The Life of Robert N. Tate, Written by Himself," Deere Archives). This opinionated but articulate personal view of the firm and its participants is a unique well of information about both the company and the town.

In the main, the records of the company in the twentieth century are excellent. William Butterworth, as chief executive officer from 1907 to 1928, left most of his papers intact. Charles Wiman's papers were largely destroyed, but a wide range of collateral documentation allows the Wiman period to come alive. Since the 1950s the company's records have been managed very well indeed; just about everything remains that a serious historian might desire. The company maintains a sophisticated, computerized corporate archives, which is effectively supervised by a professional archivist.

There are several modest books on John Deere and Deere & Company, all of which are cited in the notes. A number of dissertations at both the master's and doctoral levels have been done; one of the most valuable is William J. Kirkpatrick, "John Deere and His Steel Plow: Their Contribution to the Midwest and the World" (University of Minnesota, 1971). One of the best analytical sources of overall company history was done by P. C. Simmon, the corporate secretary of the first two decades of the twentieth century; his unpublished manuscript is dated 1920. Two company sources relating specifically to tractor history are worth mentioning: the book by Theo Brown, *Deere & Company's Early Tractor Development* (privately printed, 1953) and the promotional brochure "John Deere Tractors, 1918–1976" (1976), by Will McCracken.

Individual business histories of the many other important firms in the industry have been few in number and spotty in quality. Despite the superb raw material available on International Harvester's history—a first-rate archives at the company's headquarters, and additional important McCormick family papers at the Wisconsin Historical Society—there had yet to be a full-scale business history of this other giant of the industry by the early 1980s. Cyrus McCormick (the son of the founder), authored an interesting and surprisingly analytical view of the corporation in his *Century of the Reaper, An Account of Cyrus Hall McCormick, the Inventor of the Reaper; of the McCormick Harvesting Company, the Business He Created and of the International Harvester Company, His Heirs and Chief Memorial* (Boston: Houghton Mifflin Company, 1931). A sound biography of the founder himself is William T. Hutchinson, *Cyrus Hall McCormick,* 2 vols. (New York: D. Appleton-Century Company, 1935). Two comprehensive books on International Harvester's employee and labor relations are by Robert Ozanne—*A Century of Labor-Management Relations at McCormick and International Harvester* (1967) and *Wages in Practice and Theory: McCormick and International Harvester, 1860–1960* (1968), both published in Madison, Wisconsin, by the University of Wisconsin Press. Also helpful in understanding International Harvester's role is Helen M. Kramer, "Harvesters and High Finance: Formation of the International Harvester Company," *Business History Review* 38 (1964): 285. There is excellent documentation of International Harvester's early years in Department of Commerce, Bureau of Corporations, *The International Harvester Company* (Washington: Government Printing Office, 1913) and in the antitrust suits against the company in the 1910s and '20s (*US v. International Harvester Co., et al.*, 214 Federal Reports 987 and *US v. International Harvester*, 274 US 693). A pictorial history of International Harvester of merit is C. H. Wendel, *150 Years of International Harvester* (Sarasota, FL: Crestline Publishing Co., 1981).

For the history of the J. I. Case Company, see Stewart H. Holbrook, *Machines of Plenty* (New York: The Macmillan Company, 1955); for the history of the Oliver Chilled Plow Company, see Joan Romine, *Copshaholm: The Oliver Story* (South Bend, IN: Northern Indiana Historical Society, Inc., 1978). For histories of Massey-Ferguson and Massey-Harris, see Edward P. Neufeld, *A Global Corporation: A History of the International Development of Massey-Ferguson Limited* (Toronto: University of Toronto Press, 1969); James S. Duncan, *Not a One-Way Street: The Autobiography of James S. Duncan* (Toronto: Clarke, Irwin and Company, 1971); and Peter Cook, *Massey at the Brink: The Story of Canada's Greatest Multinational and its Struggle to Survive* (Toronto: Collin Publishers, 1981). Several biographical studies that have added significant information on the industry are: Reynold M. Wik, *Henry Ford and Grass Roots America* (Ann Arbor: University of Michigan Press, 1972); Gilbert C. Fite, *George M. Peek and the Fight for Farm Parity* (Norman: University of Oklahoma Press, 1954); Collin Fraser, *Tractor Pioneer: The Life of Harry Ferguson* (Athens: Ohio University Press, 1973); Edith Sklovsky Covich, *Max* (Chicago: Stuart Brent, publishers, 1974). One particularly unique source of information about the companies is the credit correspondence in the R. G. Dun & Company collection in the Baker Library, Harvard Business School; the material on Deere & Company, for example, extends from the 1850s to the late 1880s.

Five doctoral dissertations on the industry are particularly helpful: Michael Conant, "Aspects of Monopoly and Price Policies in the Farm Machinery Industry since 1902" (University of Chicago, 1949); Warren Wright Shearer, "Competition through Merger: An Economic Analysis of the Farm Machinery Industry" (Harvard University, 1951); Elvis Luverne Eckles, "The Development of Oligopoly in the Farm Industry" (University of Illinois, 1953); Arlyn John Melcher, "Collective Bargaining in the

Agricultural Implement Industry: The Impact of Company and Union Structure in Three Firms" (University of Chicago, 1964); and Harvey Schwartz, "The Changes in the Location of the American Agricultural Implement Industry, 1850 to 1900" (University of Illinois, 1966).

There is a wide range of sources on agricultural machinery. For the British experience, the books by George Edwin Fussell are a useful starting point, especially *The Farmer's Tools, AD 1500–1900: A History of British Farm Implements, Tools and Machinery before the Tractor Came* (London: Andrew Melrose, 1952). Also excellent is John B. Passamore, *The English Plough* (London: Oxford University Press, Humphrey Milford, 1930). The following four sources are the key pieces for the American experience in agricultural machinery in the nineteenth century: Robert L. Ardrey, *American Agricultural Implements: A Review of Invention and Development in the Agricultural Implement Industry of the United States* (1894; reprinted, Arno Press, Inc., New York, 1972); Leo Rogin, *The Introduction of Farm Machinery in Its Relation to the Productivity of Labor in the Agriculture of the United States during the Nineteenth Century* (1931; reprinted, Johnson Reprint Corp., New York, 1966); Clarence H. Danhof, *Change in Agriculture: The Northern United States, 1820–1870* (Cambridge, MA: Harvard University Press, 1969); "Report on Trials of Plows," *Transactions of the New York State Agricultural Society*, 1:27 (1867): 385–656 (often reported in the literature as the "Utica Trials"). See the notes for chapter 2 for other relevant English and American materials on the nineteenth century.

The American Society of Agricultural Engineers in St. Joseph, Michigan, has published important works on the tractor and the harvester: the classic history of the tractor in the twentieth century, R. B. Gray, *The Agricultural Tractor: 1855–1950* (1954); Lester Larson, *Farm Tractors 1950–1975* (1981); and Graeme R. Quick and Lesley F. Buchele, *The Grain Harvesters* (1978). See also C. H. Wendel, *Encyclopedia of American Farm Tractors* (Sarasota, FL: Crestline Publishing Company, 1979); Wayne Worthington, *50 Years of Agricultural Tractor Development* (St. Joseph, MI: American Society of Agricultural Engineers, 1966); and Robert Kudrle, *Agricultural Tractors: A World Industry Study* (Cambridge, MA: Ballinger Publishing Company, 1975). There are a number of books on "vintage farm tools," such as Michael Partridge, *Farm Tools through the Ages* (Boston: New York Graphic Books, 1973). Similarly, there is substantial coverage of "vintage tractors" in various largely pictorial books, such as Michael Williams, *Farm Tractors in Color* (New York: Macmillan Publishing Company, Inc., 1974) and Philip Wright, *Old Farm Tractors* (London: David & Charles, 1962, reprint, 1974). There is also a burgeoning literature on the steam plow; Reynold M. Wik, *Steam Power on the American Farm* (Philadelphia, PA: University of Pennsylvania Press, 1953), is a useful starting point.

From the mid-nineteenth century, certain key monographs, articles, and books seem to sum up agricultural machinery "at the moment." For the period before the Civil War, see the various reports of the commissioner of patents, published as executive documents of the House of Representatives for a given year; J. J. Thomas, "Farm Implements and Machinery," *Report of the Commissioner of Agriculture, US Department of Agriculture Annual Report* (1862), and *Farm Implements and the Principles of Their Construction and Use: An Elementary and Familiar Treatise on Mechanics and on Natural Philosophy Generally, As Applied to the Ordinary Practices of Agriculture* (New York: Harper and Brothers, 1854; later editions titled *Farm Implements and Farm Machinery* in 1869 and 1886, Orrin Judd & Company Publishers). For the second half of the nineteenth century, see Carroll D. Wright, *Thirteenth Annual Report of the Commissioner of Labor, 1898: Hand and Machine Labor*, 2 vols. (Washington: Government Printing Office, 1899); George K. Holmes, *The Course of Prices of Farm Implements and Machinery for a Series of Years*, US Department of Agriculture Miscellaneous Series, bulletin 18 (Washington: Government Printing Office, 1901); H. W. Quaintance, "The Influence of Farm Machinery on Production and Labor," American Economic Society, 3:5 (November 1904): 45. For the first half of the twentieth century, see M. R. D. Owings, "New Methods and New Machines for the Farm: What the Inventor Has Done for Agriculture," *Scientific American* (February 18, 1911); Barton Currie, *The Tractor and Its Influence upon The Agricultural Implement Industry* (Philadelphia, PA: Curtis Publishing Company, 1916); Lillian M. Church and H. R. Tolley, "The Manufacture and Sale of Farm Equipment in 1920: A Summary of Reports from 583 Manufacturers," US Department of Agriculture, circular 212 (April 1922); C. D. Kinsman, "An Appraisal of Power Used on Farms in the United States," US Department of Agriculture, departmental bulletin 1348 (1925); Lillian M. Church, "History of the Plow," US Department of Agriculture, Bureau of Agricultural Engineering, Division of Mechanical Equipment, Information Series 48 (October 1935); W. W. Hurst, "New Types of Farm Equipment and Economic Implications," *Journal of Farm Economics* 19 (1937): 483; Harold Barger and Hans W. Landsburg, *American Agriculture, 1899–1939: A Study of Output, Employment and Productivity* (New York: National Bureau of Economic Research, Inc., 1942); Martin R. Cooper, Glenn T. Barton, and Albert P. Brodell, "Progress of Farm Mechanization," US Department of Agriculture miscellaneous publications 620 (1947). There have been several important hearings and studies of the agricultural machinery industry by the federal government: US Department of Commerce, Bureau of Corporations, "Farm-Machinery Trade Associations" (March 15, 1915); Federal Trade Commission, "Report of the Federal Trade Commission on the Causes of High Prices of Farm Implements" (May 4, 1920); Federal Trade Commission, "Report on the Agricultural Implement and Machinery Industry," 75th Congress, 3rd Session, House Documents 702 (June 6, 1938); Temporary National Economic Committee, Investigation of Concentration of Economic Power 36, "Agricultural Implement and Machinery Inquiry" (1940); Federal Trade

Commission, "Report of the Federal Trade Commission on Manufacture and Distribution of Farm Implements" (1948).

There have been two important agricultural machinery industry journals, both originating in the late nineteenth century. *Farm Implement News* was founded in 1882 in Chicago, and *Implement and Tractor* was founded in 1876 in Kansas City; the two were merged as *Implement and Tractor* in 1958, published in Overland Park, Kansas. Two professional journals of agricultural history are *Agricultural History*, the quarterly journal of the Agricultural History Society, published by the University of California Press, and *Agricultural History Review*, published by the British Agricultural History Society. *Business History Review*, published by the Harvard Business School, has contained a number of articles relating to the industry. The literature of the trade press in agriculture has been prolific; especially useful are *Country Gentleman*, published from 1853 to 1955 in Philadelphia; *Genesee Farmer*, published from 1840 to 1865 in Rochester, New York; and *Prairie Farmer*, published since 1841 in Chicago. For an excellent review, see *Agricultural Literature: Proud Heritage–Future Promise, A Bicentennial Symposium, September 24, 26, 1975* (Washington: Associates of the National Agricultural Library, Inc., and the Graduate School Press, US Department of Agriculture, 1977).

For general books on agriculture, see Fred Shannon, *The Farmer's Last Frontier, Agriculture 1860–1897* (New York: Reinhart & Company, 1945); a number of the writings of Paul W. Gates, especially *The Farmer's Age: Agriculture 1815–1860* (New York: Holt, Rinehart, and Winston, 1960) and *The Illinois Central Railroad and Its Colonization Work* (Cambridge, MA: Harvard University Press, 1930); and the writings of Wayne D. Rasmussen, especially *Agriculture in the United States: A Documentary History*, 4 vols. (New York: Random House, 1975). Also useful is D. Gale Johnson, ed., *Food and Agricultural Policy for the 1980s* (Washington: American Enterprise Institute for Public Policy Research, 1981). US Department of Agriculture, *Farmers in a Changing World*, Part 1 (Washington: Government Printing Office, 1940), contains several excellent articles on agricultural history. The agricultural history of New England is well documented in Howard S. Russell, *A Long Deep Furrow: Three Centuries of Farming in New England* (Hanover, NH: University Press of New England, 1976). For Midwestern agricultural history, useful starting points are Allan G. Bogue, *From Prairie to Cornbelt: Farming on the Illinois and Iowa Prairies in the Nineteenth Century* (Chicago: University of Chicago Press, 1963) and Percy Wells Bidwell and John I. Falconer, *History of Agriculture in the Northern United States, 1620–1860* (Washington: Carnegie Institution of Washington, 1925). The historical writing on the Westward Movement is prolific. Particularly helpful are: Lewis D. Stilwell, *Migration from Vermont* (Montpelier, VT: Vermont Historical Society, 1948); William V. Pooley, "The Settlement of Illinois from 1830 to 1850," *Bulletin of the University of Wisconsin*, History Series I

(1908): 207; Malcolm J. Rohrbough, *The Trans-Appalachian Frontier: People, Societies, and Institutions, 1775–1850* (New York: Oxford University Press, 1978); and John D. Unruh Jr., *The Plains Across: The Overland Immigrants and the Trans-Mississippi West, 1840–1860* (Urbana, IL: University of Illinois Press, 1979). The manuscript collections of state and county historical societies provide many original documents of the Westward Movement; evocative pieces can be found at the Vermont Historical Society, the Illinois State Historical Society, the State Historical Society of Iowa, the Chicago Historical Society, and the F. Hal Higgins Library of Agricultural Technology, University of California, Davis.

Finally, certain key bibliographic contributions to agriculture should be noted. For the early nineteenth century, see Alan M. Fusonie, ed., *Heritage of American Agriculture: A Bibliography of Pre-1860 Imprints* (Beltsville, MD: National Agricultural Library, 1975). For a comprehensive bibliography of American agricultural history, see Everett E. Edwards, *A Bibliography of a History of Agriculture in the United States* (Washington: US Department of Agriculture miscellaneous publication No. 84, 1930). The US Department of Agriculture, in cooperation with the Agricultural History Center of the University of California, Davis, has supplemented this basic bibliography with a score of specialized and updated monographs; see, particularly, "A List of References for the History of Agriculture in the Midwest, 1840–1900" (1973); "A List . . . for the History of Agriculture in California" (1974); "A List . . . for the History of Agriculture in the Great Plains" (1976); "A List . . . for the History of Agricultural Technology" (1979); and "A List . . . for the History of Agricultural Science in America" (1980). See also John T. Schlebecker, *Bibliography of Books and Pamphlets on the History of Agriculture in the US, 1607–1967* (Santa Barbara, CA: American Bibliographic Center, 1969). For an annotated bibliography of traders' views of Illinois, see Solon Justus Buck, *Travel and Description, 1765–1865*, Illinois Historical Society 9, Bibliographic Sources 2 (Springfield, IL: Illinois State Historical Society Library, 1914).

INDEX

Page numbers in italics indicate illustrations.

advisory committee, 162
AFL. *See* American Federation of Labor
Agar, Cross & Company, 47, 142, 198
Agnelli, Giovanni, 222–23, 224–25, 227, 228
Agricultural Adjustment Act, 65
agricultural implements
    and depression, 57–60
    market shares, 1944, 126–27
    pricing of, 31–32
    product improvement, 127–34
    and recession, 267–69
    and selling terms, 304–5
agriculture
    and US practices, 264–65
    and world markets, 264–65
All Crop combine, 80
Alldog tractor, 172–73, 188
Allis-Chalmers Manufacturing Company, 25, 65–66, 72, 136, 228, 230, 234, 267
Allround Tool Carrier tractor. *See* Alldog tractor
American Federation of Labor (AFL), 67, 94
antitrust problems
    and (J. I.) Case Co., 124–27
    and Deere & Co., 121–24
apartheid, and South Africa, 201, 242
Argentina, as market, 46–48, 198–99, 234–38
Army-Navy E award, 96
Asia, as market, 202
assets, of Deere Company
    1930, 38–39
    1948, 111–12
    1960, 183, 185, 209–10
    overseas, 231, 258
Australia, as market, 240–42

balers, 286
Bartlett, Boyd C., 295
benefits, employee, 51–52
Bennett, Prime Minister R. B., 51
Berry, Harold J., 223
Berry combine, 8

Best combine, 8
Block, Maurice, 164
Boeke, Robert W., 279–80, 295, 321, 323
Booz, Allen & Hamilton, 155, 161–66, 206, 207, 217, 218
Boyle, Harold, 93, 104–5, 143–44
Bozeman, Virgil, 23, 143, 164
Brazil, as market, 240
Brezhnev, Premier Leonid, 250
Brown, Delno W., 219, 293–94, 295, 321, 323
Brown, Theo, 15, 19, 20, 22–23, 56–58, 80, 143, 147, 287
Burnham, Donald C., 252
Butterworth, Katherine Deere, 61, 62, 63
Butterworth, William, 3–5, 15, 32, 49, 52, 53, 54, 60–62, 69, 325, 327

Cade, J. E., 15
California, as territory, 76–79
Campins, President Luis Herrera, 237
Canada
    branch in, 48–51
    factory in, 49–50
    import tariffs in, 49, 50
Carlson, C. R., Jr., 51, 104, 165, 206, 208, 211, 218
Carlson, Edward E., 326
Carlson, Robert J., 292, 293, 295, 321, 323
Carter, President Jimmy, 266
(J. I.) Case Company, Inc., 25, 268, 270
(J. I.) Case Plow Works, 23
(J. I.) Case Threshing Machine Company, 25
CASTER/ACTION, for tractors, 276, *277*
Caterpillar tractor, 75
Caterpillar Tractor Company, 8, 11, 72–79, 270, 301, 303
Chamberlain-John Deere, Pty. Ltd., 240
China, mainland, 251–58
China Council for the Promotion of International Trade (CCPIT), 251–52
Chrysler Corporation, 302
Cindelmet, 235

361

INDEX

Clausen, F. H., 23, 98
Clausen, Leon, 15, 50, 69, 124–25
Clayton Antitrust Act, 1914, 121
Cleveland Tractor Company, 88
"Cold War," 135
collections, for company, 40–42
combines. *See also individual companies by name*
   and China, 254
   and Deere, 9–13, 80–81
   early development of, 7–9
   innovations in, 7–9, 10, 134–35, 285–86
comfort, of operator, 182, 186
Committee for Industrial Organization (CIO), 67, 95, 117
Compagnie Continentale de Motoculture (CCM), 194–95
company housing, 52
company store, 27–28, 30–32
compensation, for management, 163–64
competitive practices, 25–26, 121–24, 316
computers, use of, 297–98, 299–300
consumer products division, 272
Continental Illinois National Bank and Trust Company, 109
contracts, employee, 67, 68
Cook, Edmund "Budge," 99, 100–102, 104–5, 111, 121–22, 124, 144, 145, 162, 165, 179
Coolidge, President Calvin, 32
cooperatives, European, 174
corn head, for combines, 135
corn pickers, 11
Crampton, George, 15, 52, 53
credit, problems with, 40–42, 306–8
Cukurova Makina Imalat ve Ticaret A. S. (Cumitas), 239
Curtis, Ellwood "Woody," 141, 157, 158, 164, 165, 166, 169, 192, 198, 200–201, 206, 221, 222, 227, 228, 230–31, 242, 321, 323
cylinders, for tractors, 181–82, 186, 187

Dain, Joseph, Jr., 206, 218, 287, 295
Dain Manufacturing Company, 49–50
dealers, for company products
   overseas, 173–74, 188–89
   US, 27, 30–32, 122–23, 306–7, 316–17
debts, and collections, 40–42
decentralization, of company
   issues in, 99–101, 103–8, 288–89, 317–19, 323
   reduction of, 162–63, 293–96
Deere, Charles Henry, 53, 327
Deere, John, 52, 327
Deere, Katherine. *See* Butterworth, Katherine Deere
Deere, South Africa, 243
(John) Deere, C. A., 167
Deere Administrative Center, *156,* 165, 175–80, *176–77,* 205, 210–11, 319, *320*

(John) Deere-Bobaas (Proprietary) Limited. *See* (John) Deere (Proprietary) Limited
(John) Deere & Company
   collaboration with Caterpillar, 72–76
   and company loyalty, 308
   and competition, 25–26, 121–24, 316
   consumer products division, 272
   and credit, 40–42, 306–8
   and depression, 37–39, 51–52, 56–57
   employees of (*See* employees)
   and excellence, 319–21
   expansion of (*See* expansion, of company)
   financial performance, 303–5 (*See also* assets, of Deere Company)
   financial services units, 274
   headquarters of, 5
   industrial equipment division, 269–70, 272–74
   insurance subsidiaries, 304
   and labor relations, 67–69, 94–97, 115–21, 317
   management of (*See* management, of company)
   market shares, 187, 274–75
   merchandise unit, 272–73
   as multinational, 153, 157–59, 166–74, 188–96
   product development, 57–60, 79–81, 127–34
   rescue of bank, 52–55
   research and development, 22–24, 286–91
   and sense of history, 314–15
   war efforts, 88–89
(John) Deere Credit Company (US), 304
(John) Deere Engine Works, 294
(John) Deere Finance Limited (Canada), 304
(John) Deere Foundation, 246, 325
(John) Deere Intercontinental, Limited, 207
(John) Deere-Lanz A. G., 192, 204
(John) Deere Leasing Company (US), 304
(John) Deere (Proprietary) Limited, 200
(John) Deere Lindeman tractor, 110
Deere & Mansur Company, 95
"Deere Picnic Day," 5
(John) Deere Plow Company, Ltd., 49
Deffenbaugh, J. L., 164–65
Depression, Great, 37–39, 51, 53–57, 60
depressions, economic, 51–52, 53–57
design, of implements, 297–98
Deutz. *See* Klockner-Humboldt Deutz A. G.
Diaz Ordaz, President, 198
Dickey, Frank M., 157, 165, 218, 226
diesel, as fuel, 181–82
diesel engine, uses of, 128
Dinkeloo, John, 210
Dreyfuss, Henry, 69–71, 175, 189, *190,* 205, 223
Duncan, James S., 159–61
Dunell, Edden & Company, 47–48

362

INDEX

Echeverria Alvarez, President Luis, 232
Edwards, Richard, 157
Eisenhower, Dwight D., *156*
employees
    benefits plans for, 51–52
    contracts for work, 67
    housing for, 52
    and labor relations, 67–69, 94–97, 115–21, 317
    pensions for, 118
English, E. C., 157
Europe, as market, 135–39
European Economic Community (EEC), 193, 194, 202, 204
expansion, of company
    foreign (*See* foreign expansion)
    US, 25–27, 108–12

Fabrica Nacional de Tractores Motores, S. A. (Fanatracto), 236
factories, modern, 296–301
Farmall tractor, 14, 15
Farm Equipment Workers, 95, 96, 113, 114, 116–18
Farm Equipment Workers Organizing Committee, 68
Farm Implement News, 8, 9
Federal Trade Commission, 26, 30–31, 282
fertilizer, as product, 139–42
Fiat, 223–28, 235
financial services unit, 274
Fischer, V. C., 142
Fleming, Robben W., 326
Ford-Ferguson, 86–87
Ford-Ferguson tractor, 131
Ford Motor Company, 196, 197, 198, 200, 201–2, 241, 247, 268, 302
Fordson tractor, 14, 42
foreign expansion
    Argentina, 46–48, 198–99, 234–38
    Brazil, 240
    of company, 157–59, 166–74, 188–96
    difficulties in, 215–28
    division by regions, 207–8
    France, 193–96
    Germany, 168–69, 172–74, 188–93, 216–17, 219–23
    Iran, 238–39
    Italy, 223–28
    mainland China, 251–58
    management reorganization, 217–18
    and market share, 269
    Mexico, 166–67, 196–98, 231–34
    1970s, 258–59
    South Africa, 47–48, 200–202, 242–48
    Spain, 202–3
    Turkey, 239–40
    Venezuela, 236–38
Fraher, Maurice A., 133, 157, 159, 165, 166–67, 190

France, and Deere, 193–96
French, George, 141, 157, 164, 198, 206, 208, 218, 226, 229
"Friendship Farm" collaboration, 251–58

Gamble, Carl H., 23
General Motors Corporation, 67, 245, 302
Germany, and expansion, 168–69, 172–74, 188–93, 216–17, 219–23
Gerstenberger, Robert J., 219
Gildehaus, Thomas A., 295, 323–24
"give backs," 302–3
Glover, D. C., 206, 226, 229
Good, John, 136
Gould, John M., 52
GP tractors, 19–21, 24, 57
GP Wide-Tread tractor, 21, 57
Graflund, John H., 206, 208, 218
Graham, William B., 326
Gregerson, Nils, 200–201
group technology (GT), 298
Gyllenhammar, Pehr G., 230

Haasl, C. M., *92*
Hall, L. N. (Neel), 230, 295
Hanson, Charles A., 295
Hanson, Robert A., 167, 236, *281,* 300, 301, 323, 326, 327, *328*
harrows. See *individual harrows by name*
harvesters. See *individual harvesters by name*
hay rakes. See *individual rakes by name*
Head, A. H., 23, 67
headquarters, of company, 175–80, 210–11
Healey, M. J., 80, 99
Heinrich Lanz, 158, 159, 168–69, *170,* 172–74
Hewitt, William, 141, 144, 145–46, 153, 155–57, 164, 166, 168, 175, *176,* 178, 179–80, *184,* 187, 190, 203, 205–11, 218, 219, 222, 227, 229–30, 248–49, 250, 252, *254,* 274, *281,* 288, 307, *314,* 319, 321, 323, 324, 325, 327
hillside combine, 134–35
Hitachi, Ltd., cooperation with, 202
hitches, for tractors, 131
Holt combine, 8
Hoover, President Herbert, 52, 60
Hosford, Schiller, 55
Hosford, Willard D., Jr., 157
Hosford, Willard Deere, 54, 99, 143
housing, for workers, 52
Hutchinson, H. W., 49

IBH Holding AG, 270
industrial equipment, and Deere, 219, 269–70, 272–74
innovations, introduction of, 22–24, 57–60, 79–81, 127–34, 286–91
International Association of Machinists, 116

363

International Harvester
  and competition, 65–66, 267, 268
  foreign operations, 136, 166, 174, 196, 197, 200, 201–2, 233, 241, 280
  and labor relations, 301–2
  negotiations with Deere, 49–50
  and tractors, 14, 72
International Harvester crawler, 78
Investor Responsibility Research Center (IRRC), 245, 246
Iran, as market, 238–39
Italy, as market, 223–28

Johnson, Samuel C., 326

Keator, Benjamin, 143, 145–46, 154
Keisling, T. S., 203
Kennedy, Lloyd, 98, 101, 104, 138–39, 144, 145, 158, 159, 162, 164, 166, 168–69, 179, 188
Khruschchev, Nikita S., 248–49, *249*, 250
Kissinger, Henry, 251
Klingberg, William R., 218
Klockner-Humboldt-Deutz A. G., 221, 235
Komatsu, 270
Korean War, 135, 143
Kough, Benjamin J., 23
Kreps, Juanita, 326
Kubota, Ltd., 268

labor relations
  1930s, 66–69
  postwar, 112–21
  wartime, 94–97
labor unions. *See* unions
Lanz. *See* Heinrich Lanz
Lanz, Iberica, S. A., 172, 202
Lardner, James F., 298
lawsuits, antitrust, 121–24
legislation, agricultural, 65
Lindeman crawler tractor, 110
Lindvall, Dr. Frederick C., 288–90, 324
Locomobiles threshers, 169
Lopez Mateos, President, 198
Lopez Portillo, President José, 233
Lourie, Bruce, *92*, 138, 162, 165
Lourie, Ralph (R. B.), 28, 78, 99
Lundahl, A. B., 206, 218, 295

management, of company
  advisory committee, 162
  compensation for, 163–64
  decentralization issues, 99–101, 103–8, 288–89, 317–19, 323
  and foreign divisions, 207–8, 323
  new directions, 1955, 153–57
  organization of, 103–8
  organization study, 1966, 217–19
  policy committee, 162

  study of, 1955, 155, 161–66
  and unions, 67–69, 94–97, 115–21, 317
manufacturing, and computers, 299–300
manure spreaders, 286
marketing
  and dealers, 27, 30–32, 122–23, 306–7, 317
  new directions, 1955, 155–56
Marshall Plan, 135
Massey-Ferguson, 136, 159–61, 166, 174, 196, 197, 200, 201–2, 233, 235, 241, 247–48, 267, 268, 269
Massey-Harris Company, 8, 25, 37, 159–61
Massey-Harris-Ferguson, 159–61
Max-Emerge planter, 286, *287*
McCardell, Archie R., 268, 302
McCormick, Cyrus H., 49
McCormick, Elmer, 69–70
McCray, H. E., 15, 18
McGuire, Frank T., 203–4, 206, 209, 217, 218, 220–21, 223, 228, 242, 243, 288, 290, 324
McKahin, H. B., 15, 23, 219
McNamara, Robert, 175
merchandise unit, 272–73
Mexico, as branch, 166–67, 196–98, 231–34
Millar, Gordon H., 288, 290, 292, 295
Mixter, George, 4, 136–37
Model 40C tractor, 111
Model 40 tractor, 133
Model 50 tractor, 130, 133
Model 55H combine, 134
Model 60 tractor, 130, 133
Model 70 tractor, 132
Model 80 diesel tractor, 133
Model 300 tractor, 192
Model 310 tractor, 204
Model 500 tractor, 192
Model 510 tractor, 204
Model 630 tractor, 196
Model 710 tractor, 204
Model 730 tractor, 196
Model 830 tractor, 196
Model 850 tractor, 282
Model 950 tractor, 282
Model 1010 crawler tractor, 192
Model 1010 tractor, 186
Model 1250 tractor, 283
Model 2010 tractor, 186
Model 2040 tractor, 278, 283
Model 2150 tractor, 283
Model 3010 tractor, 186, 192–93
Model 3020 tractor, 186
Model 4010 tractor, 186
Model 4020 tractor, 186
Model 4430 tractor, 250, 275
Model 4440 tractor, 275–76
Model 8010 tractor, 182
Model 8450 tractor, 276
Model 8650 tractor, 276

INDEX

Model 8850 tractor, 276
Model A, *91*
Model A tractor, 58, 59, 60, 70–71, 79, 129, 130
Model B tractor, 19, 58, 59, 60, *70,* 70–71, 79, 129, 130
Model C tractor, 16, *16,* 17, *17*
Model DI tractor, 72
Model D tractor, 15, 16, 17–18, 19–20, *21,* 43, 57, 59, 72, 129
Model G tractor, 79
Model H tractor, 79
Model L tractor, 79
Model MC tractor, 110–11, 133
Model M tractor, 110, 130, 133
Model MT tractor, 110, 133
Model R diesel tractor, 133
Model R tractor, 129–30
Model X-21 tractor, 203, 204
Model X-22 tractor, 203, 204
Model XR-50 tractor, 250
Model XR-80 tractor, 279
Moline Implement Company, 25.
 See also Moline Plow Works
Moline National Bank, 55
Moline Plow Works, 131–32
Morgan, W. R., 9
Moschel, Herman, 23
Murphy, Governor Frank, 67
Murphy, L. A. "Pat," 23, 98–99, 100, 101, 104, 107, 132, 137–38, 141, 144, 145, 157–58, 162, 164, 179, 206

National Council for United States-China Trade, 251
National Industrial Recovery Act, 66
National Labor Relations Board (NLRB), 68, 116, 117
National War Labor Board, 95
New Deal legislation, 65
New Holland, 268
Nixon, President Richard, 250
No. 1 combine, 11–13
No. 2 combine, 11–13
No. 3 combine, 13
No. 5 combine, 13
No. 6 combine, 81
No. 11A combine, 80–81
No. 12A combine, 80–81, 134
No. 25 combine, 134
No. 45 combine, 135
No. 55 combine, 134, 135
No. 65 combine, 134

Office of Production Management, 85
Oliver Chilled Plow Company, 25
Olson, Gust, 133
Organization for Economic Cooperation and Development (OECD), 193

paint, for implements, 189
Paradise, L. A., 23
Peek, Burton, *92*
Peek, Burton F., 31–32, 62, 63, 68, 88, 98, 102, 104, 106–7, 113, 146, 165, 179, 326
Peek, George, 4, 65
Pence, Harry, 167, 169, 191, 207
People's Savings Bank of Moline, 52, 53–54
Pérez Rodriguez, Carlos Andrés, 236
Peterson, Clifford L., 203, 206, 218, 295, 306
Peterson, Comart M., 206, 208
Phillips, W. E., 160, 161
planters, 286
plows, and today's practices, 265, 267.
 See also individual plows by name
"pocket" markets, 203, 231, 240
policy committee, 162
"Poppin Johnnie" tractor, 192, 278
"power farming," 24
Power Farming Conference, 24
power steering, for tractors, 133
production quotas, for wartime, 86–87, 89
products
 development of, 57–60, 79–81, 127–34
 research on, 22–24, 286–91
PTO (power take-off) for tractor, 130–31, 182

Quad-City Industrial Union Council, 113
Quik-Tatch cultivator, 135

rakes. *See* hay rakes
reapers. *See individual reapers by name*
recession, economic, 267–69, 305–8
research and development, 22–24, 286–91
Reuther, Walter, 95, 114
robots, use of, 300
Roche, Kevin, 210, 319, 320
Roosevelt, President Franklin, 60, 65, 94
Rowland, L. A. "Duke," 23, 100, 103, 130, 133, 137, 140, 141, 165
Russia, as market, 42–43, 44, 45–46

Saarinen, Eero, *176,* 178, 179, 210, 319
Schneider, Logemann & Cia, Ltd., 240
Scotland project, 137–39, 142
self-propelled combine, 134–35
shellers. *See individual shellers by name*
Sherman Antitrust Act, 121
Shippee combine, 8
Silloway, Frank, 13, 18–19, 20, 38, 40, 41, 42, 43, 47, 53, 55, 72, 73–74, 75, 77, 98, 99, 104, 209
Sklovsky, Max, 56–57, 109, 147
Smith, Lloyd B., 326
soil culture studies, 57
Sonne, Dr. Karl-Heinz, 221
South Africa, as market, 47–48, 200–202, 242–48
South African Cultivators (Proprietary) Limited, 200

Soviet Union, 248–50. *See also* Russia, as market
Spain, as market, 202–3, 204
Stalin, Joseph, 42, 45, 248
Steel Workers' Organizing Committee (SWOC), 67, 68
Stone, Charles N., 9, 10–11, 13, 22, 23, 50, 69, 98, 104, 105–6, 137, 143, 157–58, 288
strikes
  1943, 96
  1944, 97
  1970s and '80s, 302
  and contract agreement, 67
  Germany, 191–92
  postwar, 112–13, 112–14, 115, 119–21
Stuart, Robert D., Jr., 326
Sullivan, Reverend Leon, 245

Taylor, Dr. W. E., 57
technology, sale of, 291–93
Thomas, R. J., 114, 116
threshers, 13. *See also individual threshers by name*
Thrift Plan, 51–52
tractors
  and centralization, 293–96
  and depression, 57–60
  development of, 15–21, 57–60
  diesel, 129–130, 132
  four-cylinder, 134, 182
  four-wheel drive, 276, 278
  future plans, 284–285
  general purpose, 13–15, 25
  and hitches, 131
  and horsepower, 19–21, 180–81
  industrial types, 72–76
  innovations (1950s), 180–83
  innovations (1960s), 186–87
  innovations (1970s & '80s), 275–76, 278
  military, 88, *91*
  models of (*See individual model by name*)
  and power steering, 133
  promotion of, 7
  re-styling of, 69–71
  small horsepower, 278–84
  standardization of, 202–5
  statistics, 1940, 79–81
  two-cylinder, 127–28, 187
  and wartime, 86–87
Trailer Patrol road scraper, 73
Travell, Dr. Janet, 186
Truman, President Harry, 112
Turkey, as market, 239–240

UAW. *See* United Automobile, Aircraft and Agricultural Implement Workers (UAW)
Ukkelberg, Ed W., 206, 218
Union Malleable Iron Works, 31–32, 95
unions. *See also individual unions by name*
  1930s, 66–69
  and Deere (*See* management, of company)
  and "give backs," 302–3
  postwar, 112–21
United Automobile, Aircraft and Agricultural Implement Workers (UAW), 67, 95, 97, 113, 114, 116–19, 120, 300, 302, 303
United Electrical Workers, 117
United Electrical Workers/Farm Equipment Workers, 117–18
Universal combine, 10
Universal Harvester Company, 10
US-USSR Trade and Economic Council, 250

VanSant, Robert W., 324
Veech, C. C., 3
Velie, S. H., Jr., 54
Velie, Willard, 3
Venezuela, as market, 236–38
Vogel, Walter, 323

wages, cuts in, 51. *See also* management, of company
Wagner, Senator, 68
Wagner Act, 66, 68
Wagner-Langemo Company, 13
War Food Administration, 87, 94
War Labor Board, 112, 114
War Production Board, 87, 89, 93, 101
Waterloo Boy tractor, 15, 42, *44*
Waterloo Tractor Works, 23, 69, 131–32, 133
Webber, Charles C. (C. C.), 4, 9–10, 18, 19, 37, 54, 55, 98–99
Wharton, T. F., 53, 55, 98, 227
White, Harold, 23, 143
White Motor Corporation, 268–69
Wilson, George, 67, 97
Wilson, Lewis D., 164, 218
Wiman, Charles, 85, *92*
Wiman, Charles Deere, 4, 6–7, 9, 10–11, 15–16, 17, 19, 20, 21, 22–23, 24, 32–33, 38, 51, 52–57, 54, 55, 57–58, 60, 62, 63–64, 67, 69, 80, 88, 89, 93–94, 99, 101–12, 113, 137, 138, 139–42, 143–44, 146–47, 159, 162, 326, 327
Wiman, Dwight Deere, 54, 61, 62, 143
World War II, 85, 172–73
Wormley, James, 173, 188, 191

Yamaoka, Tadao, 283–84
Yanmar, 280–84
Yanmar Diesel Engine Company, Ltd., 280–81

ABOUT THE AUTHOR

WAYNE G. BROEHL JR. (1922–2006), historian and faculty member of Dartmouth College's Amos Tuck School of Business Administration, authored or co-authored several books in the field of business history, management theory, and economic development. His special interest was the corporation's role in society, and he studied and wrote extensively on the role corporate management teams play in the American business scene.

His award-winning book *The Molly Maguires* chronicled the saga of an Irish secret society in the Pennsylvania coal fields in the nineteenth century, as early unions interacted in traumatic ways with large-scale rail and coal companies. Another book, *International Basic Economy Corporation*, was a management analysis of the Rockefeller development company in Latin America. It recorded the Rockefeller family's effort to demonstrate social responsibility under private enterprise programs in the less developed world.

Made in United States
Troutdale, OR
08/09/2024